Abstraction and Specification in Program Development

The MIT Electrical Engineering and Computer Science Series

Harold Abelson and Gerald Jay Sussman with Julie Sussman, *Structure and Interpretation of Computer Programs*, 1985

William McC. Siebert, *Circuits, Signals, and Systems*, 1986

Berthold Klaus Paul Horn, *Robot Vision*, 1986

Barbara Liskov and John Guttag, *Abstraction and Specification in Program Development*, 1986

Abstraction and Specification in Program Development

Barbara Liskov and John Guttag

The MIT Press

Cambridge, Massachusetts   London, England

McGraw-Hill Book Company

New York   St. Louis   San Francisco   Montreal   Toronto

This book is one of a series of texts written by faculty of the Electrical Engineering and Computer Science Department at the Massachusetts Institute of Technology. It was edited and produced by The MIT Press under a joint production-distribution arrangement with the McGraw-Hill Book Company.

Ordering Information:

North America

Text orders should be addressed to the McGraw-Hill Book Company.
All other orders should be addressed to The MIT Press.

Outside North America

All orders should be addressed to The MIT Press or its local distributor.

This book was printed and bound by Halliday Lithograph in the United States of America.

Library of Congress Cataloging-in-Publication Data

Liskov, B.
    Abstraction and specification in program development.

    (The MIT electrical engineering and computer science series)
    Bibliography: p.
    Includes index.
    1. Electronic digital computers—Programming.
I. Guttag, John. II. Title. III. Series.
QA76.6.L5655 1986   005.1'2   85-23078
ISBN 0-262-12112-3 (MIT Press)
ISBN 0-07-037996-3 (McGraw-Hill)

To Nate and Olga

Contents

Preface

Good programming involves the systematic mastery of complexity. It is not an easy subject to teach. Students must be convinced that programming is not an arcane art, but an engineering discipline. There are useful principles that can and should be applied. In this book we attempt to shape the way students think about programming by presenting, justifying, and illustrating a cohesive doctrine. Our principal tenet is that abstraction and specification must be the linchpins of any effective approach to programming. We place particular emphasis on the use of data abstraction to produce highly modular programs.

We have followed these principles for over a decade, and have taught them at MIT and elsewhere since 1976. We believe strongly that their application has helped to make us and our students better programmers. They are not, of course, magic. Building good programs remains challenging. A systematic approach is important, but it is no substitute for cleverness, diligence, and good taste.

Draft manuscripts of this book have been used in two different courses: an undergraduate laboratory course and a graduate-level course for professional programmers and analysts. In the undergraduate laboratory course, students undertake programming projects of varying size, culminating with a monthlong team project. Students in this course have had one course in programming and consequently know something about discovering algorithms and are familiar with at least one higher-level programming language, such as Pascal, C, or LISP. The emphasis in this course is on constructing moderately large programs rather than on inventing or implementing algorithms. The students in the graduate-level course have substantial programming experience and are able to understand and appreciate our approach to programming without doing the larger programming exercises. This leaves us time to concentrate on advanced material on formal specifications and program verification.

Chapter 1 discusses the problems that arise in constructing programs and provides an outline of our approach to program development. The remainder of the book is organized in a bottom-up manner. We begin by discussing the various kinds of abstractions that are the basic building blocks of our method and then go on to discuss some of the issues that arise in putting these blocks together. The order of presentation is that used in the laboratory course, where it is important to get the students working on projects early in the term.

Chapter 2 contains an overview of the programming language CLU, one of a growing number of languages that contain facilities supporting data abstraction. Most of the sample implementations in the text are written in CLU. We use CLU in the laboratory course we teach at MIT, and appendix

A contains a complete CLU reference manual.[1] A laboratory course based on this book can be easily taught using any language that supports data abstraction, such as Modula 2 or Ada. With some care, the course can be taught using almost any higher-level programming language. We have taught it using PL/1, and the book includes material designed to facilitate teaching the course using Pascal.

Chapters 3–6 focus on three kinds of abstractions: procedures, types, and iterators. We discuss what these abstractions are and how to specify, use, and implement them. Some of the discussion about implementation is CLU-specific, but most of it is not.

Chapter 7 discusses how to use our programming method in conjunction with Pascal. The location of this chapter in the book is a bit arbitrary. When we teach a laboratory course using CLU, we cover the material in this chapter at the end of the term. An instructor teaching a course in which the students will be programming in Pascal may want to present this material in conjunction with the material in Chapters 3–6.

Chapter 8 reemphasizes the role of specification in our programming method. It supplements the material in Chapters 3–6 with a more careful discussion of the meaning of specifications and presents some hints about writing good specifications.

Chapter 9 deals with testing and debugging. Much of the material in this chapter will be familiar to experienced programmers. Instructors will probably want to supplement the material on debugging with their own anecdotes and aphorisms.

Chapter 10 formalizes the specification language used throughout the book, and chapter 11 introduces the key concepts used in program verification. The emphasis in both chapters is on using formalism to deepen the reader's intuition about topics covered informally elsewhere in the book. These chapters assume a slightly better mathematical background than does the rest of the book. In particular, we assume that students have been exposed to simple first-order predicate calculus and mathematical induction. The location of these chapters in the book is based on their use in the graduate-level course. In the undergraduate laboratory course we defer them until the end of the term. For this reason, the rest of the book does not depend upon these chapters.

Chapters 12–14 provide an overview of various aspects of the program production process. Chapter 12 discusses requirements analysis, chapter 13 program design, and chapter 14 design reviews and program development strategies. Chapter 13 is the most detailed of the three. It emphasizes modular decomposition based on the recognition of useful abstractions. At

---

[1]For a nominal fee universities can obtain implementations of CLU by writing to the authors. At present there are implementations for the DEC20 and DEC Vax and Motorola 68000-based Unix systems.

the heart of the chapter is an extended example of program design. An implementation of this design is given in appendix B. Chapters 12 and 14 are rather superficial. They deal with problems that raise managerial as well as technical issues, many of which are beyond the scope of this book.

Chapter 15 deals with the relationship of our programming method to programming languages. It covers much the same ground as chapter 7, but in a more general context.

Most chapters of the book conclude with suggestions for further reading, and each chapter contains a set of exercises. We have kept the reading lists short. References were selected not on the basis of their scientific contribution, but because we thought that students taking a course based on this book would find them accessible and useful. Some of the exercises involve programming, but many do not. Appendix C contains a sequence of five programming problems, including a monthlong team project. These represent a typical sequence of assignments in our undergraduate laboratory course.

Acknowledgements

It is impossible to acknowledge explicitly all those who have contributed to this book. Hundreds of students have used various drafts, and many of them have contributed useful comments. Scores of graduate students have been teaching assistants in courses based on the material in this book. Many of them contributed to examples and exercises that have found their way into this text.

The publishers supplied us with a number of helpful critiques. The most comprehensive of these was by Jim Horning. His copious and perceptive comments and criticisms had a major effect on the final form of this book. Other readers whose comments were particularly helpful include Craig Chambers, Mark Day, Susan Gerhart, Gary Leavens, Nate Liskov, Brian Oki, Sharon Perl, Bob Scheifler, Gerry Sussman, Bill Weihl, Jeannette Wing, and Joe Zachary.

MIT's Department of Electrical Engineering and Computer Science and its Laboratory for Computer Science have supported this project in important ways. By reducing our teaching load, the department has given us time to write. The laboratory has provided a research environment that gave rise to many of the ideas presented in this book.

Material in appendix A is reprinted with permission from *CLU Reference Manual* by Barbara Liskov, Russell Atkinson, Toby Bloom, Eliot Moss, J. Craig Schaffert, Robert Scheifler, and Alan Snyder (New York: Springer-Verlag, 1981).

Finally, we owe a special debt of gratitude to Anne Rubin. She typed, edited, and kept track of countless versions of the manuscript. Her efficiency and good humor through it all were a big help.

1

Introduction

This book will develop a methodology for program construction. Our goal is to help programmers construct programs of high quality—programs that are reliable and reasonably easy to understand, modify, and maintain.

A very small program, consisting of no more than a few hundred lines, can be implemented as a single monolithic unit. As the size of the program increases, however, such a monolithic structure is no longer reasonable because the code becomes difficult to understand. Instead the program must be decomposed into a number of independent small programs, called *modules*, that together provide the desired function. We shall focus on this decomposition process: how to decompose large programming problems into small ones, what kinds of modules are most useful in this process, and what techniques increase the likelihood that modules can be combined to solve the original problem.

Doing decomposition properly becomes more and more important as the size of the program increases, for a number of reasons. First, many people must be involved in the construction of a large program. If just a few people are working on a program, they will naturally interact regularly. Such contact reduces the possibility of misunderstandings about who is doing what and lessens the seriousness of the consequences should such misunderstandings occur. If many people work on a project, regular communication becomes impossible because it would consume too much time. Instead the program must be decomposed into pieces that the individuals can work on independently with a minimum of contact.

The useful life of a program (its *production* phase) begins when it is delivered to the customer. Work on the program is not over at this point, however. There are likely to be residual errors in the code that will need attention, and program modifications will often be required to upgrade the program's serviceability or to provide services better matched to the user's needs. This activity of program *modification* and *maintenance* is likely to consume more than half of the total effort put into the project.

For modification and maintenance, it is rarely practical to start from scratch and reimplement the entire program. Instead, one must retrofit modifications within the existing structure, and it is therefore important that the structure accommodate change. In particular, the pieces of the program must be independent, so a change to one piece can be made without requiring changes to all pieces.

Finally, most programs have a long lifetime. People will often have to deal with programs long after they have first worked on them. Moreover, there is likely to be substantial turnover of personnel over the life of any project, and program modification and maintenance are typically done by people other than the original implementers. All of these factors suggest a requirement that programs be structured in such a way that they can be understood easily.

In the methodology we shall describe in this book, programs will be developed by means of problem decomposition based on a recognition of useful abstractions. *Decomposition* and *abstraction*, the two key concepts in this book, form our next subject.

1.1 Decomposition and Abstraction

The basic paradigm for tackling any large problem is clear—we must "divide and rule." Unfortunately, merely deciding to follow Machiavelli's dictum still leaves us a long way from solving the problem at hand. Exactly how we choose to divide the problem is of overriding importance.

Our goal in decomposing a program is to create modules that are themselves small programs that interact with one another in simple, well-defined ways. If we achieve this goal, different people will be able to work on different modules independently, without needing much communication among themselves, and yet the modules will work together. In addition, during program modification and maintenance it will be possible to modify some of the modules without affecting all of the others.

When we decompose a problem, we factor it into separable subproblems in such a way that

1. each subproblem is at the same level of detail;

2. each subproblem can be solved independently; and

3. the solutions to the subproblems can be combined to solve the original problem.

Sorting using merge sort is an elegant example of problem solving by decomposition. It breaks the problem of sorting a list of arbitrary size into the two simpler problems of sorting a list of size two and merging two sorted lists of arbitrary size.

Decomposition is a time-honored and useful technique in many disciplines. From Babbage's day onward people have recognized the utility of such things as macros and subroutines as decomposition devices for programmers. It is important to recognize, however, that decomposition is not a panacea, and that when used improperly it can have a harmful effect. Furthermore, large or poorly understood problems are difficult to decompose properly. The most common problem is creating individual components that solve the stated subproblems but do not combine to solve the original problem. This is one of the reasons why system integration is often difficult.

For example, imagine creating a play by assembling a group of writers, giving each a list of characters and a general plot outline, and asking each of them to write a single character's lines. The authors might accomplish their individual tasks admirably, but it is highly unlikely that their combined efforts will be an admirable play. It might be artistic, but it would lack any sort of coherence or sense. Individually acceptable solutions simply cannot be expected to combine properly if the original task has been divided in a counterproductive way.

Abstraction is a way to do decomposition productively by changing the level of detail to be considered. When we abstract from a problem we agree to ignore certain details in an effort to convert the original problem to a simpler one. We might, for example, abstract from the problem of writing a play to the problem of deciding how many acts it should have, or what its plot will be, or even the sense (but not the wording) of individual pieces of dialogue. After this has been done, the original problem (of writing all of the dialogue) remains, but it has been considerably simplified—perhaps even to the point where it could be turned over to another or even several others. (Dumas, père, churned out novels in this way.)

The paradigm of abstracting and then decomposing is typical of the program design process: Decomposition is used to break software into components that can be combined to solve the original problem; abstractions assist in making a good choice of components. We alternate between the two processes until we have reduced the original problem to a set of problems we already know how to solve.

1.2 Abstraction

The process of abstraction can be seen as an application of many-to-one mapping. It allows us to forget information and consequently to treat things that are different as if they were the same. We do this in the hope of simplifying our analysis by separating attributes that are relevant from those that are not. It is crucial to remember, however, that relevance often

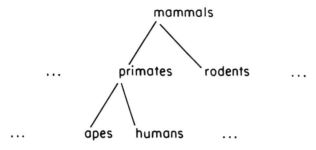

Figure 1.1 An abstraction hierarchy.

depends upon context. In the context of an elementary school classroom we learn to abstract both $(8/3) * 3$ and $5 + 3$ to the concept we represent by the numeral 8. Much later we learn, often under unpleasant circumstances, that on many computing machines this abstraction can get us into a world of trouble.

For example, consider the structure shown in figure 1.1. Here the concept is "mammal." All mammals share certain characteristics, such as the fact that females produce milk. At this level of abstraction, we focus on these common characteristics and ignore the differences between the various types of mammals.

At a lower level of abstraction, we might be interested in particular instances of mammals. However, even here we can abstract by considering not individuals, or even species, but groups of related species. At this level we would have groupings such as primates or rodents. Here again we are interested in common characteristics, such as the fact that all primates bear live young, rather than the differences between, say, humans and chimpanzees. Such differences will be relevant at a still lower level of abstraction.

The abstraction hierarchy of figure 1.1 comes from the field of zoology, but it might well appear in a program that implemented some zoological application. A more specifically computer-oriented example that is useful in many programs is the concept of a "file." Files abstract from raw storage and provide long-term, online storage of named entities. Operating systems differ in their realizations of files; for example, the structure of the file names differs from system to system, as does the way in which the files are stored on secondary storage devices.

In this book we are interested in abstraction as it is used in programs in general. The most significant development to date in this area is the development of high-level languages. By dealing directly with the constructs

```
found := false                      found := false
i := lowbound(a)                    i := highbound(a)
while i < highbound(a) + 1 do       while i > lowbound(a) − 1 do
    if a[i] = e                         if a[i] = e
        then z := i                         then z := i
        found := true                       found := true
    end                                 end
    i := i + 1                          i := i − 1
end                                 end
```

Figure 1.2 Two program fragments.

of a high-level language rather than with the many possible sequences of machine instructions into which they can be translated, the programmer achieves a significant simplification.

In recent years, however, programmers have become dissatisfied with the level of abstraction generally achieved even in high-level language programs. Consider, for example, the program fragments in figure 1.2. At the level of abstraction defined by the programming language these fragments are clearly different: If there is an occurrence of e in a, one fragment finds the index of the first occurrence, and the other the index of the last. One sets i to $highbound(a) + 1$ and the other to $lowbound(a) − 1$. It is not improbable, however, that both were written to accomplish the same goal: to set *found* to false if there is no occurrence of e in a, and otherwise to set *found* to true and z to the index of some occurrence of e in a. If this is what we want, these program fragments by themselves clearly do not achieve the desired level of abstraction.

One approach to dealing with this problem lies in the invention of "very-high-level languages," built around some fixed set of relatively general data structures and a powerful set of primitives that can be used to manipulate them. For example, suppose a language provided *is_in* and *index_of* as primitive operations on arrays. Then we could accomplish the task outlined above simply by writing

```
found := is_in(a, e)
if found then z := index_of(a, e) end
```

The flaw in this approach is that it presumes that the designer of the programming language will build into the language most of the abstractions that users of the language will want. Such foresight is not given to many; and even if it were, a language containing so many built-in abstractions might well be so unwieldy as to be unusable.

A preferable alternative is to design into the language mechanisms that allow programmers to construct their own abstractions as they need them. The most common such mechanism is the use of *procedures*. By separating procedure definition and invocation, a programming language makes two important methods of abstraction possible: *abstraction by parameterization* and *abstraction by specification*.

1.2.1 Abstraction by Parameterization

Abstraction by parameterization allows us, through the introduction of parameters, to represent a potentially infinite set of different computations with a single program text that is an abstraction of all of them. Consider the program text

x * x + y * y

This describes a computation that adds the square of the value stored in a particular variable x to the square of the value stored in another particular variable y. The *lambda expression*

λx, y: int.(x * x + y * y)

on the other hand, describes the set of computations that square the value stored in some integer variable, which we shall temporarily refer to as x, and add to it the square of the value stored in another integer variable, which we shall temporarily call y. In such a lambda expression we refer to x and y as the *formal parameters* and $x * x + y * y$ as the *body* of the expression. We invoke a computation by binding the formal parameters to arguments and then evaluating the body. For example,

λx, y: int.(x * x + y * y) (w, z)

is identical in meaning to

w * w + z * z

In more familiar notation, we might denote the above lambda expression by

squares = **proc** (x, y: int) **returns** (int)
     return (x * x + y * y)
     end

and the binding of actual to formal parameters and evaluation of the body by the procedure call

squares(w, z)

Programmers often use abstraction by parameterization without even noticing that they are doing so. For example, suppose we need a procedure that sorts an array of integers a. At some time in the future we shall probably have to sort some other array, perhaps even somewhere else in this same program. It is highly unlikely, however, that every array we need to sort will be named a; we therefore invoke abstraction by parameterization to generalize the procedure and thus make it more useful.

Abstraction by parameterization is an important means of achieving generality in programs. A *sort* routine that works on any array of integers is much more generally useful than one that works only on a particular array of integers. By further abstraction we can achieve even more generality. For example, we might define a *sort* abstraction that works on arrays of reals as well as arrays of integers, or even one that works on arraylike structures in general.

Abstraction by parameterization is an extremely powerful mechanism. Not only does it allow us to describe a large (even infinite) number of computations relatively simply, but it is easily and efficiently realizable in programming languages. Nonetheless, it is not a sufficiently powerful mechanism to describe conveniently and fully the abstraction that the careful use of procedures can provide.

1.2.2 Abstraction by Specification

Abstraction by specification allows us to abstract from the computation (or computations) described by the body of a procedure to the end that procedure was designed to accomplish. We do this by associating with each procedure a *specification* of its intended effect and then considering the meaning of a procedure call to be based on this specification rather than on the procedure's body.

We are making use of abstraction by specification whenever we associate with a procedure a comment that is sufficiently informative as to allow others to use that procedure without looking at its body. A good way to write such comments is to use pairs of *assertions*. The *requires assertion* (or *pre-condition*) of a procedure specifies something that is assumed to be true on entry to the procedure. In practice what is most often asserted is a set of conditions sufficient to ensure the proper operation of the procedure.

```
sqrt = proc (coef: real) returns (real)
    % requires coef > 0
    % effects returns an approximation to the square root of coef
        ans: real = coef/2.0;
        i: int := 1
        while i < 7 do
            ans := ans − ((ans * ans − coef)/(2.0 * ans))
            i := i + 1
            end
        return (ans)
    end sqrt
```

Figure 1.3 The *sqrt* procedure.

(This is often simply the vacuous assertion "true.") The *effects assertion* (or *post-condition*) specifies something that is supposed to be true at the completion of any invocation of the procedure for which the pre-condition was satisfied.

Consider, for example, the *sqrt* procedure in figure 1.3. Because a specification is provided, we can ignore the body of the procedure and take the meaning of the procedure call $y := sqrt(x)$ to be "If x is greater than 0 when the procedure is invoked, then after the execution of the procedure, y is an approximation to the square root of x." Notice that the requires and effects assertions permit us to say nothing about the value of y if x is not greater than 0.

In using a specification to reason about the meaning of a procedure call, we follow two distinct rules:

1. After the execution of the procedure we can assume that the post-condition holds.

2. We can assume *only* those properties that can be inferred from the post-condition.

The two rules mirror the two benefits of abstraction by specification. The first asserts that users of the procedure need not bother looking at the body of the procedure in order to use it. They are thus spared the effort of first understanding the details of the computations described by the body and then abstracting from these details to discover that the procedure really does compute an approximation to the square root of its argument. For complicated procedures, or even simple ones using unfamiliar algorithms, this is a nontrivial benefit.

The second rule makes it clear that we are indeed abstracting from the procedure body, that is, forgetting some supposedly irrelevant information. This insistence on forgetting information is what distinguishes abstraction from decomposition. By examining the body of *sqrt*, users of the procedure could gain a considerable amount of information that cannot be gleaned from the post-condition, for example, that *sqrt*(4) will return +2. In the specification, however, we are saying that this information about the returned result is to be ignored. We are thus saying that the procedure *sqrt* is an abstraction representing the set of all computations that return "an approximation to the square root of *x*."

In this book, abstraction by specification will be the major method used in program construction. Abstraction by parameterization will be taken almost for granted; abstractions will have parameters as a matter of course.

1.2.3 Kinds of Abstractions

Abstraction by parameterization and by specification are powerful methods for program construction. They enable us to define three different kinds of abstractions: procedural abstraction, data abstraction, and iteration abstraction. In general, each procedural, data, and iteration abstraction will incorporate both methods within it.

For example, *sqrt* is like an operation: It abstracts a single event or task. We shall refer to abstractions that are operationlike as *procedural abstractions*. Note that *sqrt* incorporates both abstraction by parameterization and abstraction by specification.

Procedural abstraction is a powerful tool. It allows us to extend the virtual machine defined by a programming language by adding a new operation. This kind of extension is most useful when we are dealing with problems that are conveniently decomposable into independent functional units. Often, however, it is more fruitful to think of adding new kinds of data objects to the virtual machine.

The behavior of the data objects is expressed most naturally in terms of a set of operations that are meaningful for those objects. This set includes operations to create objects, to obtain information from them, and possibly to modify them. For example, *push* and *pop* are among the meaningful operations for stacks, while integers need the usual arithmetic operations. Thus a *data abstraction* (or *data type*) consists of a set of objects and a set of operations characterizing the behavior of the objects.

As an example, consider *multi_sets*. Multi_sets are like ordinary sets except that elements can occur more than once in a multi_set. Multi_set operations include *empty*, *insert*, *delete*, *number_of*, and *size*. These operations create an empty multi_set, add and delete elements from a multi_set, tell how many times a particular element occurs in a multi_set, and tell how many total elements are in a multi_set, respectively. The operations might be implemented within the runtime environment of the programming language by calls to various procedures. Programmers using multi_sets, however, need not worry about how these procedures are implemented. To them *empty*, *insert*, *delete*, *number_of*, and *size* are abstractions defined by such statements as

The *size* of the multi_set *insert*(*s*, *e*) is equal to *size*(*s*) + 1.

For all *e*, the *number_of* times *e* occurs in the multi_set *empty*( ) is 0.

The key thing to notice is that each of these statements deals with more than one operation. We do not present independent definitions of each operation, but rather define them by showing how they relate to one another. The emphasis on the relationships among operations is what makes a data abstraction something more than just a set of procedures. The importance of this distinction is discussed throughout this book.

In addition to procedural and data abstraction, we shall deal in this book with *iteration abstraction*. Iteration abstraction is used to avoid having to say more than is relevant about the flow of control in a loop. A typical iteration abstraction might allow us to iterate over all the elements of a multi_set without constraining the order in which the elements are to be processed.

The remainder of this book is concerned with how to do program decomposition based on abstraction. Our emphasis will be on data abstraction. We believe that while procedural and iteration abstraction have valuable roles to play, it is data abstraction that most often provides the primary organizational tool in the programming process.

The next few chapters are concerned with the three kinds of abstractions—what they are, how to specify their behavior, and how to implement them. We shall begin by describing a programming language called CLU, which was designed to support the methodology of this book and which provides a mechanism for each kind of abstraction. It is, therefore, particularly easy to implement abstractions in CLU, and chapters 3–6 will use CLU as the implementation language. In chapter 7 we discuss how to implement abstractions in Pascal.

The latter portion of the book focuses on the use of abstractions in program construction. We discuss the phases of program construction, how to do program design, and how to carry on into implementation. The book concludes with a discussion of a number of practical issues that arise in using the methodology in constructing real programs.

Further Reading

Boehm, Barry W., 1976. Software engineering. *IEEE Transactions on Computers* C-25(12): 1226–1241.

Brooks, Frederick P., Jr., 1975. *The Mythical Man-Month*. Reading, Mass.: Addison-Wesley Publishing Co.

Buxton, John M., Peter Naur, and Brian Randell (eds.), 1976. Software engineering concepts and techniques. In *Proceedings of the NATO Conference* (New York: Petrocelli/Charter).

Gries, David (ed.), 1978. Part 1: viewpoints on programming. In *Programming Methodology, A Collection of Articles by Members of IFIP WG2.3* (New York: Springer-Verlag), pp. 7–74.

Exercises

1.1   Describe an abstraction hierarchy with which you are familiar.

1.2   Select a procedure that you have written or used and discuss how it supports abstraction by specification and by parameterization.

2

An Overview of CLU

This chapter provides an overview of CLU. We shall concentrate primarily on semantics but shall also describe briefly the language's built-in types and control structures. We shall discuss how CLU provides compile-time type checking and separate compilation and its support for object-oriented programming. Appendix A contains a detailed description of the language.

2.1 Program Structure

A CLU program is made up of one or more modules. Each module is a *procedure*, *cluster*, or *iterator*. Procedures are used for implementing procedural abstractions, clusters for data abstractions, and iterators for iteration abstractions. In this chapter we shall discuss only procedures, although much of what we have to say applies to clusters and iterators as well.

The modules in a CLU program do not have a nested structure relative to one another, except insofar as procedures and iterators can be used inside clusters to implement the data abstraction's operations. It is not possible in CLU to nest one procedure definition inside another to obtain an "internal" procedure that can be called only within a limited context.

Figure 2.1 shows an example of a procedure definition. Each such definition has a *header* that defines the name of the procedure; the number, order, and types of the arguments that must be passed to the procedure when it is called; and the number, order, and types of the results returned from a call. In the example,

gcd = **proc** (n, d: int) **returns** (int)

is the header of the procedure *gcd*, which takes two integers as arguments and returns one integer as a result. The formal names of the arguments are *n* and *d*; these names can be used inside the body of *gcd* to refer to the actual arguments.

The *body* of a procedure definition consists of a sequence of statements and can also contain declarations of local variables. The statements in the body can refer only to these local variables and the formal names. Since CLU modules cannot be nested, they cannot use global variables, that is,

```
gcd = proc (n, d: int) returns (int)
    % the gcd is computed by repeated subtraction
    while n ~= d do   % ~ is "not" in CLU
        if n > d then n := n − d else d := d − n end
        end       % while
    return (n)
    end gcd
```

Figure 2.1 A procedure definition.

variables that are declared outside the module; all communication between modules must consist of explicitly passed arguments and results.

2.1.1 Statements

Within a module, CLU uses fairly conventional control structures. The basic units of computation are assignment statements and procedure call statements, which can be combined into larger units in three ways: sequencing, conditionals, and loops. Comments are introduced by the symbol % and continue from there to the end of the line.

Sequencing is achieved simply by writing the statements one after another; it is not necessary to separate or terminate statements by means of semicolons. **If** statements are used for conditionals and **while** statements for loops; like all statements in CLU, these are self-terminating, and we use the reserved word **end** in each case, as in

if *expr* **then** *body* **else** *body* **end**

The **begin** statement can be used to group a number of statements into a single statement. This is rarely done, however, since all the above statement forms provide new bodies, each consisting of a sequence of statements and declarations.

CLU does not contain any **goto** statement, but it does contain a number of alternatives to the **goto**. A procedure can return control to its caller from anywhere in its body by executing the **return** statement. This statement must be used if the procedure returns results, but it is optional if the procedure returns no results.

The **break** and **continue** statements can be used in conjunction with loops. The **continue** statement terminates the current iteration of the smallest containing loop; control continues at the next iteration of the loop. The **break** statement terminates the smallest containing loop; control continues at the statement immediately following that loop.

In addition to the control structures mentioned above, there are the **except**, **signal**, and **exit** statements, which will be discussed in chapter 5; the **for** statement, which will be discussed in chapter 6; and the **tagcase** statement, which will be discussed in section 2.7.

2.1.2 Type Checking and Separate Compilation

Each CLU module is compiled separately. When a procedure definition is compiled, the compiler checks that only results of the proper types are returned. When a procedure call is compiled (possibly as part of compiling a different module), the compiler checks that arguments of the proper types are supplied and that results of the proper types are expected. CLU is a *type-safe* language: At runtime one cannot operate on an object of one type as if it belonged to another type. Type safety is achieved by means of compile-time type checking.

There is no notion of a "main" program in CLU. After compilation, modules can be linked to form an executable program, and at this time one of the procedures can be chosen as the main procedure. If no main procedure is specified, any procedure among the modules can be called from the CLU runtime system.

2.2 Integers, Booleans, and Arrays

Each data type in CLU can be thought of as a data abstraction. As mentioned in chapter 1, a data abstraction consists of a set of objects and a set of operations. The operations permit the user of the type to create objects of the type and to manipulate those objects. CLU provides a rich set of built-in types and a mechanism that allows users to define new, abstract data types. This section describes three built-in types: integers, booleans, and arrays. Others will be described later in this chapter, and appendix A contains a complete presentation of built-in types.

CLU integers and booleans are like those in other languages. The name for the integer type is *int*; the boolean type is called *bool*. Int provides the usual arithmetic operations ($+$, $-$, $*$, $/$), relational operations ($=$, $<$, $>$, $<=$, $>=$), and literals (1, 2). The bool literals are **true** and **false**; operations include the relational operation $=$ and the predicates & (and), | (or), and $\sim$ (not).

CLU arrays are unlike those in most languages because they are dynamic; an array can grow or shrink at either end after being created. CLU arrays are one-dimensional and are always indexed by integers. (Arrays of arrays are used for multidimensional arrays.) An array has a low and a high bound; all integers between the bounds (inclusive) are legal indexes into that array. Since arrays are dynamic, the bounds may vary during program execution.

All elements of an array must be of the same type, and this type determines the type of the array. For example,

array[int]

is the type of an array containing integer elements, while

array[array[int]]

is the type of an array containing arrays of integers as elements. Note that the array bounds are not part of the type.

Array types provide a number of operations. First, there are operations that create new arrays. For example, the operation

new = **proc** ( ) **returns** (array[T])

returns a new, empty array. The low bound of this array is 1, and the high bound is 0. Since there are no integers in the range [1, 0], this new array contains no elements. Here T stands for an arbitrary type; if T were int, for example, we would obtain an array[int].

Arrays can also be created by means of a special array constructor. For example,

array[int]$[3: 6, 17, 24]

creates an array of integers with low bound 3, high bound 5, and elements 6, 17, and 24 at indexes 3, 4, and 5, respectively.

Next, there is a group of operations that return information about the array. The operations *size, low,* and *high* return the current size, low bound, and high bound of the array, respectively. The headers of these operations are

size = **proc** (a: array[T]) **returns** (int)
low = **proc** (a: array[T]) **returns** (int)
high = **proc** (a: array[T]) **returns** (int)

For example, if these operations were applied to an array returned by a call of the *new* operation, then *size* and *high* would return 0 and *low* would return 1. Another information-returning operation is

fetch = **proc** (a: array[T], i: int) **returns** (T)

which returns the ith element of a if i is within bounds, that is, if $low(a) \leq i \leq high(a)$. If i is not within bounds, the call of *fetch* terminates by signaling an exception (see chapter 5). All built-in operations signal exceptions if they cannot perform their intended function, but we shall ignore these exceptions in the rest of this chapter.

Finally, there are operations that modify an array. The operation

store = **proc** (a: array[T], i: int, e: T)

stores the element e at index i if i is within bounds. In addition, the array can be extended or shrunk. The operations

addh = **proc** (a: array[T], e: T)
addl = **proc** (a: array[T], e: T)

make the array a grow. *Addh* extends a by 1 position in the high direction and stores e in this new position. For example, if x were a new array of integers created by a call of the *new* operation, then

addh (x, 7)

would cause x to have low bound 1 (the same as before), high bound 1, and x_1 would have the value 7. *Addl* is like *addh* except it extends the array by 1 position in the low direction.

The operations

remh = **proc** (a: array[T]) **returns** (T)
reml = **proc** (a: array[T]) **returns** (T)

shrink the array and return the element that has been removed. *Remh* shrinks the array from the high end, while *reml* shrinks it from the low end. For example, if

reml(x)

is executed after performing *addh* above, it returns 7, and x now has low bound 2 and high bound 1.

Because of the way they grow and shrink, there is always an element associated with any integer between the low and high bounds in CLU arrays. For example, if array a has low bound 1 and high bound 10, then a_1, a_2, ..., a_{10} all have defined values. Thus fewer kinds of runtime errors are possible with CLU arrays than with arrays in other languages.

2.2.1 Naming Operations

Different types are likely to have operations with the same name. For example, ints and bools both have an operation named *equal*. Arrays and sequences (another built-in CLU type that will be defined in section 2.7) both have operations *new, fetch,* and *addh* (among others). To avoid ambiguity about the operation being named, we use the compound form

T$op

This form specifies the operation named *op* of the type named *T*. For example,

array[int]$new

distinguishes the array[int] *new* operation from the *new* operation for array[bool] or any other array or sequence type.

The compound form for naming operations must be used except in cases where special short forms are provided. For example, there are short forms for fetching and storing array elements. If *x* is an integer and *a* is an array[int], then

x := a[i]

is short for

x := array[int]$fetch(a, i)

and

a[i] := x

is short for

array[int]$store(a, i, x)

Short forms are discussed further in section 2.6.

2.2.2 Example: Removing Duplicates

Suppose we want to remove all duplicates from an array of integers, so that, for example, the array [1: 1, 13, 3, 1, 3] becomes [1: 1, 13, 3]. We can accomplish this by starting with the first element of the array and removing all duplicates of it from the rest of the array, then repeating the process with the second element of the array and so on. The *remove_dupls*

```
remove_dupls = proc (a: array[int])
    i: int := array[int]$low(a)
    while i < array[int]$high(a) do
        x: int := a[i]      % fetch the ith element of a
        j: int := i + 1
        % remove all duplicates of x from the rest of a
        while j <= array[int]$high(a) do
            if x = a[j]
                then t: int := array[int]$remh(a)
                    if j > array[int]$high(a) then break end
                    a[j] := t
                else j := j + 1
                end  % if
            end  % inner while
        i := i + 1
        end  % outer while
    end remove_dupls
```

Figure 2.2 Removing duplicates.

procedure shown in figure 2.2 makes use of this strategy. Note the use of
the short forms for the array *fetch* and *store* operations. Note also that
remove_dupls returns simply by reaching the end; no **return** statement is
needed since no result is returned.

The method used in removing duplicates requires roughly $n^2/2$ comparisons in the worst case (when there are no duplicates). Thus the algorithm
is Order(n^2). We shall show a faster algorithm in chapter 4.

2.3 Objects

CLU programs perform computations on *objects*, which are containers for
data. The data contained in an object are referred to collectively as that
object's *value*. For example, the value of an integer object might be 3. The
value of an array object is its low and high bounds and its elements. For
example, the value of the array of integers created by a call of the *new*
operation has low bound 1, high bound 0, and no elements.

Each object in CLU has a data type that determines the kind of value
the object contains and the operations that can be performed on it. For
example, *add* is a meaningful operation on integer objects but not on array
objects.

Objects exist as part of a *universe*. They are created by calling particular
operations that have this function. For example, the array *new* operation
causes a new array object to come into existence in the object universe.
In theory, objects exist forever. In practice, the space used by inaccessible

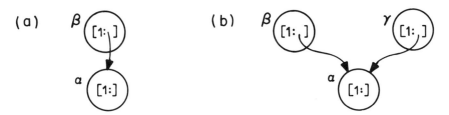

Figure 2.3 Objects and references: (a) object β refers to object α; (b) objects β and γ share object α.

objects is reclaimed by *garbage collection*. An object is *accessible* if it is referred to by one of the program's variables or by another accessible object.

Each object is distinct from all other objects and has a unique name that can be used for reference. The value of an object cannot physically contain another object, but it can contain the names of other objects; when this happens, we say that the containing object *refers to* the contained object. For example, figure 2.3a shows two objects, α and β: α is a newly created array[int]; β is an array[array[int]] that has a single element, namely the array[int] α. Because objects can refer to other objects, it is possible to have cyclic objects (which refer to themselves) and recursive data structures without the use of explicit pointers. (Examples of recursive data structures will be given in chapter 4.) Also, it is possible for several objects to refer to the same object. For example, a second array[array[int]] object, γ, might also refer to α as its single element, as shown in figure 2.3b. When this happens, we say that the objects *share* the object referred to.

There are two kinds of objects in CLU, those that always have the same value and those whose value can change. Objects that always have the same value are called *immutable*. For example, integer objects are immutable: An integer object containing 3 as its value will always have 3 as its value. Objects whose value can change are called *mutable*. For example, arrays are mutable: When we change the first element of an array of integers to be a different integer object, or extend the array by means of the *addh* operation, the value of that array has changed.

Whether an object is mutable or not depends on its type and in particular on whether any of the type's operations modify the type's objects. Integers and booleans are immutable because none of their operations do modifications. However, there are several array operations that modify array objects (including *addh* and *store*); thus arrays are mutable.

When two or more objects share a mutable object, changes to the shared object are visible through either of the sharing objects. For example, any change to the shared array α in figure 2.3b will be visible through arrays β and γ. However, when an object is immutable, it does not matter how many other objects share it; since it can never be modified, we cannot observe the effect of the sharing. For example, suppose the shared object in figure 2.3 were an int instead of an array. The sharing of this object could never be observed by any CLU program since ints are immutable.

2.4 Declarations and Assignment

To operate on objects, programs need some way to name them, and to do this they use *identifiers*. Identifiers may be *constants*, referring to just one object, or *variables*, referring to different objects as time goes by. Both constants and variables are created by means of *declarations*, and both have types that are defined by their declarations. The type checking of CLU guarantees that a variable or constant always refers to an object of its type.

2.4.1 Constants

Constants are created by means of a special kind of declaration called an *equate*. An equate defines the name and type of the constant and causes the constant to refer to an object of that type. For example,

    x = 3

is an equate that creates a constant x of type int that refers to the integer object 3. It is not necessary to state the type of x in the equate, since this is the same as the type of the object that the constant refers to.

CLU limits constants to built-in immutable types. There can be int and bool constants, as well as real, char, and string constants. (These types will be defined in section 2.7.) However, there are no array constants, since arrays are mutable. This constraint ensures that constants are truly unvarying; not only do they always refer to the same object, but that object's value can never change.

Another constraint on the use of equates is that the object the constant refers to must be computable statically (for example, at compile time). Such a *manifest expression* consists of built-in operators applied to literals or constants. For example,

    6 * (4 + c)

is a manifest expression if c is a constant declared in another equate. However,

   p (4)

where p is a user-defined procedure, is not a manifest expression.

   In addition to int constants, bool constants, and so on, CLU also permits type constants. For example,

   ai = array[int]

introduces the constant *ai*. As with other constants, a type constant serves simply as an abbreviation for the object it refers to. For example, given the above equate, we could say

   x: ai

This means exactly the same thing as the declaration

   x: array[int]

Similarly, if we had

   pi = 3.1416

then a later expression

   z * pi

has exactly the same meaning as

   z * 3.1416

In either case, we use the constant as an abbreviation for the expression occurring on the right-hand side of the equate.

2.4.2 Variables and Assignment

A variable is created by a declaration that defines its name and type. For example,

   x: int
   a, b: array[int]

are two declarations creating three new variables, x, a, and b. Variable x has type int, while a and b have type array[int]. These variables are

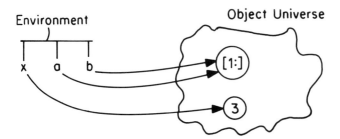

Figure 2.4 Variables and assignment.

uninitialized since they do not refer to any objects. (Declarations with initialization are also possible, as will be discussed.) As a program runs, a variable can refer to different objects, but all objects referred to must be of the variable's type. For example, only integer objects can be referred to by x.

Assignment causes a variable to refer to an object. Specifically, it causes the variable named on the left-hand side of the := to refer to the object obtained by evaluating the expression occurring on the right-hand side of the :=. For example,

x := 3

causes x to refer to an integer object whose value is 3. Similarly,

a := array[int]$new( )

causes a to refer to a new, empty array object. The assignment

b: array[int] := a

causes variable b to refer to the same object that a refers to. In other words, assignment can cause two variables to *share* the same object. Note that assignment never requires copying an object.

The effect of the declarations of a, b, and x and the three assignments is shown in figure 2.4. Variables and objects are indicated in an abstract way. Objects are shown as existing in the object universe; variables are not in this universe but are instead contained in the *environment*. The variables refer to objects in the universe.

Assignment can be combined with declaration, as in

y: int := (z * 3)/7

where z is an already existing integer variable or constant. This form is short for a declaration followed by an assignment:

y: int
y := (z * 3)/7

In addition to the simple assignment statement, there is a multiple assignment statement. For example,

x, y := y, x

can be used to swap the values of x and y.

2.4.3 Scopes

Each body defines a new scope in CLU, within which local constants and variables can be declared. Variable declarations can appear anywhere within a body. The *scope* of a variable is from immediately after its declaration to the end of the body that contains the declaration. If, for example, we had the body

S1
i: int
S2

i could be used in S2 but not in S1; that is, the scope of i is S2. CLU allows declarations to appear anywhere, so that new variables can be declared when needed rather than in advance.

Equates must appear in a group at the beginning of a body. The scope of the constant defined by an equate is the entire containing body. However, an equate may not be defined in terms of itself. For example,

t = array[s]
s = array[t]

is illegal. Therefore, equates cannot be used to define recursive types. (We shall discuss the definition of recursive types in chapter 4.)

Since bodies can be nested, scopes can be nested. However, an identifier cannot be redefined in an inner scope. For example,

```
while p do
   x: int
   . . .
   if q
      then x: int
         . . .
      end
   end
```

is illegal, since x is redefined in the inner scope. By ruling out such redefinition, CLU avoids the "hole in the scope" problem that can lead to program errors. (The scope of the variable x in the **while** statement extends to the **end** of the **while** statement. However, the inner declaration of x makes the outer x inaccessible within the **if** statement. This region of inaccessibility is referred to as a *hole in the scope*.)

2.5 Procedures

A procedure can take any number of arguments and return any number of results. For example, *gcd* (figure 2.1) takes two integer arguments and returns a single integer result, while *remove_dupls* (figure 2.2) takes a single argument of type array of integers and returns no result. Note that CLU treats functions simply as procedures that return results. CLU procedures can return more than one result; for example,

   div = **proc** (num, denom: int) **returns** (int, int)

returns two integer results.

A procedure returning a single result can be called in an expression. For example,

   x: int := gcd(10, 3) + 7

is a legal call of *gcd*, since a single result is returned, but *remove_dupls* or *div* cannot be called in such a context. Similarly,

   a: int := array[int]$new( )

is a legal call of the *new* operation; note that the parentheses must be present in the call even though no arguments are passed. A procedure returning multiple results can be called as the right-hand side of a multiple assignment. For example,

x, y: int := div(3, 4)

is a legal call of *div*; the two results are assigned to the variables x and y. A procedure returning no results can be called only as a statement, as in

remove_dupls(a)

Procedures that return results can also be called as statements; in this case the returned results are simply discarded. For example, *remh* shrinks the array by removing and returning the highest element. If the caller is interested only in shrinking the array, but not in the removed element, then *remh* can be called in a statement and the result will be discarded automatically.

When a procedure call is evaluated, the following things happen:

1. The argument expressions are evaluated from left to right to obtain the argument objects.

2. New variables are created for the procedure's formals. For example, in a call of *div*, two new variables are created for the formals *num* and *denom*.

3. The argument objects are assigned to the formals.

4. The procedure body is evaluated. When it terminates, its result objects, if any, are retained, but its formals, and also any local variables that were created for it, are discarded.

Procedure call evaluation is illustrated in figure 2.5, where we trace the call *remove_dupls(b)*. (*Remove_dupls* was defined in figure 2.2.) Figure 2.5a shows the situation just before the call; at this point the environment contains variable b, but not the formal of *remove_dupls*. Figure 2.5b shows the situation just after the call. Now there is a subenvironment containing the new variable created for the formal of *remove_dupls*; this variable refers to the object obtained by evaluating the expression b. Note that the new variable is distinct from the variables of the caller, but that objects are shared between variables of the caller and the called procedure. Finally, in figure 2.5c, we show the situation after *remove_dupls* returns. Note that the new subenvironment and its variable have disappeared.

We call the CLU method of parameter passing *call by sharing* to emphasize the sharing of the argument objects between the caller and the called procedure. If those shared objects are mutable and the called procedure changes them, the changes are visible to the caller, as shown in figure 2.5. However, the variables of a module are strictly local to that module and cannot be accessed by any other module. For example, in figure 2.5, the

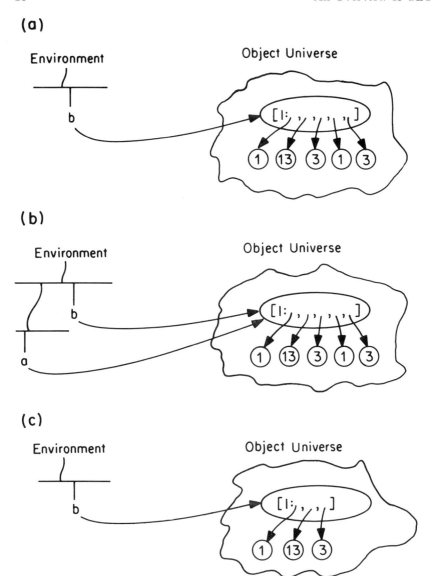

Figure 2.5 Procedure call and return: (a) before the call of *remove_dupls*; (b) immediately after starting the call *remove_dupls(b)*; (c) after *remove_dupls* returns.

variable *b* is local to the caller of *remove_dupls*, and *remove_dupls*, or any procedure it calls, cannot access *b*. Similarly, *a* is local to this call of *remove_dupls* and is not accessible either to the caller or to any procedure called by *remove_dupls*. If we decide to assign some other array to *a*, the assignment will have no effect on the caller's variable, *b*. For example, the *gcd* procedure in figure 2.1 does assignments to its formals *n* and *d*. These assignments can be made safely because they have no effect on *gcd*'s caller.

Call by sharing is the same as the parameter-passing method used in Lisp. It is similar to call by value when the arguments are pointers. However, as noted before, there are no explicit pointers in CLU.

2.6 Expressions

An expression evaluates to an object in the universe of objects. The simplest forms of expressions are literals, variables, and constants. These forms name their result object directly. For example, 3 evaluates to an integer object whose value is 3, while the variable *x* evaluates to the object it refers to. More complex expressions are built up out of procedure calls. The result of such an expression is the value returned by the outermost call.

Like most languages, CLU has prefix and infix operators for the common arithmetic and comparison operations and uses the familiar syntax for array indexing and record component selection (for example, $a[i]$ and $r.s$). However, in CLU these notations are considered to be simply abbreviations for procedure calls. For example, an expression

$3 + 2$

is an abbreviation for the procedure call

int$add(3, 2)

Obviously the form 3+2 is more convenient than the form int$$add$(3, 2). We refer to such forms as *syntactic sugar*, since they are are not a fundamental part of the semantics of the language but simply make it "sweeter" to use.

Almost all infix and prefix expression forms in CLU are abbreviations for procedural forms. Each operator symbol has a related operation name; for example, + has the related name *add* and = has the related name *equal*. When the compiler encounters one of these symbols, it "desugars" the containing expression by expanding it into the underlying procedural form. The general rule for a binary operator is that

arg1 op arg2

is translated into

T$opname(arg1, arg2)

where T is the type of *arg1* and *opname* is the related name of operation
op. Once it has obtained the expanded form, the compiler simply uses it in
place of the short form. The use of the short form is legal if the expanded
form is legal.

Note that the expansion rule is defined in a way that works for any type;
no rule was given that restricts the use of short forms to the built-in types.
Therefore, short forms can be used for user-defined types as well as for
built-in types, although the operator symbols that can be used in this way
are predefined. A full list of expression forms and their expansions is given
in appendix A.

As was mentioned earlier, short forms can be used for fetching and storing
arrays. The translation of E1[E2] depends on the context of its use: If it
appears in an expression, it is expanded into a call to the *fetch* operation.
If it appears on the left-hand side of the := in an assignment statement, it
is expanded into a call to the *store* operation, as in

a[i] := 3   % short for array[int]$store(a, i, 3)

As was the case for expressions, these abbreviations can be used for user-
defined types as well as built-in types.

CLU provides two special expression forms, the conditional predicates
cand and **cor**, which are neither abbreviations nor syntactically like pro-
cedure calls. **Cand** and **cor** are similar to the boolean operations & and |,
but their second argument is evaluated only if necessary. They are useful
for avoiding runtime errors. For example, in

i >= ai$low(a) **cand** i <= ai$high(a) **cand** a[i] = x

a[i] = x is evaluated only if *i* is within the bounds of the array.

2.7 Data Types

We have already described three built-in CLU types—ints, bools, and
arrays. This section describes briefly most remaining built-in types (see
appendix A for a complete description).

CLU provides an approximation to real numbers in the *real* data type.
Reals have literals, for example,

3.14   3.14E0   314e − 2   .0314e + 2

and the usual arithmetic (+, −, *, /) and relational (>, <, =, >=, <=)

operations. These operations take reals as arguments and provide real or boolean results. There are no operations that operate on both integers and reals, but operation *r2i* rounds a real to an integer, operation *trunc* truncates a real to an integer, and operation *i2r* returns a real that is a close approximation to its integer argument. Like integers, reals are immutable.

The type *char* provides characters, which are immutable, form an ordered set, and have the usual relational operations. Character literals are surrounded with single quotes:

‘a’   ‘7’   ‘Z’

The type *string* provides immutable sequences of characters. String literals are surrounded by double quotes:

“ ”   “a string literal”

Strings have the usual relational operations and can be indexed sequentially from 1 using the *fetch* operation. Like the array *fetch* operation, a call of the string *fetch* operation can be abbreviated, as in

s[i]   % short for string$fetch(s, i)

Strings can be concatenated with the *concat* operation, whose operator symbol is ‖:

s1 ‖ s2   % short for string$concat(s1, s2)

Since strings are immutable, *s1* ‖ *s2* produces a new string containing the elements of *s1* followed by the elements of *s2*; it does not modify *s1* or *s2*.

Procedures (and iterators) are CLU objects; they can be passed as arguments and returned as results, assigned to variables, and stored as components of composite objects. The type of a procedure (or iterator) is derived in a straightforward way from its header. For example, the type of

gcd = **proc** (n, d: int) **returns** (int)

is

proctype (int, int) **returns** (int)

while the type of the array of integers *addh* operation is

proctype (array[int], int)

For example, a procedure that computed

$$\sum_{i\,=\,low}^{high} f(a[i])$$

```
ai = array[int]
summation = proc (a: ai, f: proctype (int) returns (int)) returns (int)
    sum: int := 0
    i: int := ai$low(a)
    while i <= ai$high(a) do
        sum := sum + f(a[i])
        i := i + 1
        end
    return (sum)
    end summation
```

Figure 2.6 The summation procedure.

Table 2.1 Composite types

	Mutable	Immutable
Homogeneous	array	sequence
Heterogeneous	record	struct
Tagged union	variant	oneof

for an arbitrary function f would be generally useful. The *summation* procedure shown in figure 2.6 provides this service. The call

   summation (b, square)

computes the sum of the squares of the elements in the array b, assuming that

   square = **proc** (x: int) **returns** (int)

returns the square of its input.

   CLU provides a number of *composite* types that are used to construct collections of objects of other types. Three kinds of collections are provided: homogeneous (for example, arrays), heterogeneous, and tagged unions. In each case, both a mutable and an immutable form are provided (see table 2.1). So far we have described only arrays; now we describe the other composite types.

   Sequences are immutable arrays. Like arrays, they are one-dimensional and can be indexed by integers. Unlike arrays, they cannot be modified once created, and they always have a low bound of 1. The type of a sequence depends on the type of its elements; for example, objects of type

   sequence[string]

contain strings as elements.

Sequences have many of the same operations as arrays. They can be created by calling the *new* operation. The special constructor form can also be used, but in this case the lower bound is not specified since it is always 1. For example,

sequence[int]$[2, 6]

constructs a 2-element sequence with 2 at index 1 and 6 at index 2. Sequences have a *fetch* operation, which can be abbreviated as usual to $s[i]$. They also have *addh*, *addl*, *remh*, and *reml* operations, but these operations now return new sequences instead of modifying their inputs. For example, the call

sequence[int]$addl(s, 7)

produces a new sequence with 7 at index 1 and the elements of s, in order, as the remaining elements. Sequences do not have a *store* operation but do have a *replace* operation,

replace = **proc** (s: sequence[T], i: int, x: T) **returns** (sequence[T])

which returns a new sequence with the same elements as the original sequence except that the ith element is x. Finally, sequences have a *concat* operation, which is like that of strings and is used with the operator symbol ‖.

Records are used for building heterogeneous collections of objects, in which the contained objects may be of different types. A record consists of one or more named components; all the component names must be distinct. The names and types of the components (but not their order) determine the record type. For example,

RT = record[size: int, status: bool]

is the type of a two-component record; one component is of type int and has the name *size*, while the other is of type bool and has the name *status*.

The only way to create new record objects is to use a special constructor operation. The form of this constructor is

T${n_1: expr_1, \ldots, n_k: expr_k}$

where T is a record type, $n_1, n_2, \ldots, n_k$ are the component names of T, and each $expr_i$ is an expression of the type of that component. Between the braces each component name must be listed with an associated expression of that component's type. For example,

RT${size: 3, status: **true**}

creates a new record of type *RT*; the *size* component of this record is the integer object 3, and the *status* component is the boolean object **true**.

Once a record object has been created, we can decompose it to obtain its components or modify its components. Decomposition is accomplished by means of the *get_n* operations; there is one such operation for every component name *n* of the record type. Each *get_n* operation returns the component named *n*. For example, *RT* provides two *get_n* operations:

get_size = **proc** (r: RT) **returns** (int)
get_status = **proc** (r: RT) **returns** (bool)

If *get_size* and *get_status* are applied to the newly created record object above, they will return 3 and **true**, respectively.

Records can be modified by means of the *set_n* operations. There is a *set_n* operation for each component name *n* of the record type; this operation modifies its argument object by changing the component named *n*. Thus *RT* provides two *set_n* operations:

set_size = **proc** (r: RT, s: int)
set_status = **proc** (r: RT, b: bool)

If *x* is the above record object, then

RT$set_size(x, 7)

changes the *size* component of *x* to be the integer object 7.

There are short forms for calling the *get_* and *set_* operations. For example, in

y: int := x.size

x.size is short for *RT$get_size(x)*, while

x.size := 18

is short for *RT$set_size(x, 18)*. As was the case for arrays, the desugaring depends on whether the abbreviated form appears to the right or the left of the assignment symbol.

Records cannot be created by calls of *set_n* operations. Instead they must be created by means of the special constructor operation discussed previously. Furthermore, in using the constructor, values must be defined for all components; the compiler will detect a violation of this constraint. Therefore records, like arrays, have defined values for all components.

Structs are immutable records. Just like a record type, a struct type lists the names and types of the components. Structs have a constructor like the record constructor, and they provide *get_n* operations, one for each field name *n* of the struct type. They do not provide any *set_* operations, but they do provide *replace_n* operations, one for each field name *n*; these operations construct a new struct with the same components as the original struct, except for field *n*, which contains the new component.

Variants are mutable tagged unions, somewhat like variant records in Pascal. They are used to form objects that are one of a set of alternatives. A variant type lists the alternatives, each of which consists of a field name, called the *tag*, and a type. Unlike a record, a variant object does not contain a component for each field. Instead it contains a component for just one of the fields, which are the alternative possibilities.

A variant object consists of a *tag part* and a *data part*. The tag part contains one of the tags of the variant type; it specifies the current alternative of the variant object. The data part contains an object of the type associated with the tag in the variant type.

For example, the following variant type could be used to represent the status of a bank account:

status = variant[overdrawn: info, ok: int]

where

info = record[amount: int, authority: string]

Each status object is either in the *overdrawn* alternative or the *ok* alternative. If the alternative is *ok*, the data part of the variant object shows the account balance as a positive integer. If the alternative is *overdrawn*, the data part shows the amount of the overdraft and the identity of the person who authorized the overdraft.

A variant type provides operations to construct new variant objects and to modify and decompose existing objects. New variant objects are created using the *make_* operations; there is a *make_n* operation for each tag *n* of the variant type. An existing variant object can be modified by using the *change_* operations; again, there is one *change_n* operation for each tag *n* of the variant type. For example, the variant type status shown above provides two *make_* operations,

make_ok = **proc** (b: int) **returns** (status)
make_overdrawn = **proc** (i: info) **returns** (status)

to construct new status objects in the two alternatives, and two *change_* operations,

change_ok = **proc** (s: status, b: int)
change_overdrawn = **proc** (s: status, i: info)

to take an existing status object *s* and change it to contain the tag *n* of the *change_n* operation as its tag part and the argument of the operation as its data part. Thus

s: status := status\$make_ok(100)

causes *s* to refer to a new status object in the *ok* alternative, containing a balance of 100 dollars. Later,

status\$change_overdrawn(s, info\${50, "Fred"})

changes this status object to be in the *overdrawn* alternative, showing an overdraft of 50 dollars that was authorized by Fred.

Note that the *change_* operations change both the tag and the data parts, thus ensuring that the data part is always of the type that is associated with the tag. There is no operation on variants that changes the data part without changing the tag as well. Similarly, there is no variant operation that allows access to the data part without checking the tag part, since in the absence of such a check, a type error could occur. Instead variants provide *value_n* operations, one for each tag *n*, which check the tag of the argument object and return the data part of the object only if the tag is *n*. Thus status provides two *value_* operations:

value_ok = **proc** (s: status) **returns** (int)
value_overdrawn = **proc** (s: status) **returns** (info)

It is also possible to check the tag using the *is_n* operations; for example, status provides two *is_* operations:

is_ok = **proc** (s: status) **returns** (bool)
is_overdrawn = **proc** (s: status) **returns** (bool)

Although variant objects can be decomposed by means of the *is_* and *value_* operations, it is more convenient to use the special **tagcase** statement, which provides an arm for each alternative of the variant type, as in

```
tagcase s
   tag ok: body
   tag overdrawn (i: info): body
   end
```

When the **tagcase** statement is executed, the arm for the current alternative is selected, the data part of the variant object is assigned to the arm variable (if one is declared), and the body is executed. Thus if *s* is in the *ok* alternative, the first arm is selected; within the body of this arm, there is no access to the data part of *s*, since no arm variable is declared. Similarly, the second arm is selected if *s* is in the *overdrawn* alternative; in this case, arm variable *i* can be used to access information about the overdraft.

Just as sequences are immutable arrays and structs are immutable records, *oneofs* are immutable variants. Oneofs provide *make_* and *value_* operations just like variants, but they do not provide any *change_* operations. Like variants, oneofs are usually decomposed using a **tagcase** statement.

Type *null* is often used as a component of a oneof or variant. It is immutable and has a single object, **nil**, and a single operation, *equal*. For example,

color = oneof[red, green, yellow: null]

might be used to represent the current state of a traffic light. (It is similar to an enumeration type in Pascal.) Since only the tag and not the data part is of interest, the type for each data part is null.

2.8 Input/Output

CLU provides for input/output to both devices and files by means of a special data type called *stream*. Some streams are read-only, while others are write-only; to read and write to the same device, two streams, one read-only and the other write-only, must be connected to the device.

There are three operations for creating streams connected to "default" devices. The operation

primary_input = **proc** ( ) **returns** (stream)

returns the "primary" input stream, suitable for reading. This is usually a stream to the user's terminal, but may be set by the operating system.

The operation

primary_output = **proc** ( ) **returns** (stream)

returns the "primary" output stream, suitable for writing. Again, this is usually a stream to the user's terminal. Finally, the operation

error_output = **proc** ( ) **returns** (stream)

returns the "primary" output stream for error messages, suitable for writing. This, too, is usually a stream to the user's terminal.

To use streams to read and write files, we need some way to name files in the file system. CLU fills this need by means of type *file_name*. The exact format of file_names is system-dependent. Details about file_names and their operations can be found in appendix A.

A stream to a file is created by calling the *open* operation,

open = **proc** (fn: file_name, mode: string) **returns** (stream)

Fn names the file to which the new stream should be connected, while *mode* specifies the type of access desired: read, write, or append. In all three cases, a new stream is created; for write and append modes, a new file will also be created if one named *fn* does not already exist. The stream created in read mode is read-only; in write or append mode it is write-only. For read and write modes, the stream is positioned at the first character of the file. In append mode, the stream is positioned at the end of the file; in this mode, new information can be added to the file, but its current contents cannot be overwritten.

The mode of a stream can be determined by calling either of the following:

can_read = **proc** (s: stream) **returns** (bool)
can_write = **proc** (s: stream) **returns** (bool)

Can_read returns **true** if *s* can be read and **false** otherwise, while *can_write* returns **true** if *s* can be written and **false** otherwise.

The basic way to read or write a stream is character by character. Two operations read characters: The operation

getc = **proc** (s: stream) **returns** (char)

removes the next character from the stream and returns it; the operation

peekc = **proc** (s: stream) **returns** (char)

```
center = proc (s: stream, t: string)
    cnt: int := (80 − string$size(t))/2      % integer division returns the integer
                                             % quotient
    stream$putspace(s, cnt)      % output cnt spaces
    stream$putl (s, t)       % output t and an end-of-line character
    end center

get_int = proc (s: stream)  returns (int)
    w: string := ""   % the empty string
    while ∼stream$empty(s)  do      % get the digits
        c: char := stream$peekc(s)      % look at next char but don't remove it
        if c < '0' cor c > '9' then break end     % have found all digits
        w := w ‖ string$c2s(c)      % convert c to a string and concatenate
        stream$getc(s)      % remove the char from the stream
        end
    if w = "" then return (0) end
    return (int$parse(w))      % convert the string to an int
    end get_int
```

Figure 2.7 Using streams.

returns the next character but does not remove it. Writable files have the operation

   putc = **proc** (s: stream, c: char)

which appends c to s. In addition, there are operations to read and write more than a character at a time—for example, to read a line or write a string. Details can be found in appendix A.

   For a stream connected to a file, the operation

   close = **proc** (s: stream)

can be used to terminate the connection between the stream and the file.

   Figure 2.7 shows some procedures that use streams. *Center* centers a string in the middle of an 80-character line on a given stream; it assumes that the string is less than 80 characters, and that its stream argument is writable. *Get_int* assumes that its stream argument s is readable. If s is positioned at the start of a sequence of digits, *get_int* removes the digits from s and returns the integer corresponding to those digits, leaving s positioned immediately after the last digit scanned; otherwise it leaves s unchanged and returns zero.

   Get_int illustrates the use of the int$*parse* operation. To make input and output more convenient, ints and reals provide a *parse* operation to convert a string to an int or real, respectively, and an operation *unparse* to do the reverse transformation.

2.9 Object-Oriented Programs

All languages provide their users with a model of computation. The CLU model differs from that of "Algol-like" languages such as Pascal and PL/I in that CLU is *object oriented* rather than variable oriented.

Objects are the important entities in CLU programs. All data are contained in objects. Expressions evaluate to objects. Procedures receive objects as arguments and return objects as results. (In fact call by object might be a better name than call by sharing for our method of passing arguments.) Procedures share objects and communicate not only by the particular objects shared but by modifying the shared objects.

Variables are simply identifiers used to refer to objects in programs. In contrast to objects, variables are never shared between procedures; they cannot be passed as arguments or results, nor can they be shared directly, since procedures can refer only to local variables. Variables are needed because programs need to refer to objects. But it is the objects themselves that really matter.

This emphasis on objects has two significant ramifications. First, it leads to a simple and uniform semantics. It is not necessary to have more than one parameter-passing method in CLU because there is no need to distinguish between passing a value (call by value in conventional languages) and passing a container for a value (call by reference).

Even more important is the effect on programming style. In an object-oriented language, attention is focused on objects, what they mean and how they behave. This focus is well matched to the programming methodology to be developed in this book. In this methodology, data abstractions, consisting of data objects and the operations used to manipulate them, will be the most important means of organizing programs.

Exercises

2.1 Computing the greatest common divisor by repeated subtraction (figure 2.1) is not very efficient. Reimplement *gcd* to use division instead.

2.2 Implement a procedure with the header

max = **proc** (a: array[int]) **returns** (int)

Max should return the maximum value in *a*. You may assume that *a* is nonempty.

2.3 Illustrate calling the *gcd* procedure of figure 2.1 with a diagram like that in figure 2.5.

2.4 The array passed as an argument to *remove_dupls* in figure 2.5 contains duplicate elements 1 and 3. Why are these elements not shown in figure 2.5c?

2.5 Implement a procedure that determines whether or not a string is a palindrome. (A palindrome reads the same backward and forward; "deed" is an example.)

2.6 Implement a procedure with the header

combine = **proc** (a: array[int], comb: **proctype** (int, int) **returns** (int))
    returns (int)

It should combine all elements of *a* using procedure argument *comb*. For example,

combine(a, int$add)

adds all elements of *a*. You may assume that *a* has at least 2 elements.

2.7 Implement a procedure to scan and return the next word from a stream, where a word is any combination of characters not including spaces and line separators. You may assume that the stream is positioned at the first character of the word.

2.8 Implement a procedure to skip blanks in a stream. It should return with the stream positioned at the first nonblank character, or at the end-of-file if there are no nonblank characters.

3

Procedural Abstraction

In this chapter we discuss the most familiar kind of abstraction used in programming, the *procedural abstraction*, or *procedure* for short. Anyone who has introduced a subroutine to provide a function that can be used in other programs has used procedural abstraction. Procedures combine the methods of abstraction by parameterization and specification in a way that allows us to abstract a single operation or event, such as computing the gcd of two integers or sorting an array.

A procedure provides a transformation from input arguments to output arguments. More precisely, it is a mapping from a set of input arguments to a set of output results, with possible modifications of the inputs. The set of inputs or outputs or both might be empty. For example, *gcd* has two inputs and one output, and it does not modify its inputs. By contrast, *remove_dupls* has one input and no output, and it does modify its input.

We begin with the benefits of abstraction and, in particular, of abstraction by specification. Next we discuss specifications and why they are needed. Then we discuss how to specify and implement procedures, and we conclude with some general remarks about their design.

3.1 The Benefits of Abstraction

An abstraction is a many-to-one map. It "abstracts" from "irrelevant" details, describing only those details that are relevant to the problem at hand. Its *realizations* or *instances* must all agree in the relevant details, but can differ in the irrelevant ones. Of course, distinguishing what is relevant from what is irrelevant is relative to the task at hand. A major portion of this book will be concerned with how this is done.

In abstraction by parameterization, we abstract from the identity of the data being used. The abstraction is defined in terms of formal parameters; the actual data are bound to these formals when the abstraction is used. Thus the identity of the actual data is irrelevant, but the presence, number, and types of the actuals are relevant. Parameterization generalizes modules, making them useful in more situations. A virtue of such generalizations is that they decrease the amount of code that needs to be written and, thus, modified and maintained.

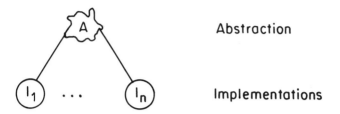

Figure 3.1 The general structure of abstraction by specification.

In abstraction by specification, we focus on the behavior that the user can depend on and abstract from the details of implementing that behavior. Thus the behavior—"what" is done—is relevant, while the method of realizing that behavior—"how" it is done—is irrelevant. For example, in a *sort* procedure, the fact that an array is sorted is relevant, while the sorting algorithm—for example, merge sort—is irrelevant.

A key advantage of abstraction by specification is that the implementation is irrelevant, and we can therefore change to another implementation without affecting any program that uses the abstraction (figure 3.1). For example, we could reimplement *sort* using heap sort, and programs using *sort* would continue to run correctly with this new implementation (although some improvement in performance might be noticed). The implementations could even be written in different programming languages, provided that the data types of the arguments are treated the same in these languages. For example, in the CLU compiler and runtime system, most of which is implemented in CLU, some abstractions are implemented in assembly language to improve performance.

Abstraction by specification provides a method for achieving a program structure with two advantageous properties. The first property is *locality*, which means that the implementation of one abstraction can be read or written without our needing to examine the implementation of any other abstraction. To write a program that uses an abstraction, a programmer need understand only its behavior, not the details of its implementation.

Locality is beneficial both when a program is being written and later when someone wants to understand it or reason about its behavior. Because of locality, different abstractions that make up a program can be implemented by people working independently. One person can implement an abstraction that uses another abstraction being implemented by someone else. As long as both people agree on what the used abstraction is, they can work independently and still produce programs that work together properly. Also, understanding a program can be accomplished one abstraction at a time. To understand the code that implements one abstraction, it is necessary to understand what the used abstractions are, but not the code that implements them. In a large program, the amount of information that is not needed can be enormous; we can ignore not only the code of the used abstractions but also the code of any abstractions they use, and so on.

The second property is *modifiability*. Abstraction by specification helps to bound the effects of program modification and maintenance. If the implementation of an abstraction changes but its specification does not, the rest of the program will not be affected by the change. Of course, if the number of abstractions that must be reimplemented is large, making a modification will still be a lot of work. As will be discussed later, the workload can be reduced by identifying potential modifications while designing the program and then trying to limit their effects to a small number of abstractions. For example, if the effects of machine dependencies can be limited to just a few modules, the result will be software that can be transported readily to another machine.

Modifiability leads to a sensible method of tuning performance. Programmers are notoriously bad at predicting where time will actually be spent in a complex system, probably because it is difficult to anticipate where bottlenecks will arise. Since it is unwise to invest effort in inventing structures that avoid nonexistent bottlenecks, a better method is to start with a simple set of abstractions, run the system to discover where the bottlenecks are, and then reimplement the abstractions that are bottlenecks.

3.2 Specifications

It is essential that abstractions be given precise definitions; otherwise, the advantages discussed in section 3.1 cannot be achieved. For example, we can replace one implementation of an abstraction by another only if everything that was depended on in the old implementation is supported by the new one. The entity depended on and supported is the abstraction. Therefore we must know what the abstraction is.

We shall define abstractions by means of *specifications*, which are written in a *specification language* that can be either formal or informal. The advantage of formal specifications is that they have a precise meaning. Informal specifications are easier to read and write than formal ones, but giving them a precise meaning is difficult because the informal specification language is not precise. Despite this, informal specifications can be very informative and can be written in such a way that readers will have little

```
pname = proc (...) returns (...)
      requires    % this clause states any constraints on use
      modifies    % this clause identifies all modified inputs
      effects     % this clause defines the behavior
```

Figure 3.2 Specification template for procedural abstractions.

trouble understanding their intended meaning. We shall use both formal and informal specifications in this book, but shall concentrate on informal specifications. Formal specifications will be described in chapter 10.

A specification is distinct from any implementation of the abstraction it defines. The implementations are all similar because they implement the same abstraction; they differ because they implement it in different ways. The specification defines their commonality.

3.3 Specifications of Procedural Abstractions

The specification of a procedure consists of a header and a description of effects. The *header* gives the name of the procedure and the number, order, and types of its inputs and outputs. In addition, names must be given for the inputs and may be given for the outputs. For example, the header for *remove_dupls* is

remove_dupls = **proc** (a: array[int])

while the header of *sqrt* is

sqrt = **proc** (x: real) **returns** (rt: real)

The information in the header is just syntactic; it describes the "form" of the procedure. It is similar to a description of the "form" of a mathematical function, as in

f: integer → integer

In neither case is the meaning—what the procedure or the function does—described. The meaning is captured in the semantic part of the specification, in which the behavior of the procedure is described in English, possibly extended with convenient mathematical notation. This description makes use of the names of the inputs and outputs.

Figure 3.2 shows a template of a procedure specification. The semantic part of a specification consists of three parts: the requires, modifies, and effects clauses. These clauses should appear in the order shown, although the requires and modifies clauses are optional.

concat = **proc** (a, b: string) **returns** (ab: string)
  effects On return, *ab* is a new string containing the characters of *a* (in
   the order they occur in *a*) followed by the characters of *b* (in the
   order they occur in *b*).

remove_dupls = **proc** (a: array [int])
  modifies *a*
  effects Removes all duplicate elements from *a*. The low bound of *a*
   remains the same, but the order of the remaining elements may
   change. E.g., if $a = [1{:}\,3, 13, 3, 6]$ before the call, then on return *a* has
   low bound 1 and contains three elements, $3, 13$ and 6, in some order.

search = **proc** (a: array [int], x: int) **returns** (i: int)
  requires *a* is sorted in ascending order.
  effects If *x* is in *a*, returns *i* such that $a[i] = x$; otherwise *i* is one greater
   than the high bound of *a*.

Figure 3.3 Procedural abstraction specifications.

  The *requires clause* states the constraints under which the abstraction
is defined. The requires clause is needed if the procedure is partial, that
is, if its behavior is not constrained for some inputs. If the procedure is
total, that is, if its behavior is constrained for all inputs, the requires clause
can be omitted. In this case, the only restrictions on a legal call are those
implied by the header, that is, the number and types of the arguments.
The *modifies clause* lists the names of any inputs that are modified by the
procedure; it can be omitted when no inputs are modified. The absence of
the modifies clause means that none of the inputs are modified. Finally,
the *effects clause* describes the behavior of the procedure for all inputs not
ruled out by the requires clause. It must define what outputs are produced
and also what modifications are made to the inputs listed in the modifies
clause. The effects clause is written under the assumption that the requires
clause is satisfied, and it says nothing about the procedure's behavior when
the requires clause is not satisfied.

  Several specifications are given in figure 3.3. *Concat* does not modify
its inputs, but *remove_dupls* modifies its input array, as indicated in the
modifies clause. Note the use of an example in the *remove_dupls* specification. Examples can clarify a specification and should be used whenever
convenient.

  Concat and *remove_dupls* are total, since neither specification contains
a requires clause. *Search*, however, is partial; it only does its job if the
input array is sorted. Note that the effects clause does not state what
search does if the input array is not sorted. In this case the implementer
can do whatever is convenient; for example, the implementation of *search*

could return an index out of bounds, or an index in bounds that does not contain x, or even run forever.

Specifications are written in a specification language, not a programming language. Thus, the specifications shown previously are not written in CLU. Furthermore, specifications are usually quite different from programs in that they focus on describing what the abstraction is rather than how it is implemented. We shall discuss specifications further in chapter 8.

3.4 Implementing Procedures

The implementation of a procedure should produce the behavior defined by its specification. In particular, it should modify only those inputs that appear in the modifies clause, and for all inputs that satisfy the requires clause, it should produce the outputs in accordance with the effects clause.

Every programming language includes some mechanism for implementing procedural abstractions. In CLU, these abstractions are implemented by means of CLU procedures, or *procs* for short.

Figure 3.4 shows two CLU procs that implement *search*; one uses linear search, while the other uses binary search. These two implementations differ in many details. For example, for all but very small arrays, binary search is faster than linear search. Moreover, if x appears in a more than once, the two procs may return different indexes. Finally, if x is contained in a but a is not sorted, the proc using binary search may return $high(a)+1$ when the other proc finds x or vice versa (consider $a = [1: 1, 7, 6, 4, 9]$ and $x = 7$, for example). Nevertheless, both procs are correct realizations of the *search* abstraction since both provide behavior that is consistent with the specification.

In the example procs we have followed some conventions to enhance program readability. First, we have used the same names for the formals as were used in the specification. This convention makes it easier to relate an implementation of an abstraction to the specification. Following the header, we have included a comment explaining the algorithm in use. Finally, we have adopted formatting conventions to make the code easy to read.

As a second example, consider sorting an array. One possible method is merge sort, which reduces the problem to that of merging two arrays that have already been sorted. We begin by dividing the array in half, then sort each half and merge the results:

```
% sort the first half of the array
% sort the second half of the array
% merge the two sorted halves
```

ai = array[int] % an abbreviation

search = **proc** (a: ai, x: int) **returns** (int)
     % implemented using linear search
     i: int := ai$low(a)
     while i <= ai$high(a) **do**
          if a[i] = x **then return** (i) **end**
          if a[i] < x **then** i := i + 1 **else return** (ai$high(a) + 1) **end**
          end % while
          return (i)
     end search

search = **proc** (a: ai, x: int) **returns** (int)
     % implemented using binary search
     low: int := ai$low(a)
     high: int := ai$high(a)
     while (low <= high) **do**
          mid: int := (low + high)/2
          val: int := a[mid]
          if x < val **then** high := mid − 1
               elseif x = val **then return** (mid)
               else low := mid + 1
               end
          end
     return (ai$high(a) + 1)
     end search

Figure 3.4 Two implementations of *search*.

To carry out these steps we use two subsidiary procedures: *merge*, to do
the merging, and *merge_sort*, to carry out the sorting of a subpart of the
array. *Merge_sort* itself will carry out the same three steps on the subpart
of the array, giving a recursive algorithm.

     Figure 3.5 shows the specifications of *sort* and the two subsidiary abstrac-
tions. Note that *sort* does not modify its input array and that both *merge*
and *merge_sort* are partial. Figure 3.6 shows CLU procs that implement
the abstractions.

3.5 More General Procedures

The *sort* procedure discussed previously will work for any array of integers.
If it applied to other kinds of arrays, such as arrays of characters, strings,
or reals, it would be more generally useful. We can achieve this extra generality, which comes from carrying abstraction by parameterization further
than we have done so far, by using data types as parameters.

     Types are clearly useful as parameters. Evidence for this is the fact

sort = **proc** (a: array[int]) **returns** (b: array[int])
    effects Returns a new array with the same bounds as a and containing
    the elements of a arranged in ascending order.

merge_sort = **proc** (a: array[int], low, high: int) **returns** (array[int])
    requires $low(a) \leq low \leq high \leq high(a)$.
    effects Returns a new array with the same low bound as a and contain-
    ing elements $a[low], \ldots, a[high]$ arranged in ascending order.

merge = **proc** (a, b: array[int]) **returns** (array[int])
    requires a and b are sorted in ascending order.
    effects Returns a new array with the same low bound as a and contain-
    ing the elements of a and b arranged in ascending order.

Figure 3.5 Specification for merge sort.

that many built-in data types, such as arrays, records, and procedures, are
parameterized by types. It is true in general that whatever is useful at the
level of a programming language is probably also useful for its users, and
there is no doubt that type parameters fall into this category.

When types are used as parameters, some parameter values may not be
meaningful. For example, arrays can be sorted only if the elements belong
to a type that is totally ordered. Constraints on type parameters take the
form of requiring the parameter type to have certain operations that must
behave in certain ways. The specification of an abstraction must state such
constraints in the requires clause.

Some specifications for parameterized procedures are shown in figure 3.7.
Note that these specifications differ from those of nonparameterized procedures only in the header, which now has an extra part that defines the
parameters. The requires clause states constraints on parameters in addition to any other constraints; if there are no constraints, the requires
clause can be omitted as usual. Thus the *sort* abstraction requires that
parameter *t* have an operation *lt* that defines a total order on *t*. Similarly,
search requires that its parameter provide both an *lt* operation and an *equal*
operation. (*Lt* is the name used for operations for which the short form $<$
is permitted; *equal* is the name associated with the short form $=$.)

A parameterized abstraction is really a class of related abstractions defined by a single specification. For example, the parameterized *sort* abstraction is not a single procedure but a class of many procedures, one
taking as an argument an array of integers, another an array of strings,
and so on. We shall name such procedures by combining the name of the
class with the name of the type, as in *sort*[int] and *sort*[string]. Similarly we
have *search*[int] and *search*[real]. We have already been using such names
for the built-in abstraction classes of CLU—for example, array[int].

ai = array[int]

sort = **proc** (a: ai) **returns** (ai)
 % sort using merge sort
 if ai$empty(a) **then**
  % create empty array with low bound low(a)
  return (ai$create(ai$low(a)))
  end
 return (merge_sort(a, ai$low(a), ai$high(a)))
 end sort

merge_sort = **proc** (a: ai, low, high: int) **returns** (ai)
 if low < high
  then   % sort the two halves of a and merge the result
   mid: int := (low + high)/2
   return (merge(merge_sort(a, low, mid),
    merge_sort(a, mid + 1, high)))
  else   % a is already sorted, but we must return new array
   b: ai := ai$create(low)
   ai$addh(b, a[low])
   return (b)
  end
 end merge_sort

merge = **proc** (a, b: ai) **returns** (ai)
 a_low: int := ai$low(a)
 b_low: int := ai$low(b)
 a_high: int := ai$high(a)
 b_high: int := ai$high(b)
 c: ai := ai$create(a_low)   % start off with low bound a_low
 % merge a and b
 while a_low <= a_high **cor** b_low <= b_high **do**
  if a_low > a_high **cor** (b_low <= b_high **cand** b[b_low] < a[a_low])
   then % move element from b, either because all elements of a have
    % been moved, or because the next element of b is less than the
    % next element of a
    ai$addh(c, b[b_low])
    b_low := b_low + 1
   else % move element from a
    ai$addh(c, a[a_low])
    a_low := a_low + 1
   end
  end   % while
 return (c)
 end merge

Figure 3.6 An implementation of merge sort.

sort = **proc** [t: **type**] (a: array[t]) **returns** (array[t])
    requires t has an operation
        lt: **proctype** (t, t) **returns** (bool)
        that totally orders t.
    modifies a
    effects Returns a new array with the same bounds as a and
        containing the elements of a arranged in ascending order, where
        the order is determined by $t\$lt$.

search = **proc** [t: **type**] (a: array[t], x: t) **returns** (int)
    requires t has operations
        equal, lt: **proctype** (t, t) **returns** (bool)
        such that t is totally ordered by lt, and a is sorted in ascending
        order based on lt.
    effects If x is in a, returns i such that $a[i] = x$; otherwise returns
        $high(a) + 1$.

Figure 3.7 Specifications of parameterized procedures.

A class contains a member for each type that satisfies the constraints in the requires clause. For example, *sort*[int] is in the *sort* class, while *sort*[bool] is not (since CLU booleans have no *lt* operation). The specification of a procedure is derived in the obvious way from the class specification, by replacing the name of the parameter with its value and removing constraints on the type parameter(s) from the requires clause (since these constraints have been satisfied). Thus the specification of *search*[int] is the same as the specification of *search* in figure 3.3.

It is desirable to implement a parameterized abstraction with a single program, for reasons of convenience in writing, debugging, testing, and maintenance. CLU provides this ability. For example, a parameterized proc is used to implement a procedure class. Such a proc has a header similar to the header of the specification, as in

    sort = **proc** [t: **type**] (a: array[t]) **returns** (array[t])

Typically, the operations listed in the requires clause of the specification will be used in the implementation. For example, to implement *sort* we use the *lt* operation. To type-check the use of such operations, the compiler must know their names and the number and types of their arguments and results. We provide such information in the **where** clause, which appears right after the header, as in

    sort = **proc** [t: **type**] (a: array[t]) **returns** (array[int])
        where t **has** lt: **proctype** (t, t) **returns** (bool)

```
search = proc [t: type] (a: array[t], x: t) returns (int)
        where t has lt, equal: proctype (t, t) returns (bool)
    % implemented using linear search
    at = array[t]
    i: int := at$low(a)
    while i <= at$high(a) do
        if a[i] = x then return (i) end   % use of t$equal
        if a[i] < x   % use of t$lt
            then i := i + 1 else return (at$high(a) + 1) end
        end   % while
    return (i)
    end search
```

Figure 3.8 The parameterized *search* abstraction.

The only operations of the parameter type that can be used in a parameterized module are those listed in the **where** clause.

The **where** clause includes a portion of the information given in the requires clause of the specification. The requires clause includes both syntactic information (the names and types of required operations) and semantic information (the meanings of the operations). In the associated **where** clause we list only syntactic information, which is sufficient for type checking.

The code in a parameterized procedure is similar to that in a nonparameterized procedure. The implementation of *search* given in figure 3.8, for example, differs from the one in figure 3.4 in that wherever we assumed previously that the array elements were of type integer, we now assume they are of type *t*, the parameter type. Thus *search* compares array elements using *t$lt* and *t$equal*. These calls are abbreviated using their associated short forms.

To use a parameterized proc, we must first supply values for the parameters—for example, *search*[int]. Thus

search[int](a)

is a legal call if *a* is an array[int], but

search(a)

is not, since no value has been supplied for *t*.

When a parameter value is supplied, the compiler checks that the type has the operations listed in the **where** clause. For example,

search[array[int]]

would not be permitted, since array[int] does not have an *lt* operation. This checking guarantees that operations used in the parameterized module are provided by the actual parameter and that the numbers and types of arguments and results are as expected.

A parameterized procedure must list an operation in the **where** clause even if it only uses the operation indirectly, by calling another parameterized procedure that requires the operation. Suppose, for example, that we provide parameterized versions of the procedures in the merge sort implementation shown in figure 3.5. Only the *merge* proc actually calls *lt*. However, *sort* and *merge_sort* must both list *lt* in their **where** clauses, so that the compiler can check that the call on *merge*[*t*] made in *merge_sort* is legal.

3.6 Designing Procedural Abstractions

In this section we discuss a number of issues that arise in designing procedures. Procedures, as well as the other kinds of abstractions that we shall discuss later, should be designed to be minimal; care should be taken to constrain details of the procedure's behavior only to the extent necessary. In this way we leave more freedom to the implementer, who may be able to provide a more efficient implementation as a result. For example, if *sort* in figure 3.5 were permitted to modify its array argument, less space could be used to implement it. However, details that matter to users must be constrained, or the procedure will not be what is needed.

One kind of detail that is almost certainly left undefined is the method to be used in the implementation. Generally, users do not depend on such details. (There are exceptions, however; for example, a numerical procedure may be constrained to use a well-known numerical method so that its behavior with respect to rounding errors will be well-defined.) Some details of what the procedure does may also be left undefined, leading to a procedure that is *underdetermined.* This means that for certain inputs, instead of a single correct output, there is a set of acceptable outputs. An implementation is constrained to produce some member of that set, but any member will do.

The *search* procedure is underdetermined because we did not state exactly what index should be returned if *x* occurs in the array more than once. As mentioned earlier, the two implementations of *search* in figure 3.4 differ in their behavior in this regard. Yet each was a correct implementation.

Remove_dupls (see figure 3.3) is also underdetermined, since it does not necessarily preserve the order of elements in its input array. This lack of constraint may be a mistake, because users may care about the order; if the input array is sorted, for example, it might be desirable to preserve the order. The important point is that what matter depends on what users need. Every detail that matters to users must be specified, while all others should be left undefined.

An underdetermined abstraction can have a deterministic implementation, that is, one that, if called twice with identical inputs, behaves identically on the two calls. Both implementations of *search* in figure 3.4 are deterministic. (Nondeterministic implementations require the use of own variables, which are described in appendix A but not used in this book.)

In addition to minimality, another important property of procedures is generality, which is achieved by using parameters instead of specific variables or assumptions. For example, a procedure that works on arrays of any size is more general than one that works only on arrays of some fixed size. Similarly, a procedure that works on arrays of elements of arbitrary type is more general than one that requires the elements to be integers. Generalizing a procedure is only worthwhile, however, if doing so increases its usefulness. This is almost always true when size assumptions are eliminated, since by doing so we ensure that a minor change in the context of use (for example, doubling the size of an array) requires little, if any, program modification.

Another important property of procedures is simplicity. A procedure should have a well-defined and easily explained purpose that is independent of its context of use. A good check for simplicity is to give the procedure a name that describes its purpose. If it is difficult to think of a name, there may be a problem with the procedure.

Finally, the implementation of a procedure should really do something. Procedures are introduced during program design to shorten the caller and make its structure clearer. In this way the caller becomes easier to understand and to reason about. However, it is possible to introduce too many procedures. For example, the *merge* procedure in figure 3.6 is worth introducing because it has a well-defined purpose and because it allows us to separate the details of sorting from those of merging, thus making *merge_sort* easier to understand. Further decomposition is probably counterproductive, however. For example, the loop body in *merge* could be made into a procedure, but its purpose would be difficult to state, and neither *merge* itself nor the new procedure would do much.

Some of the procedures discussed earlier are partial, while others are total. This dichotomy leads to the question of when it is appropriate to define a partial abstraction. Partial procedures are not as safe as total ones, since they leave it to the user to satisfy the constraints in the requires clause. When the requires clause is not satisfied, the behavior of a partial procedure is completely unconstrained, and this can cause the using program to fail in mysterious ways. For example, *search* might return the wrong index when its input array is not sorted. This error may not be noticed until long after *search* returns. By then, the reason for the error may be obscure, and data objects may have been damaged.

On the other hand, partial procedures can be more efficient to implement than total ones. For example, if *search* had to work even when the input array was not sorted, then neither implementation in figure 3.4 would be correct; only a less efficient implementation that examined all elements of the array could be used.

In choosing between a partial and a total procedure, we have to make a trade-off. On the one hand is efficiency; on the other is safe behavior, with fewer potential surprises at runtime. How is such a choice to be made? An important consideration is the expected context of use. If the procedure is intended for general use (for example, if it is to be made available as part of a program library), safety considerations should be given great weight. In such a situation, it is impossible to perform a static analysis of all calls to ensure that they satisfy the constraints. Therefore it is wise not to rely on such an analysis.

Alternatively, some procedures are intended to be used only in a limited context. This was the situation with *merge* and *merge_sort*, which are supposed to be used only within the implementation of *sort* using merge sort. In a limited context, it is easy to establish that constraints are satisfied. For example, the two arrays passed to *merge* as arguments have just been sorted. Therefore we have safe behavior without sacrificing efficiency.

Finally, it is worth noting that a specification is the only record of its abstraction. Therefore, it is crucial that the specification be clear and precise. How to write good specifications is the subject of chapter 8.

3.7 Summary

This chapter has been concerned primarily with procedures: what they are, how to describe their behavior, and how to implement them. We also discussed two important benefits of abstraction and the need for specifications.

Abstraction provides the two key benefits of locality and modifiability. Both are based on the distinction between an abstraction and its implementations. Locality means that each implementation can be understood in isolation. An abstraction can be used without our having to understand how it is implemented, and it can be implemented without our having to understand how it is used. Modifiability means that one implementation can be substituted for another without disturbing the using programs.

To obtain these benefits, we must have a description of the abstraction that is distinct from any implementation. To this end we introduced the specification, which describes the behavior of an abstraction using a special specification language. This language can be formal or informal. Users can assume the behavior described by the specification, and implementers must provide this behavior. Thus the specification serves as a contract between users and implementers.

A procedure is a mapping from inputs to outputs, with possible modifications of some of the inputs. Its behavior, like that of any other kind of abstraction, is described by a specification, and we presented a form for informal specifications of procedures. A procedure is implemented in CLU by a proc; in other languages it would be implemented by a similar mechanism.

Since we are interested in design and how to invent good abstractions, we concluded the chapter with a discussion of what procedures should be like. Desirable properties include minimality, simplicity, and generality. Minimality often gives rise to underdetermined abstractions. We also discussed the pros and cons of partial and total procedures. We shall continue to discuss desirable properties in the following chapters as we introduce additional kinds of abstractions.

Further Reading

Abelson, Harold, and Gerald J. Sussman, with Julie Sussman, 1985. *Structure and Interpretation of Computer Programs*, chapter 1. Cambridge, Mass.: MIT Press and New York: McGraw-Hill.

Dijkstra, Edgser W., 1972. Notes on structured programming. In *Structured Programming* (New York: Academic Press), pp. 1–82.

Wirth, Niklaus, 1971. Program development by stepwise refinement. *Communications of the ACM* 14(4): 221–227. Reprinted in *Programming Methodology, A Collection of Articles by Members of IFIP WG2.3*, edited by David Gries (New York: Springer-Verlag), 1978.

Exercises

3.1   Implement

maxt = **proc** [t: **type**] (a: array[t]) **returns** (t)
       requires *a* is not empty and *t* has operation
            gt: **proctype** (t, t) **returns** (bool)
          that defines a total order on *t*.
       effects Returns the largest element of *a* as determined by *t$gt*.

3.2   Specify and implement a procedure *is_prime* that determines whether an integer is prime.

3.3   Generalize the *remove_dupls* procedure of figure 3.3 so that it works on an array of arbitrary type. Give a specification and an implementation.

3.4   Define a procedure

key_sort = **proc** [etype: type] (a: ar)
ar = array[record[key1, key2: int, elem: etype]]

Key_sort modifies its input array *a* by sorting *a*'s elements into increasing order according to the values of the two key fields. *Key1* is the primary key; it is used for a preliminary sort of *a*. Then *key2* is used as a secondary key to sort those elements that have the same *key1* value (that is, elements *x* and *y* such that *x.key1* = *y.key1*). Give a specification and implementation of this procedure.

3.5   The specification of *key_sort* may or may not require the sort to be *stable*. A sort is stable if all elements that are the same for all keys retain their relative positions in the input array. Does your specification in exercise 4 imply anything about stability?

3.6   You are to choose between two procedures both of which compute the minimum value in an array of integers. One procedure returns the smallest integer if its array argument is empty. The other requires a nonempty array. Which procedure should you choose and why?

3.7   Suppose that the implementation of sorting by merge sort shown in figure 3.6 were changed as follows: Procedure *merge* is retained, but *merge_sort* is eliminated, so that its work is done directly in *sort*. Is this change a good idea? What purpose does *merge_sort* have? Discuss.

4

Data Abstraction

In chapter 3 we discussed one of the major kinds of abstractions used in programming: the procedural abstraction. Procedures allow us to extend the base language with new operations. In addition to operations, however, the base level provides various types of data—for example, integers, reals, booleans, strings, and arrays. As discussed in chapter 1, we need to extend the base level with new types of data just as much as with new operations. This need is satisfied by *data abstraction*.

What new types are needed depends on the application domain of the program. For example, in implementing a compiler or interpreter, stacks and symbol tables are useful, while accounts are a natural abstraction in a banking system. Polynomials arise in a symbolic manipulation system, and matrices are useful in defining a package of numeric functions. In each case, the data abstraction consists of a set of objects—for example, stacks or polynomials—plus a set of operations. For example, matrix operations include addition, multiplication, and so on, and deposit and withdraw are operations on accounts.

The new data types should incorporate abstraction both by parameterization and by specification. Abstraction by parameterization can be achieved in the same way as for procedures—by using parameters wherever it is sensible to do so. We achieve abstraction by specification by making the operations part of the type. To understand why the operations are needed, consider what happens if we view a type as just a set of objects. Then all that is needed to implement the type is to select a storage representation for the objects; all the using programs can be implemented in terms of this representation. However, if the representation changes, or even if its interpretation changes, all programs that use the type must be changed. (An example of different interpretations of the same representation is given in section 4.5.1.)

On the other hand, suppose we include operations in the type, obtaining

data abstraction = ⟨objects, operations⟩

and we require users to call the operations instead of accessing the representation directly. Then to implement the type, we implement the operations in terms of the chosen representation, and we must reimplement the operations if we change the representation. However, we need not reimplement any using programs, because they depend only on the operations and not on the representation. Therefore we have achieved abstraction by specification.

If enough operations are provided, lack of access to the representation will not cause users any difficulty—anything they need to do to the objects can be done, and done efficiently, by calls on the operations. In general, there will be operations to create and modify objects and to obtain information about their values. Of course, users can augment the set of operations by defining procedures, but such procedures would not use the representation.

Data abstraction is the most important method in program design, as will be discussed further in chapter 13. Choosing the right data structures is crucial to achieving an efficient program. In the absence of data abstraction, data structures must be defined too early—they must be specified before the implementations of using modules can be designed. At this point, however, the uses of the data are typically not well understood. Therefore the chosen structure may lack needed information or be organized in an inefficient way.

Data abstraction allows us to defer decisions about data structures until the uses of the data are fully understood. Instead of defining the structure immediately, we introduce the abstract type with its objects and operations. Implementations of using modules can then be designed in terms of the abstract type. Decisions about how to implement the type are made later, when all its uses are understood.

Data abstraction is also valuable during program modification and maintenance. In this phase, data structures are particularly likely to change, either to improve performance or to accommodate changing requirements. Data abstraction limits the changes to just the implementation of the type; none of the using modules need be changed.

In this chapter we describe how to specify and implement data abstractions. We also discuss ways to reason about the correctness of programs that use and implement types, and some issues that arise in designing new types.

4.1 Specifications for Data Abstractions

Just as was the case for procedures, the meaning of a type should not be given by any of its implementations. Instead there should be a specification that defines its behavior. Since objects of the type are used only by calling

dname = **data type is**   % list of operations

Overview
   % an overview of the data abstraction goes here

Operations
   % a specification for each operation goes here

end dname

Figure 4.1 Template for a specification of a data abstraction.

the operations, most of the specification consists of explaining what the operations do. A specification template is given in figure 4.1. It consists of a header that names the type and its operations and two main sections, the overview section and the operations section.

The *overview section* describes the type as a whole. Sometimes it presents a model for the objects; that is, it describes the objects in terms of other objects that the reader of the specification can be expected to understand. For example, stacks might be defined in terms of mathematical sequences. The overview section should also say whether the type is mutable or immutable.

The *operations section* contains a specification for each operation. If the operation is a procedure, its specification will be a procedure specification. (The operation might also be an iteration abstraction; these are discussed in chapter 5.) In these specifications, the concepts introduced in the overview section can be used.

Figure 4.2 gives a specification of the *intset* data abstraction. Intsets are unbounded sets of integers with operations to create a new, empty intset, test whether a given integer is an element of an intset, and add or remove elements. Note that in the overview section we choose to describe intsets in terms of mathematical sets. Also, we mention that intsets are mutable and identify all the mutating operations. In the operations section we specify each operation in terms of mathematical sets. Note that the specifications of *insert* and *delete* use the notation s_{post} to indicate the value of s when the operation returns. An input argument name without the *post* qualifier always means the value when the operation is called.

In this specification, we are relying on the reader knowing what mathematical sets are; otherwise, the specification would not be understandable. In general, this reliance on informal description is a weakness of informal specifications. It is reasonable to expect the reader to understand a number of mathematical concepts, such as sets, sequences, and integers. However, not all types can be described nicely in terms of such concepts. If the concepts are inadequate, we must describe the type as best we can, even by using pictures, but of course there is always the danger that the reader

intset = **data type is** create, insert, delete, member, size, choose

Overview

Intsets are unbounded mathematical sets of integers. Intsets are mutable: *insert* and *delete* add and remove integers from the set.

Operations

create = **proc** ( ) **returns** (intset)
effects Returns a new, empty intset.

insert = **proc** (s: intset, x: int)
modifies s
effects Adds x to the elements of s; after insert returns, $s_{post} = s \cup \{x\}$, where s_{post} is the set of values in s when *insert* returns.

delete = **proc** (s: intset, x: int)
modifies s
effects Removes x from s (i.e., $s_{post} = s - \{x\}$).

member = **proc** (s: intset, x: int) **returns** (bool)
effects Returns ($x \in s$).

size = **proc** (s: intset) **returns** (int)
effects Returns the number of elements in s.

choose = **proc** (s: intset) **returns** (int)
requires s is not empty.
effects Returns an arbitrary member of s.

end intset

Figure 4.2 Specification of intset.

will not understand the description or will interpret it differently than we intended. As we shall see in chapter 10, this weakness is avoided if we use formal specifications.

A specification of the *poly* data type is shown in figure 4.3. Polys are immutable polynomials with integer coefficients. Operations are provided to create a one-term poly, to add, subtract, and multiply polys, and to test two polys for equality.

4.2 Implementing Data Abstractions

To implement a data type, we select a representation for the objects and implement the operations in terms of that representation. The chosen representation must permit all operations to be implemented in a reasonably simple and efficient manner. In addition, if some of the operations must run quickly, the representation must make this possible. Often a represen-

poly = **data type is** create, degree, coeff, add, mul, sub, minus, equal

Overview

Polys are immutable polynomials with integer coefficients.

Operations

create = **proc** (c, n: int) **returns** (poly)
    requires $n \geq 0$.
    effects Returns the poly cx^n. For example,
        poly\$create(6, 3) = $6x^3$
        poly\$create(3, 0) = 3
        poly\$create(0, 0) = 0

degree = **proc** (p: poly) **returns** (int)
    effects Returns the degree of p, i.e., the largest exponent in p with
        a nonzero coefficient. The degree of the zero polynomial is zero.
        For example,
        poly\$degree(x^2+1) = 2
        poly\$degree(17) = 0

coeff = **proc** (p: poly, n: int) **returns** (int)
    requires $n \geq 0$
    effects Returns the coefficient of the term of p whose exponent is n.
        Returns 0 if n is greater than the degree of p. For example,
        poly\$coeff(x^3 + 2x + 1, 4) = 0
        poly\$coeff(x^3 + 2x + 1, 1) = 2

add = **proc** (p, q: poly) **returns** (poly)
    effects Returns the poly that is the result of adding p and q.

mul = **proc** (p, q: poly) **returns** (poly)
    effects Returns the poly that is the result of multiplying p and q.

sub = **proc** (p, q: poly) **returns** (poly)
    effects Returns the poly that is the result of subtracting q from p.

minus = **proc** (p: poly) **returns** (poly)
    effects Returns $z - p$ where z is the zero polynomial.

equal = **proc** (p, q: poly) **returns** (bool)
    effects Returns true if p and q have the same coefficients for
        corresponding terms; otherwise returns false.

end poly

Figure 4.3 Informal specification of the polynomial data abstraction.

tation that is fast for some operations will be slower for others. We might
therefore require multiple implementations of the same type.

For example, a plausible representation for an intset object is an array of integers, where each integer in the intset occurs as an element of the array. We could choose to have each element of the set occur exactly once in the array or allow it to occur many times. The latter choice makes the implementation of *insert* run faster, but slows down *delete* and *member*. If *member* is likely to be called frequently, we would make the former choice.

Note that there are two types under discussion here: the new abstract type, intset, which we are implementing, and the type array of integers, which we are using as the representation. Every implementation will have two such types: the *abstract type* and the *rep type*. Outside the implementation, the rep type is supposed to be invisible; all that can be done with the objects of the abstract type is to apply the type's operations to them. This restriction is enforced in CLU by type checking.

We now discuss how to implement data abstractions in CLU. We shall discuss how to implement types in other languages in chapters 7 and 15.

4.2.1 Implementations in CLU

Although it is certainly possible to implement types in a language that provides no special support for them, it is more convenient if a type can be implemented by a single program module. CLU provides such a program unit, called a *cluster*. (The name CLU comes from the first three letters of "cluster.")

A cluster contains the following:

1. A header, which gives the name of the type being implemented and the names of the operations. This header is like the header in a specification.

2. A definition of the representation chosen in this implementation.

3. Implementations of the primitive operations. Additional routines (procedures and iterators) can also be included in the cluster, but only those named in the header can be called from outside.

A cluster template showing these parts is given in figure 4.4.

To represent intsets by array, for example, we would say

rep = array[int]

This line serves two functions: It informs the CLU compiler that in this cluster the representation is an array of integers, and it is an equate, so that it permits the reserved word **rep** to be used as an abbreviation for this array type throughout the body of the cluster.

Inside a cluster both the abstract type and the rep type must be visible. Furthermore, it must be possible to go back and forth between them. For example, *insert* receives an intset as an argument, but to implement *insert*

dname = **cluster is**   % list of names of operations

    rep =   % description of representation goes here

  % implementations of the operations go here.
  % some helping routines may also be provided if desired.

end dname

Figure 4.4 Cluster template.

it must be possible to make use of the array that represents the intset.

CLU provides the ability to convert back and forth between the abstract and rep types by means of two special operations: **up** and **down**. **Up** takes a rep object as an argument and produces an abstract object as a result, while **down** performs the reverse transformation. Each cluster has its own versions of **up** and **down** to allow it to convert between its abstract and rep types. **Up** and **down** are defined automatically by the CLU compiler. For example, in the intset cluster, the compiler provides **up** and **down** operations with the following headers:

  up = **proc** (a: array[int]) **returns** (intset)
  down = **proc** (s: intset) **returns** (array[int])

Up and **down** may only be used in clusters, and always convert between the abstract and rep types of the cluster in which they appear. Thus it is not possible to use **up** and **down** to undermine the type checking of CLU.

Although **up** and **down** may appear anywhere in cluster operations, they are used most often in the following two ways. First, when an operation takes an abstract object as an argument, it often uses **down** to convert that object to the rep type; second, when an operation returns a newly created abstract object, it often uses **up** just before returning to convert a rep object to the abstract type. To facilitate these two cases, CLU provides special syntax, the reserved word **cvt**. **Cvt** may be used as the type of an argument or result in the headers of operations in clusters. Its use as the type of an argument indicates that the actual argument is of the abstract type but the formal is of the rep type, so that **down** should be applied implicitly to the actual argument immediately after the call, and the resulting rep object should be assigned to the formal. When **cvt** is used as the type of a result, it indicates that the result object is of the abstract type but the object being returned is of the rep type, so that **up** should be applied implicitly to the returned object just before it is returned. **Cvt** does not add any functionality that was not already available from **up** and **down**. It provides convenience only, by making common cases easy to write.

intset = **cluster is** create, insert, delete, member, size, choose

    rep = array[int]

    create = **proc** ( ) **returns** (**cvt**)
      return (**rep**$new( ))
      end create

    insert = **proc** (s: intset, x: int)
      if ~member(s, x) **then rep**$addh(**down**(s), x) **end**
      end insert

    delete = **proc** (s: **cvt**, x: int)
      j: int := getind(s, x)
      if j <= **rep**$high(s)
        then s[j] := **rep**$top(s)   % top returns s[high(s)]
          rep$remh(s)
      end
      end delete

    member = **proc** (s: **cvt**, x: int) **returns** (bool)
      return (getind(s, x) <= **rep**$high(s))
      end member

    size = **proc** (s: **cvt**) **returns** (int)
      return (**rep**$size(s))
      end size

    choose = **proc** (s: **cvt**) **returns** (int)
      return (**rep**$bottom(s))   % bottom returns s[low(s)]
      end choose

    getind = **proc** (s: **rep**, x: int) **returns** (int)
      i: int := **rep**$low(s)
      while i <= **rep**$high(s) **do**
        if x = s[i] **then return** (i) **end**
        i := i + 1
      end
      return (i)
      end getind

end intset

Figure 4.5 Implementation of intset.

Now we are ready to look at the implementation of intset (see figure 4.5). First note the implementation of the *create* operation. Its header is

    create = **proc** ( ) **returns** (**cvt**)

Therefore it is returning an abstract object—that is, an intset—to its caller.

The **return** statement is apparently returning an array, but this object is **up**'d just before being returned.

Next, look at the implementation of *insert*. In this case, **cvt** is not used; the header of *insert* is

insert = **proc** (s: intset, x: int)

Therefore the abstract, intset object provided by the caller is not **down**'d.

Insert does not return any results; instead it modifies its intset argument. In this implementation, each element of the set occurs only once in the representing array; therefore *insert* cannot simply add the integer x to the array. Instead it must check first whether x is already in the array. *Insert* uses *member* to do this check. *Member* requires an intset argument; this is why we did not use **cvt** in the header of *insert*. Note that in the call to *member* the "intset$" prefix is omitted; this prefix is not required when an operation is called within its own cluster.

If it is determined that x is not a member of s, s is **down**'d explicitly and x is added to the array by using the *addh* operation. If x is already in s, the array is not modified.

The header of *delete* is

delete = **proc** (s: **cvt**, x: int)

In this case the abstract intset object passed as an argument is **down**'d immediately, and inside *delete* the type of s is array[int].

Delete must first determine whether x is in s. To do this it uses the internal procedure *getind*. *Getind* returns the index at which an element resides if the element is in the array; otherwise, it returns one greater than the high bound of the array. Since *getind* is not listed in the cluster header, it is internal to this cluster; that is, it can only be called from inside this cluster.

If x is in s, *delete* simply stores the top element of the array in x's position and then deletes the top element by using *remh*. *Remh* actually returns the removed element as a result. In this case, however, the unwanted result is simply discarded.

Choose is allowed to return an arbitrary element of s; therefore in implementing *choose* we can do whatever is convenient. One convenient choice is the first element of the array; another, equally convenient, choice would have been the last element of the array. Of course, that would have resulted in a different object being returned to the caller of *choose*, but the caller should not be relying on a particular object being returned, since the specification of *choose* says that the choice is arbitrary. Note that, in accordance with its specification, *choose* does not modify the intset passed to it as an argument.

4.2.2 Remarks on Up and Down

It is helpful to think of **up** as taking an object of the rep type and surrounding it with an impenetrable shield. The user of the **up**'d object cannot see through this shield to the representation. Instead the value of the object can be observed only by calling the primitive operations of the object's type. For example, a user cannot see the array that represents an intset, but can tell what elements are in the set and can add and remove elements by calling the intset operations.

Down does the opposite of **up**—it melts the shield. If it were possible for the user to call **down**, then the way in which the abstract object is represented would be visible, and the benefits of data abstraction would be lost. Therefore CLU limits the use of **down** to the cluster that implements the object.

The use of **up** must be limited too, to prevent counterfeit objects. If **up** could be used outside the defining cluster, it would be possible to create objects that masquerade as objects of a type. If these objects were later passed as arguments to the operations of the cluster, their rep might be different than expected and all sorts of nasty errors could occur. By limiting the use of **up** to the cluster, we eliminate these errors. Limitations on both **up** and **down** are needed to obtain the benefits of data abstraction.

As mentioned, when **down** is used inside a cluster, it melts the shield and makes the rep of the object visible. However, this rep is simply another abstract object; it is protected by a shield constructed by its own implementation, which may be either a user-defined cluster or part of the CLU implementation. Again, the real rep of the object cannot be seen, but information about the object can be obtained by applying the operations. Thus inside the intset cluster, we cannot see inside the array that implements the intset, but we can observe interesting information about it and we can modify it by applying the array operations.

Up and **down** (and hence **cvt**) do not cause any code to be generated. They are used to inform the compiler of a change in the way an object is to be viewed. The compiler uses this information for type checking. Since type checking is done at compile time, the information is not needed at runtime, although it is useful for debugging.

remove_dupls = **proc** (a: array[int])
    modifies a.
    effects Removes all duplicate elements from a. The low bound of a
    remains the same but the order of the remaining elements may
    change. E.g., if a = [1: 3, 13, 3, 6] before the call, on return a has
    low bound 1 and contains the three elements 3, 13, and 6 in some
    order.

ai = array[int]
remove_dupls = **proc** (a: ai)
    a_low: int := ai$low(a)
    i: int := a_low
    j: int := a_low
    s: intset := intset$create( )
    while i <= ai$high(a) **do**
      if ~intset$member(s, a[i])
        then intset$insert(s, a[i]) % a[i] is not a duplicate, so keep it
          a[j] := a[i]
          j := j + 1
      end
      i := i + 1
    end
    ai$trim(a, a_low, j − a_low) % remove trailing elements of a
    end remove_dupls

Figure 4.6 Using intset.

4.3 Using Data Abstractions

Once a data abstraction has been defined, it is no different from a built-in type, and its objects and operations should be usable in the same way as are those of a built-in type. The extent to which this is possible depends on the programming language in use. In CLU, variables can be declared of the new type, as in

    i: intset

and objects can be created and manipulated, as in

    i := intset$create( )
    intset$insert(i, 3)

Figure 4.6 gives a specification and implementation of *remove_dupls*. This implementation uses an intset to remember elements already encountered in the array.

User-defined types can also be used to represent other user-defined types. For example, if we were trying to implement an *account* data type for a banking system, we might find it desirable to keep track of the individual amounts of deposits and withdrawals as well as a cumulative balance. Thus we might have

rep = record [balance: int, deposits, withdrawals: intset]

Here we are using intsets to represent accounts.

4.4 Implementing Polynomials

As a second example, we implement the poly type specified in figure 4.3. Polys can be represented by arrays in which the ith element of the array holds the coefficient of the ith term. The array stores coefficients only for terms up to the degree; we need not store the coefficients of terms above the degree, since these are all zero. Also, we need not store leading zero coefficients, and the zero poly can be represented by the empty array; if we make the high bound of this array zero, then the high bound of the representing array will always be the degree of the poly. This representation is good if polys are not sparse; if most polys were sparse, like $x + x^{20}$, the representation would not be good, since there would be many zeros stored in the array.

A cluster using this representation is shown in figure 4.7. The operations *create*, *degree*, and *coeff* are easy given this representation. *Create* returns either the empty array to represent the zero poly or a one-element array containing the given coefficient. *Degree* returns the high bound of the array; thus it depends on this high bound being the degree of the poly. This requirement is supported in *create* and in the *add* operation, which removes any trailing zeros after it has added its inputs. It is also supported in the *mul* operation because a zero can occur in the low or high term only if one of the input polys is the zero poly, and this case is handled explicitly.

Equal simply calls the *similar* operation for arrays; this operation returns true if its inputs have the same low bounds and the same elements in the same order, which is just what we need here. Note that the array *equal* operation would not work because it returns true only if its two input arrays are the same object.

The names of the poly operations were chosen to take advantage of short forms. For example, $p + q$ can be used for *add* and $-p$ for *minus*. These short forms are used in implementing *sub*.

poly = **cluster is** create, degree, coeff, add, minus, mul, sub, equal

    rep = array[int]

    create = **proc** (c, n: int) **returns (cvt)**
       if c = 0 **then return** (rep$new( )) **end**   % the zero poly
       r: **rep** := rep$create(n)   % degree will be low and high bound
       rep$addh(r, c)   % add nonzero coeff
       return (r)
       end create

    degree = **proc** (p: **cvt**) **returns** (int)
       return (rep$high(p))
       end degree

    coeff = **proc** (p: **cvt**, n: int) **returns** (int)
       if n > rep$high(p) **cor** n < rep$low(p)
          then return (0) **else return** (p[n]) **end**
       end coeff

    add = **proc** (p, q: **cvt**) **returns (cvt)**
       % make p be the poly with the lower low bound
       if rep$low(q) < rep$low(p) **then** p, q := q, p **end**
       s: **rep** := rep$copy(p)   % initialize s to p
       qhigh: int := rep$high(q)
       % extend s with 0's up to q's high bound
       while rep$high(s) < qhigh **do**
          rep$addh(s, 0)
          end
       i: int := rep$low(q)
       while i <= qhigh **do**   % add coeffs of q to s
          s[i] := s[i] + q[i]
          i := i + 1
          end
       while ~rep$empty(s) **do**   % remove leading and trailing zeros
          if rep$bottom(s) = 0 **then** rep$reml(s)
             elseif rep$top(s) = 0 **then** rep$remh(s)
             else return (s)
             end
          end
       return (rep$new( ))   % the zero poly
       end add

Figure 4.7 Implementation of poly (continues on next page).

```
minus = proc (p: cvt) returns (cvt)
    i: int := rep$low(p)
    q: rep := rep$create(i)
    while i <= rep$high(p) do
        rep$addh(q, −p[i])
        i := i + 1
        end
    return (q)
    end minus

sub = proc (p, q: poly) returns (poly)
    return (p + (−q))   % Note use of short forms
    end sub

mul = proc (p, q: cvt) returns (cvt)
    % check for multiplication by zero
    if rep$empty(p) cor rep$empty(q) then return (rep$new( )) end
    s: rep := rep$create(rep$low(p) + rep$low(q))
                % create s with the new low bound
    shigh: int := rep$high(p) + rep$high(q)   % the new high bound
    while rep$high(s) < shigh do   % initialize s to zero
        rep$addh(s, 0)
        end
    i: int := rep$low(p)
    phigh: int := rep$high(p)
    qlow: int :=  rep$low(q)
    qhigh: int := rep$high(q)
    while i <= phigh do   % do multiplication
        j: int := qlow
        while j <= qhigh do
            s[i + j] := s[i + j] + p[i]*q[j]
            j := j + 1
            end
        i := i + 1
        end
    return (s)
    end mul

equal = proc (p, q: cvt) returns (bool)
    return (rep$similar(p, q))
    end equal

end poly
```

Figure 4.7 (continued)

4.5 Aids to Understanding Implementations

In this section, we discuss two pieces of information, the abstraction function and the representation invariant, that are particularly useful in understanding an implementation of a data abstraction. This information should be included as comments in the implementation. In addition, we discuss reasoning about the correctness of implementations and mutable representations.

4.5.1 The Abstraction Function

Any implementation of a data abstraction must define how objects belonging to the type are represented. In choosing the representation, the implementer has in mind a relationship between the rep objects and the abstract objects. Particular rep objects are expected to represent particular abstract objects. For example, in figure 4.5, intsets are represented by arrays, where the elements of the array are the elements of the set.

This relationship can be defined by a function, called the *abstraction function*, that maps rep objects to abstract objects:

A: **rep** $\rightarrow$ $\mathcal{A}$

Here $\mathcal{A}$ stands for the set of abstract objects. For each rep object r, $A(r)$ is the abstract object $a \in \mathcal{A}$ that r represents.

For example, the abstraction function for the intset implementation maps array[int] to intset. Figure 4.8 illustrates this function at some points. Note that A is many-to-one: many rep objects map into the same abstract element of $\mathcal{A}$. For example, [1: 1, 2] and [1: 2, 1] both represent the intset $\{1, 2\}$. Since the process of abstraction involves forgetting irrelevant information, it is not surprising that abstraction functions are often many-to-one. In this example, the order in which the elements appear in the array is irrelevant.

The abstraction function is a crucial piece of information about an implementation. It defines the meaning of the representation, the way in which the rep objects are supposed to implement the abstract objects. It should always be described in a comment in the implementation. In writing such a description, however, we are hampered by the fact that if the specification of the type is informal, the range of the abstraction function (the set $\mathcal{A}$) is not really defined. We shall overcome this problem in what follows by giving an informal description of a "typical" abstract object. When we discuss formal specifications in chapter 10, we shall see that they enable us to have well-defined ranges for abstraction functions.

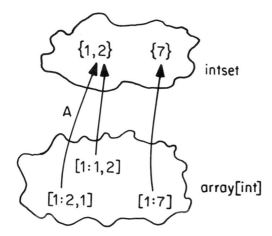

Figure 4.8 An example of an abstraction function.

To define the abstraction function, we first define a typical element of the abstract type. This provides us with a way to talk about the abstract objects. Then we can define the abstraction function in terms of this typical object. For example, for intsets we might say

% A typical intset is $\{x1, \ldots, xn\}$

Here we are using mathematical sets to denote intsets, just as we did in the specification of intset. Then we can say

% The abstraction function is
% $A(r) = \{r[i] \mid low(r) \leq i \leq high(r)\}$

where $\{x \mid p(x)\}$ is the set of all x such that $p(x)$ is true.

As a second example, consider the poly implementation. We chose to represent a poly as an array in which the ith element held the ith coefficient unless this coefficient was a leading or trailing zero. We can describe this representation as follows:

% A typical poly is $c_0 + c_1x + c_2x^2 + \ldots$
% The abstraction function produces the coefficients c_i:
% $c_i = r[i]$ if $low(i) \leq i \leq high(r)$
% $= 0$ otherwise

This abstraction function is defined only for arrays with nonnegative low bound. Abstraction functions are frequently partial like this (we shall discuss this issue further in the next section).

One virtue of the abstraction function is that it can clear up ambiguities about the interpretation of the rep. For example, suppose that stacks were implemented using arrays. Here we have a choice of which way to grow the array when we push an element on the stack. This choice will be clearly stated in the abstraction function. If we decide to grow the array at the high end, the abstraction function is

% A typical stack is a sequence [e1, ... , en] where en is the top element
% The abstraction function is
% $A(r) = [r[low(r)], \ldots, r[high(r)]]$

If we decide to grow the array from the low end, we have

% $A(r) = [r[high(r)], \ldots, r[low(r)]]$

Note that we have just discussed two implementations of stacks that use the same rep but interpret it differently.

4.5.2 The Representation Invariant

In a strongly typed language like CLU, type checking guarantees that whenever an abstract object of the type is passed as an argument to an operation of the type, that abstract object is represented by an object of the rep type. Frequently, however, not all objects of the rep type are legitimate representations of abstract objects. For example, for the intset cluster in figure 4.5, every array, no matter what its value, is an element of $\mathcal{R}$; examples are the arrays [1: 1, 7, 6], [1: 1, 6, 6], and [1: 6, 1, 7]. However, in the implementation of intset we decided that each element of the set would be entered in the array exactly once. Therefore legitimate representations of intsets in this cluster do not contain duplicate entries; [1: 1, 6, 6], for example, is not a legitimate representation.

A statement of a property that all legitimate rep objects satisfy is called a *representation invariant*, or *rep invariant*. A rep invariant I is a predicate,

I: **rep** → bool

that is true of legitimate rep objects. For example, for the intset cluster, we might state the following rep invariant:

% The rep invariant is
%   For all integers i, j such that $low(r) \leq i < j \leq high(r)$
% $r[i] \sim= r[j]$

Note that I is false for [1: 1, 6, 6] but is true for both [1: 1, 7, 6] and [1: 6, 1, 7]. Note also that we did not include in I any constraints that are true for all arrays—for example, $size(r) = high(r) - low(r) + 1$—because these are guaranteed by arrays and can be assumed in the intset implementation.

As a second example of a rep invariant, consider an alternative representation of intsets that consists of an array of 100 booleans plus an array of integers:

rep = record[els: array[bool], other_els: array[int], size: int]

The idea here is that for an integer i in the range 1–100, we shall record membership in the set by storing true in $r.els[i]$. Integers outside this range will be stored in *other_els* in the same manner as in our previous implementation of intset. Since it would be expensive to compute the size of the intset if we had to examine every part of the *els* array, we also store the size explicitly in the rep. This representation is a good one if almost all members of the set are in the range 1–100 and if we expect the set to have quite a few members in this range. (Otherwise the space required for the *els* array will be wasted.) Thus we have

```
% The abstraction function is
%     A(r) = { r.other_els[i] | low(r.other_els) ≤ i ≤high(r.other_els) }
%                 ∪
%             { j | 1 ≤ j ≤ 100 & r.els[j] }
```

Here ∪ stands for the mathematical union of the two sets. Also,

```
% The rep invariant is
%     size(r.els) = 100 & low(r.els) = 1
%     & all elements in r.other_els are not in the range 1 to 100
%     & there are no duplicates in r.other_els
%     & r.size = size(r.other_els) + (count of "true" entries in r.els)
```

Note that the *size* field of this rep is redundant: It holds information that can be computed directly from the rest of the rep. Whenever there is redundant information in the rep, the relationship of this information to the rest of the rep should be explained in the rep invariant (for example, in the last line of this rep invariant).

Sometimes it is convenient to use a helping function in the rep invariant or abstraction function. For example, the last line of the rep invariant above could be rewritten

% r.size = size(r.other_els) + cnt(r.els, low(r.els))
%   where cnt(a, i) = if i > high(a) then 0
%   else if r[i] then 1 + cnt(a, i+1)
%   else cnt(a, i + 1)

The helping function *cnt* is defined by a recurrence relation.

The implementation of poly in figure 4.7 has an interesting rep invariant. Recall that we chose to store coefficients only up to the degree, not to store leading zeros, and to represent the zero poly by the empty array. Therefore we do not expect to find a zero in the low or high element of the array. In addition, these arrays always have a nonnegative low bound. Finally, the empty array (representing the zero poly) must have high bound zero. Thus we have

% The rep invariant is
%   low(r) $\geq$ 0
%   & if empty(r) then high(r) = 0
%   else r[low(r)] $\sim=$ 0 & r[high(r)] $\sim=$ 0

Recall that the implementation of the *degree* operation depended on the high bound of the array being the degree of the poly; now we see this requirement spelled out in the rep invariant.

Sometimes all concrete objects are legal representations. Then we have simply

% The rep invariant is
%   true

The rep invariant should be stated explicitly even in this case. It may prevent the implementer from depending on a stronger, unsatisfied invariant.

A rep invariant is "invariant" because it always holds for the reps of abstract objects; that is, it holds whenever an object is used outside its implementation. The rep invariant need not hold all the time, since it can be violated while executing one of the type's operations. For example, poly$*add* may produce an array with zero in the low or high element, but such elements are removed from the array before *add* returns. The rep invariant must hold whenever operations return to their callers.

There is a relationship between the abstraction function and the rep invariant. The abstraction function is only of interest for legal representations, since only these represent the abstract objects. Therefore it need not be defined for illegal representations. For example, as was mentioned earlier, the abstraction function for poly is partial; it is defined only when the array has a nonnegative low bound. All legal poly representations satisfy this constraint, as indicated by the rep invariant.

There is an issue concerning how much to say in a rep invariant. A rep invariant should express all constraints on which the operations depend, but it need not state additional constraints. For example, all concrete objects for the intset implementation in figure 4.5 have low bound 1, but none of the operation implementations depends on this, so it need not be stated in the rep invariant. A good way to think of this is to imagine that the operations are to be implemented by different people who cannot talk to one another; the rep invariant must contain all the constraints that these various implementers depend on.

Whenever a data abstraction is implemented, the rep invariant is one of the first things the programmer thinks about. It must be chosen before any operations are implemented, or the implementations will not work together harmoniously. To ensure that it is understood, the rep invariant should be written down and included as a comment in the code. Writing down the rep invariant forces the implementer to articulate what is known and increases the chances that the operations will be implemented correctly.

All operations must be implemented in a way that preserves the rep invariant. For example, suppose we implemented *insert* by

```
insert = proc (s: cvt, x: int)
    rep$addh(s, x)
    end insert
```

This implementation can produce a rep object with duplicate elements. If we know that the rep invariant prohibits such objects, then this implementation is clearly incorrect.

The rep invariant is also useful for the reader of an implementation. For example, in an alternative implementation of intset we might have decided to keep the rep array sorted. In this case we would have

```
% The rep invariant is
%     For all i, j such that low(r) ≤ i < j ≤ high(r)
%         r[i] < r[j]
```

and the operations would be implemented differently than in figure 4.5. The rep invariant tells the reader why the operations are implemented as they are.

4.5.3 Preserving the Rep Invariant

As part of showing that a type is implemented correctly, we must show that the rep invariant holds for all legitimate rep objects. We can do this as follows. First, we show that the invariant holds for objects returned by operations, such as poly$*create*, that return objects of the type but take no arguments of the type. For all other operations, we can assume when they are called that the invariant holds for all argument objects of the type; we must show that it holds at return both for arguments of the type and for returned objects of the type.

For example, the intset implementation of figure 4.5 has invariant

% For all integers i, j such that $low(r) \leq i < j \leq high(r)$
% $r[i] \sim= r[j]$

Intset$*create* establishes this invariant because the newly created array is empty. The *member* operation preserves it because we can assume that the invariant holds for argument s and that *member* does not modify s. Operation *insert* preserves the invariant because

1. the invariant holds for its argument s at the time of the call;

2. the call to *member* by *insert* preserves the invariant, since *member* preserves the invariant; and

3. *insert* adds x to s only if $\sim member(s, x)$—therefore, since s satisfies the invariant at the time of the call, it still satisfies the invariant after x has been added to it.

As a second example, consider the poly implementation in figure 4.7 and recall that the invariant is

% $low(r) \geq 0$
% & if empty(s) then $high(r) = 0$
%   else $r[low(r)] \sim= 0$ & $r[high(r)] \sim= 0$

Poly$*create* preserves the invariant because it explicitly tests for the zero polynomial. The *mul* operation preserves the invariant because

1. the invariant holds for p and q at the time of the call;

2. if either p or q is the zero poly, this is recognized and the proper rep constructed; and

3. otherwise, neither p nor q contains a zero in its low or high term—therefore neither the low term of the returned array, $bottom(p) * bottom(q)$, nor its high term, $top(p) * top(q)$, can be zero.

We are assuming here that abstract objects cannot be modified outside their implementation. If they could be modified, we could not assume that the rep invariant holds when operations are called, because using modules could cause it to be violated.

4.5.4 Mutable Reps

Mutability (or immutability) is a property of an abstraction that must be preserved by the implementation. A mutable abstraction must have a mutable rep, or it will not be possible to provide the required mutability. However, an immutable abstraction need not have an immutable rep. For example, polys are immutable, but they have a mutable rep. A mutable rep is acceptable if modifications made to the rep cannot be observed by the abstraction's users.

Sometimes it is useful to create an object by incrementally mutating its rep, though once the object is created, its rep is never modified again. This is the way polys were created. Mutability is also useful for *benevolent side effects*, which are modifications that are not visible outside the implementation. As an example, let us represent rational numbers as a pair of integers:

rep = record[num, denom: int]

The abstraction function is

```
% A typical rational is n/d
% The abstraction function is
%     A(r) = r.num/r.denom
```

Given this rep, there are several choices to make: what to do with a zero denominator, how to store negative rationals, and whether or not to keep the rational in reduced form (that is, with the numerator and denominator reduced so that there are no common terms). We choose to rule out zero denominators, to represent negative rationals by means of negative numerators, and *not* to keep the rep in reduced form (to speed up operations like multiplication). Thus we have

```
% The rep invariant is
%     r.denom > 0
```

Let us also choose to compute reduced forms when testing whether two rationals are equal. We can do this using the *gcd* procedure:

rep = record [num, denom: int]

equal = **proc** (r1, r2: **cvt**) **returns** (bool)
   if r1.num = 0 **then return** (r2.num = 0)
     elseif r2.num = 0 **then return** (**false**)
     end
   reduce(r1)
   reduce(r2)
   return (r1.num = r2.num **cand** r1.denom = r2.denom)
   end equal

% reduce is an internal routine that changes its argument to its reduced form
reduce = **proc** (r: **rep**)
   g: int := gcd(int\$abs(r.num), r.denom)
   r.num := r.num/g
   r.denom := r.denom/g
   end reduce

Figure 4.9 A benevolent side effect.

gcd = **proc** (n, d: int) **returns** (int)
      requires n and d are positive.
      effects Returns the greatest common divisor of n and d.

The implementation of *equal* is shown in figure 4.9. Once computed, the reduced forms are stored in the rep because this will speed up the next equality test.

The modification of the rep performed by the *equal* operation is a benevolent side effect. Such side effects are often performed for reasons of efficiency. They are possible whenever the abstraction function is many-to-one, since there are then many rep objects that represent a particular abstract object. Sometimes it is useful within an implementation to switch from one of these rep objects to another. Such a switch is safe since the rep still maps to the same abstract object.

4.6 Parameterized Data Abstractions

Types are useful parameters for types, just as they are for procedures. For example, consider a general *set* abstraction in which the elements of the set are of some arbitrary type. Of course, not all types make sense as parameters to the generalized set; since sets do not store duplicate elements, there must be some way to determine whether elements are duplicates. Therefore the specification of set shown in figure 4.10 requires that the element type have an *equal* operation. The requires clause is positioned right after the header. (Requirements on individual operations are also permitted; this will be discussed in the next section.)

set = **data type** [t: **type**] **is** create, insert, delete, member, size, choose

Requires t has an operation
    equal: **proctype** (t, t) **returns** (bool)
   that is an equivalence relation on t.

Overview

   Sets are unbounded mathematical sets of integers. Sets are mutable: *insert* and *delete* add and remove elements from the set.

Operations

   create = **proc** ( ) **returns** (set[t])
      effects Returns a new empty set.

   insert = **proc** (s: set[t], x: t)
      modifies s
      effects Adds x to the elements of s; after insert returns,
       $s_{post} = s \cup \{x\}$.

   delete = **proc** (s: set[t], x: t)
      modifies s
      effects Removes x from s (i.e., $s_{post} = s - \{x\}$).

   member = **proc** (s: set[t], x: t) **returns** (bool)
      effects Returns $x \in s$.

   size = **proc** (s: set[t]) **returns** (int)
      effects Returns the number of elements in s.

   choose = **proc** (s: set[t]) **returns** (t)
      requires s is not empty.
      effects Returns an arbitrary member of s.

end set

Figure 4.10 Specification of a parameterized data abstraction.

The implementation of set is shown in figure 4.11. Note the **where** clause following the cluster header; this clause includes the syntactic part of the requires clause, which is needed for type checking. The implementation of set makes use of parameter type t as the type of the elements. Thus the representation in the set cluster is

   rep = array [t]

and the header of *insert* is

   insert = **proc** (s: **cvt**, i: t)

Also, where we previously used integer operations on the elements, we now use t operations. Thus in *getind* we use *t$equal* to check whether the

set = **cluster** [t: **type**] **is** create, insert, delete, member, size, choose
   where t **has** equal: **proctype** (t, t) **returns** (bool)

  rep = array[t]

  create = **proc** ( ) **returns** (**cvt**)
    return (rep$new( ))
    end create

  insert = **proc** (s: **cvt**, x: t)
    if ~member(**up**(s), x) **then** rep$addh(s, x) **end**
    end insert

  delete = **proc** (s: **cvt**, x: t)
    j: int := getind(s, x)
    if j <= rep$high(s)
      then s[j] := rep$top(s)
        rep$remh(s)
      end
    end delete

  member = **proc** (s: **cvt**, x: t) **returns** (bool)
    return (getind(s, x) <= rep$high(s))
    end member

  size = **proc** (s: **cvt**) **returns** (int)
    return (rep$size(s))
    end size

  choose = **proc** (s: **cvt**) **returns** (t)
    return (rep$bottom(s))
    end choose

  getind = **proc** (s: **rep**, x: t) **returns** (int)
    i: int := rep$low(s)
    while i <= rep$high(s) **do**
      if x = s[i] **then return** (i) **end**   % = is short for t$equal
      i := i + 1
      end
    return (i)
    end getind

end set

Figure 4.11 Implementation of set.

given element is in the set. Note that *equal* is the only operation of *t* used in the set cluster.

To use set, we must supply a type as a parameter, as in

pset = set[poly]

Since poly has an *equal* operation, it is a legal parameter for set. The type obtained has all the listed operations, with every occurrence of the parameter name replaced by the parameter type. For example, operations for *pset* include

create = **proc** ( ) **returns** (pset)
insert = **proc** (s: pset, x: poly)

4.7 Lists

Our next example is another parameterized type in which requirements are placed on individual operations rather than on the type as a whole. In addition, we shall show a recursive implementation.

Unbounded *lists* contain elements of some arbitrary type. They have operations to create an empty list, to form a new list consisting of a given element and the elements of an already existing list, and to decompose a list into its first element and the remaining elements. A specification is given in figure 4.12. Note that lists are immutable; operations like *cons* do not modify their argument list, but instead return a new list.

Note that lists as a whole place no constraints on the parameter type. Thus list[intset] is a legal type. However, the *equal* operation needs to test elements for equality, and it thus constrains *t* in its requires clause. Therefore list[intset] has all list operations except *equal*; since intset has no *equal* operation, list[intset] has none either. Limiting the constraint to just the *equal* operation makes it possible to form a list of elements of any type, even one (like intset) that has no *equal* operation. Defining lists in this way is appropriate, since, like arrays, lists are a very general grouping mechanism; arrays also have constraints on particular operations instead of general constraints.

A convenient way to implement a list is as linked pairs, where the first element of a pair contains an element of the list and the second refers either to another pair or to a special value indicating that there are no more pairs. An example is shown in figure 4.13a; here we are assuming that *t* is int. Note that whenever the second element of the pair points to another pair, we can view this second pair as the start of a list containing all elements

list = **data type** [t: **type**] **is** create, cons, first, rest, empty, equal

Overview

Lists are immutable, unbounded sequences of elements of type t.

Operations

create = **proc** ( ) **returns** (list[t])
    effects Returns the empty list.

cons = **proc** (i: t, x: list[t]) **returns** (list[t])
    effects Returns a list containing i as its first element and the
        elements of x as its remaining elements.

first = **proc** (x: list[t]) **returns**(t)
    requires x is not empty.
    effects Returns the first element of x.

rest = **proc** (x: list[t]) **returns** (list[t])
    requires x is not empty.
    effects Returns the list containing all but the first element of x.

empty = **proc** (x: list[t]) **returns** (bool)
    effects Returns true if x is empty; otherwise returns false.

equal = **proc** (x, y: list[t])
    requires t has an operation
            equal: **proctype** (t, t) **returns** (bool)
        that is an equivalence relation on t.
    effects Returns true iff x and y contain equal elements (as
        determined by $t\$equal$) in the same order.

end list

Figure 4.12 Specification of the list data abstraction.

but the first of the original list. In fact, even when the second element
does not point to another pair, we can still consider it as pointing to a
list, namely, the empty list. This view is illustrated in figure 4.13b. In
either case, the second element of the pair refers to another list, and the
representation of lists is thus recursive.

A pair can be represented in CLU by

pair = struct[elem: t, rest: list[t]]

The *pair* data type is sufficient for representing lists having at least one
element, but cannot represent empty lists. The empty list must be treated
as a special case: Either a list has some elements, and is represented as a
pair, or it is empty. To express the concept of a list having either some
elements or no elements, we need a data type that expresses this "either-

(a)

(b)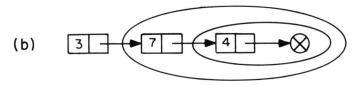

Figure 4.13 Linked pairs: (a) linked pairs representing the list (3 7 4); (b) linked pair representation viewed recursively.

or" property. CLU oneofs (see section 2.7) provide the needed facility, and we have

rep = oneof[some: pair, none: null]

An empty list is represented using type null.

The list cluster is shown in figure 4.14. As usual, we begin by defining the representation:

rep = oneof[some: pair, none: null]
pair = struct[elem: t, rest: list[t]]

These definitions are recursive, since the rep is defined (via pair) in terms of list. Recursive type definitions are limited in CLU to providing recursive representations of data abstractions: The rep of a type can be defined (directly or indirectly) in terms of that type.

We have included a description of the abstraction function and rep invariant as comments in the cluster. Here the definition of the abstraction function is given as a recurrence relation. (Recall that *value_some* is an operation of the oneof type; it returns the data part of its input if its tag is *some*.)

The *create* operation simply creates the list rep in the *none* alternative and returns it; it is *up*'d to the abstract type automatically because of the **cvt**. *Cons* creates the rep in the *some* alternative; the new pair contains the new element and, in the *rest* field, refers to the argument list. Thus the newly created list shares the argument list with the caller of *cons*, as shown in figure 4.15. This sharing is safe because lists are immutable.

Note the **where** clause following the header of the *equal* operation; it contains the syntactic part of the information in the requires clause of the

list = **cluster** [t: **type**] **is** create, cons, first, rest, empty, equal

   rep = oneof[some: pair, none: null]
   pair = struct[elem: t, rest: list[t]]

   % A typical list is a sequence [s1, ..., sn]
   % The abstraction function is
   %   A(r) = [ ]   if r's tag is none
   %   A(r) = [v.elem] || A(v.rest)   if r's tag is some
   %   where v = value_some(r)

   % The rep invariant is
   %   true

   create = **proc** ( ) **returns** (**cvt**)
      return (**rep**$make_none(**nil**))
      end create

   cons = **proc** (i: t, x: list[t]) **returns** (**cvt**)
      return (**rep**$make_some(pair${elem: i, rest: x}))
      end cons

   first = **proc** (x: **cvt**) **returns** (t)
      tagcase x
         tag some (p: pair): **return** (p.elem)
         tag none: % CLU requires all tags to be listed if there is no
                 % "others" arm
      end
      end first

   rest = **proc** (x: **cvt**) **returns** (list[t])
      tagcase x
         tag some (p: pair): **return** (p.rest)
         tag none:
      end
      end rest

   empty = **proc** (x: **cvt**) **returns** (bool)
      return (**rep**$is_none(x))
      end empty

   equal = **proc** (x, y: list[t]) **returns** (bool)
      where t **has** equal: **proctype** (t, t) **returns** (bool)
      if empty(x) **then return** (empty(y))
         elseif empty(y) **then return** (**false**)
         elseif first(x) = first(y)   % t$equal
           then return (rest(x) = rest(y))   % list[t]$equal
         else return (**false**)
      end
      end equal

   end list

Figure 4.14 Implementation of list.

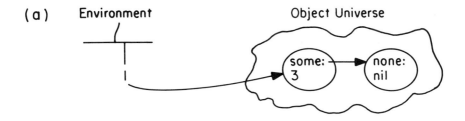

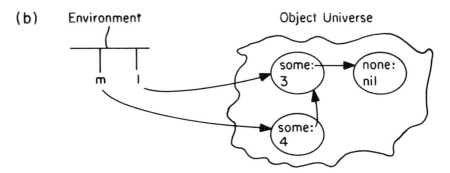

Figure 4.15 Sharing in lists: (a) the representation of the list (3), referred to by variable l; (b) the effect of executing m: list := list$cons(4, l)$.

specification. The required operation is used in

first(x) = first(y)

Equal is implemented recursively; it calls itself using the short form

rest(x) = rest(y)

4.6 Ordered Lists

As a final example, consider lists in which the elements are accessible in sorted order. A specification of such a type, *olist*, is given in figure 4.16. Olists are mutable and are parameterized by a type that must be totally ordered by its *lt* and *equal* operations. Also, olists contain no duplicates.

It is not necessary to keep the representation of an *olist* in sorted order. However, an efficient representation is a sorted binary tree. Each node of the tree contains an element and a left and a right subtree; each subtree is

olist = **data type** [t: **type**] **is** create, addel, remel, is_in, empty, least

Requires t has operations
　　　lt, equal: **proctype** (t, t) **returns** (bool)
　　that define a total order on t.

Overview

　　Olists are mutable lists of elements. *Addel* and *remel* modify the olist. Each
　　element in an olist is different from the rest (as determined by $t\$equal$).
　　Least returns the smallest element of the olist (as determined by $t\$lt$).

Operations

　　create = **proc** ( ) **returns** (olist[t])
　　　　effects Returns a new, empty olist.

　　addel = **proc** (s: olist[t], x: t)
　　　　requires $\sim is\_in(s,\ x)$.
　　　　modifies s
　　　　effects Inserts x in s.

　　remel = **proc** (s: olist[t], x: t)
　　　　requires $is\_in(s,\ x)$.
　　　　modifies s
　　　　effects Removes x from s.

　　is_in = **proc** (s: olist[t], x: t) **returns** (bool)
　　　　effects Returns true if s contains an element equal to x
　　　　　　(using $t\$equal$); otherwise returns false.

　　empty = **proc** (s: olist[t]) **returns** (bool)
　　　　effects Returns true if s contains no elements; otherwise returns
　　　　　　false.

　　least = **proc** (s: olist[t]) **returns** (t)
　　　　requires $\sim empty(s)$.
　　　　effects Returns element e of s such that no element of s is $< e$
　　　　　　(as determined by $t\$lt$).

end olist

Figure 4.16 Specification of olist.

an olist. This is expressed in CLU as

　　node = record[val: t, right, left: olist[t]]

As was the case for lists, we must treat the empty olist as a special case,
giving

　　rep = variant[empty: null, some: node]

This is another recursive representation. Note that it uses mutable types (variants and records) since we are implementing a mutable type.

At this point we can specify the abstraction function:

```
% A typical olist is [e1, ..., en]
% The abstraction function is
% A(r)  = [ ]                               if r is empty
%        = A(n.left) || [n.val] || A(n.right)   if r has some elements
%            where n = value_ some(r)
```

We keep the representation sorted by storing all elements less than a node's value in the node's left subtree, and all elements greater than the value in the right subtree. Thus we have

```
% The rep invariant is
%     if is_ some(r) then
%         all components of n.left are < n.val
%         & n.val < all components of n.right
%             where n = value_ some(r)
```

Once we have identified the abstraction function and the rep invariant, implementing the operations is relatively straightforward (see figure 4.17). The implementations are recursive (iterative implementations are also possible, but they are more complicated).

This implementation is Order($\log_2 n$) for operations *addel*, *remel*, *is_in*, and *least*, assuming that the tree is balanced—that is, that there are roughly equivalent numbers of elements in the left and right subtrees of each node. If the tree is not balanced, it can be reshuffled occasionally (as a benevolent side effect) to balance it, although we do not do so here. The implementation is not very space efficient, however, because of the need to store two pointers at each node in the tree.

If intsets were implemented using olists, then the *remove_dupls* implementation shown in figure 4.6 would require Order($n \log_2 n$) comparisons if the tree were balanced. One case in which the tree is not balanced is when the array argument of *remove_dupls* is sorted!

4.9 Discussion

In this section, we discuss aspects of data abstractions that are of interest both to the definer of the abstraction and to those who use or implement that abstraction. The issues to be discussed are mutability, adequacy,

olist = **cluster** [t: **type**] **is** create, addel, remel, is_in, least, empty
   where t **has** equal, lt: **proctype** (t, t) **returns** (bool)

  node = record[val: t, left, right: olist[t]]
  rep = variant[some: node, empty: null]

  % A typical olist is [e1, ..., en]
  % The abstraction function is
  %   A(r) = [ ]   if r is empty
  %   = A(n.left) || [n.val] || A(n.right)   if r has some elements
  %   where n = value_some(r)

  % The rep invariant is
  %   if is_some(r) then
  %   all components of n.left are < n.val
  %   & n.val < all components of n.right
  %   where n = value_some(r)

  create = **proc** ( ) **returns** (**cvt**)
    return (rep$make_empty(**nil**))
    end create

  addel = **proc** (s: **cvt**, v: t)
    tagcase s
      tag some (n: node):
        if v < n.val **then** addel(n.left, v) **else** addel(n.right, v) **end**
      tag empty:
        rep$change_some(s, node${val: v, left: create( ), right: create( )})
      end
    end addel

  remel = **proc** (s: **cvt**, v: t)
    tagcase s
      tag empty: **return**
      tag some(n: node):
        if v = n.val
          then if empty(n.right)
            then % replace this node with the left subtree
              rep$v_gets_v(s, **down**(n.left))
            else % make n.val be the value from the right subtree
              n.val := least(n.right)
              remel(n.right, n.val)
            end
          elseif v < n.val **then** remel(n.left, v)
          else remel(n.right, v)
          end
      end
    end remel

Figure 4.17 Implementation of ordered lists (continues on next page).

```
least = proc (s: cvt) returns (t)
    tagcase s
        tag empty:
        tag some (n: node):
            if empty(n.left) then return (n.val)
                else return (least(n.left)) end
    end
    end least
is_in = proc (s: cvt, v: t) returns (bool)
    tagcase s
        tag empty: return (false)
        tag some (n: node):
            if v = n.val
                then return (true)
                elseif v < n.val then return (is_in(n.left, v))
                else return (is_in(n.right, v))
                end
    end
    end is_in
empty = proc (s: cvt) returns (bool)
    return (rep$is_empty(s))
    end empty
end olist
```

Figure 4.17 (continued)

reasoning about properties of abstract objects, and the meaning of the *equal, similar,* and *copy* operations.

4.9.1 Mutability

Data abstractions are either mutable, with objects whose values can change, or immutable. Care should be taken in deciding on this aspect of a type. In general, a type should be immutable if its objects would naturally have unchanging values. This might be the case, for example, for such mathematical objects as integers, polys, and complexes. A type should usually be mutable if it is modeling something from the real world, such as a memory device or an automobile in a simulation system, where it is natural for the values of objects to change over time. For example, an automobile might be running or stopped, and contain passengers or not. However, we might still prefer to use an immutable type in such a case because of the greater safety immutability provides. This is why lists are immutable.

In deciding about mutability, it is sometimes necessary to make a trade-off between efficiency and safety. Immutable abstractions are safer than mutable ones because no problems arise if their objects are shared. However, new objects may be created and discarded frequently for immutable abstractions, which means that garbage collection is done more frequently. For example, representing intsets as lists is probably not a good choice if *insert* and *delete* are used frequently.

Note in any case that mutability or immutability is a property of the type and not of its implementation. An implementation must simply support this aspect of its abstraction's behavior.

4.9.2 Classes of Operations

The operations of a data abstraction fall into four classes:

1. *Primitive constructors.* These operations create objects of their type without taking any objects of their type as inputs. Examples are the *create* operations of intset and list.

2. *Constructors.* These operations take objects of their type as inputs and create other objects of their type. For example, *add* and *mul* are constructors for poly, and *cons* and *rest* are constructors for lists.

3. *Mutators.* These operations modify objects of their type. For example, *insert* and *delete* are mutators for intsets. Clearly, only mutable types can have mutators.

4. *Observers.* These operations take objects of their type as inputs and return results of other types. They are used to obtain information about objects. Examples are *size*, *member*, and *choose* for intsets, and *coeff*, *degree*, and *eval* for polys.

Usually, the primitive constructors produce some but not all objects; for example, poly$create only produces single-term polynomials, while intset$create only produces the empty set. The other objects are produced by constructors or mutators. Thus poly$add can be used to obtain polys with more than one term, while sets containing many elements are produced by intset$insert.

Mutators play the same role in mutable types that constructors play in immutable ones. A mutable type can have constructors as well as mutators; for example, intsets might have a *copy* operation that produces a new intset containing the same elements as its argument. Sometimes observers are combined with constructors or mutators; for example, the array *remh* operation modifies the array and returns the removed element.

4.9.3 Adequacy

A data type is adequate if it provides enough operations so that everything users need to do with its objects can be done with reasonable efficiency. It is not possible to give a precise definition of adequacy, although there are limits on how few operations a type can have and still be useful. For example, if we provide only the *create*, *insert*, and *delete* operations for intset, programs cannot find out anything about the elements in the set (because there are no observers). On the other hand, if we add to these three operations just the *size* operation, we can learn about elements in the set (for example, we could test for membership by deleting the integer and the seeing if the size changed), but the type would be costly and inconvenient.

In general, a data abstraction should have operations from at least three of the four classes discussed in the preceding section. It must have primitive constructors, observers, and either constructors (if it is immutable) or mutators (if it is mutable).

Whether a type is adequate or not depends on the context of use; that is, it must have a rich enough set of operations for its intended uses. If the type is to be used in a limited context, such as a single program, then just enough operations for that context need be provided. If the type is intended for general use, a rich set of operations is desirable. This is why the built-in types in CLU have many operations.

To decide whether a data abstraction has enough operations, identify everything users might reasonably expect to do. Next, think about how these things can be done with the given set of operations. If something seems too expensive or too cumbersome to do (or both), investigate whether the addition of an operation would help. Sometimes a substantial improvement in performance can be obtained simply by having access to the representation. For example, we could eliminate the *member* operation for intsets because this operation can be implemented outside the type by using the other operations. However, testing for membership in a set is a common use, and will be faster if done inside the implementation. Therefore, intset should provide this operation.

There can also be too many operations in a type. In this case the abstraction may be less comprehensible, and implementation and maintenance are more difficult. The desirability of extra operations must be balanced against the cost of implementing those operations. If the type is adequate, its operations can be augmented by procedures that are outside the type's implementation.

4.9.4 Reasoning about Data Types

To show that a program that uses a data type is correct, we must reason about properties of objects of the type. This reasoning is done at the abstract level; that is, we use the type's specification and ignore its implementation. Such reasoning will be discussed in detail in chapter 11.

As an example, sometimes it is useful to establish a property that holds for all objects of a type. Such a property is called an *abstract invariant*—it is the abstract analog of the rep invariant. Such a property is similar to a theorem about integers and is proved by a similar technique, namely, induction. To establish an abstract invariant, we must show that it holds for all objects produced by the type. Since only the primitive constructors, constructors, and mutators produce objects, we have the following method:

1. Show that the property holds for all objects returned by primitive constructors.

2. Show that the property holds for objects returned by constructors, assuming that the property holds for objects passed to the constructor as arguments.

3. Show that on exit from mutators the property holds for modified objects, assuming that the property held at the time the mutator was called.

For example, an interesting property of olists (see figure 4.16) is that they contain no duplicate elements, as determined by the *t$equal* operation. This invariant would be useful in determining whether an implementation of intsets using olists is correct. It can be established as follows:

1. It clearly holds for the only primitive constructor, *create*, since *create* returns a new, empty olist.

2. It holds for *addel(s, x)*. When *addel* returns, *s* contains exactly the elements it contained before the call plus *x*. By assumption, there are no duplicates among the elements *s* contained before the call. Furthermore, *addel* requires (in its requires clause) that *x* not already be in *s*. Therefore *s* with *x* added contains no duplicates.

3. It holds for *remel(s, x)* because it holds for *s* (by assumption), and *remel* only removes elements from *s*.

This kind of reasoning is called *data type induction*. The induction is on the number of procedure invocations used to produce the current value of the object. The first step of the induction is to establish the property for the primitive constructor(s); the induction step establishes the property for the constructors and mutators.

Note that we are reasoning at an abstract level, not an implementation level. We are not concerned with how olists are implemented. Instead, we work directly with the olist specification. Working at the abstract level greatly simplifies the reasoning.

Induction is also used in showing that operation implementations preserve the rep invariant (see section 4.5.3). However, this reasoning is done at the level of the implementation. As before, the first step is to establish the property for the primitive constructors. If the rep is mutable, all operations must be considered in the induction step, since any of them might modify the rep. Only the constructors need be considered in the induction step if the rep is immutable.

4.9.5 Equal, Similar, and Copy Operations

It may appear that data abstractions should get *copy* and *equal* operations "for free": Every type should be provided with these operations automatically, and their implementations should be generated by the compiler. However, such implementations cannot be generated automatically because they depend on the interpretation of the rep—that is, on the abstraction function.

Equality cannot be implemented automatically because two abstract objects can be equal even if their reps are not. Consider, for example, the rational number implementation discussed earlier, in which the reps were not in reduced form. Clearly the rational number represented by the record

{num: 1, denom: 2}

is equal to that represented by

{num: 13, denom: 26}

Copying cannot be implemented automatically because it is not always clear what an object consists of. For example, to copy an olist, do we copy just the top node, so that the copy shares the right and left subtrees with the original, or do we copy the right and left subtrees as well? The correct answer depends on the meaning of the type.

Even though the implementation of such operations as *equal* is left to the programmer, we recommend that conventions be followed to ensure a uniformity of meaning for these names. We have adopted a convention of reserving the *equal* operation for the equality relation; that is, it should be impossible to distinguish between two equal objects. Therefore, two mutable objects can be equal if and only if they are the very same object;

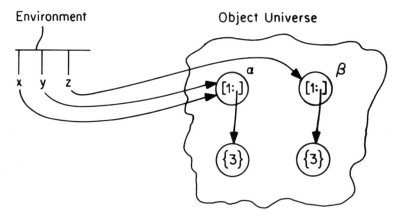

Figure 4.18 Two concepts of equality.

equal = **proc** (s1, s2: intset) **returns** (bool)
    effects Returns true if *s1* and *s2* are the same intset object.

similar = **proc** (s1, s2: intset) **returns** (bool)
    effects Returns true if *s1* and *s2* contain the same elements.

copy = **proc** (s: intset) **returns** (intset)
    effects Returns a new intset object containing the same elements as *s*.

Figure 4.19 Specifications of *equal*, *similar*, and *copy* for intsets.

otherwise we could distinguish them by applying a mutating operation to one and not the other. For example, consider the two situations shown in figure 4.18. Here, $x = y$ but $x \sim= z$.

We use the name *similar* for the operation that captures the relationship between α and β, namely, the fact that the two objects contain similar values. We require the *copy* operation to produce a result that is similar to its argument but that does not share any components with the argument. Thus, z's object could have been produced by calling $copy(x)$.

The CLU built-in types follow these conventions for the meaning of *equal*, *copy*, and *similar*. User-defined types should too. Thus if intset provides these operations, their specifications would be as in figure 4.19 and their implementations as in figure 4.20. Note that *similar* compares the values of its two arguments to see whether they contain the same elements. *Equal* and *copy* can be implemented in terms of the *equal* and *copy* operations provided by arrays.

```
equal = proc (s1, s2: cvt) returns (bool)
    return (s1 = s2) % uses rep$equal
    end equal

similar = proc (s1, s2: intset) returns (bool)
    return (subset(s1, s2) cand subset(s2, s1))
    end similar

% subset is an internal routine that returns true if s1 is a subset of s2
subset = proc (s1, s2: cvt) returns (bool)
    i: int := rep$low(s1)
    while i <= rep$high(s1) do
        if ~ member(up(s2), s1[i]) then return (false) end
        i := i + 1
        end
    end subset

copy = proc (s: cvt) returns (cvt)
    return (rep$copy(s))
    end copy
```

Figure 4.20 Implementations of *equal, similar,* and *copy* for intsets.

4.10 Summary

This chapter has discussed data abstractions: what they are, how to specify their behavior, and how to implement them, both in general and in CLU. Four examples were discussed in detail. Intsets and olists are mutable, while lists and polys are immutable. The implementations of list and olist illustrate the use of type parameters and recursive data structures.

In addition, we discussed some important aspects of data type implementations. In general, not all objects of the rep type are legal representations of the abstract objects; the rep invariant defines the legal representations. The abstraction function defines the meaning of the rep by stating the way in which the legal rep objects represent the abstract objects. Both the rep invariant and the abstraction function should be included as comments in the implementation. They are helpful in starting the implementation, since they force the implementer to be explicit about assumptions. They are also helpful to anyone who examines the implementation later, since they explain what must be understood about the rep.

Finally, we explored some issues that must be considered in designing data types. Care must be taken in deciding whether or not a type is mutable and in choosing its operations so that it serves the needs of its users adequately. We also discussed data type induction and how it is used to prove properties of objects, and we discussed the need for types to provide *equal* and *copy* operations.

The benefits of locality and modifiability apply to data abstractions as well as to procedures. These benefits can be achieved only if we have abstraction by specification. Locality requires that a representation be modifiable only within its type's implementation. If modifications can occur elsewhere, then we cannot establish the correctness of the implementation just by examining its code; for example, we cannot guarantee locally that the rep invariant holds, and we cannot use data type induction with any confidence. Modifiability requires even more—all access to a representation must occur within its type's implementation. If access occurs in some other module, we cannot replace the implementation without affecting that other module.

Thus it is crucial that access to the representation be restricted to the type's implementation. It is desirable to have the programming language enforce this restriction, since otherwise restricted access is another property that must be proved about programs. CLU provides this restriction via compile-time type checking.

Further Reading

Morris, James H., Jr., 1973. Types are not sets. In *Conference Record of ACM Symposium on Principles of Programming Languages*, pp. 120–124.

Hoare, C. A. R., 1972. Proof of correctness of data representations. *Acta Informatica* 1(1): 271–281. Reprinted in *Programming Methodology, A Collection of Articles by Members of IFIP WG2.3*, edited by David Gries (New York: Springer-Verlag), 1978.

Exercises

4.1   Implement the procedure

diff = **proc** (p: poly) **returns** (poly)
      effects Returns the poly that is the result of differentiating p, e.g.,
        diff $(x^3 + 7x + 6) = 3x^2 + 7$

(See figure 4.3 for a specification of *poly*.)

4.2   Specify and implement the procedure

reverse = **proc** (l: list) **returns** (list)

which reverses the order of the elements in a list. (Lists are defined in figure 4.12.)

Exercises

97

4.3 Suppose intsets were implemented using an array of integers as in figure 4.5, but the rep array was kept sorted in increasing size. Give the rep invariant and abstraction function for this implementation.

4.4 Suppose polys (figure 4.3) were implemented with the zero poly represented by the array [0: 0]. Give the rep invariant and abstraction function for this implementation.

4.5 Implement intsets (figure 4.2) using olists (figure 4.16). Be sure to include the rep invariant and abstraction function.

4.6 Suppose you knew that polys (figure 4.3) would usually be sparse. Give an implementation that is efficient in such a situation. (Hint: Lists may be useful.) Be sure to include the rep invariant and abstraction function.

4.7 Polys (figure 4.3) should provide *parse* and *unparse* operations similar to those for ints and reals. Invent an external form for polys, specify these operations, and then implement them as part of the implementation of figure 4.7 or exercise 6.

4.8 Specify and implement *equal*, *similar*, and *copy* operations for *olist* (figure 4.16).

4.9 Bounded queues have an upper bound, established when a queue is created, on the number of elements that can be stored in the queue. Queues are mutable and provide access to their elements in first-in/first-out order. Queue operations include

```
create = proc (n: int) returns (queue)
enq = proc (q: queue, elem: t)
deq = proc (q: queue) returns (t)
```

Create creates a new queue with maximum size *n*, *enq* adds an element to the front of the queue, and *deq* removes the element from the end of the queue; the queue elements are of some arbitrary type *t*. Provide a specification of queue, including extra operations as needed for adequacy. Implement your specification. Give the rep invariant and abstraction function.

4.10 Specify and implement a rational number type. Be sure to include *equal*, *similar*, and *copy* operations. Give the rep invariant and abstraction function.

4.11 Give an informal argument that the implementation of poly in figure 4.7 preserves the rep invariant.

4.12 Give an informal argument that the implementation of olist in figure 4.17 preserves the rep invariant.

4.13 An abstract invariant of lists (figure 4.12) is that the size of a list is never negative. Give an informal argument to establish this invariant.

4.14   Suppose we wanted to evaluate a poly (figure 4.3) at a given point:

eval = **proc** (p: poly, x: int) **returns** (int)
           effects Returns the value of p evaluated at x , e.g.,
              poly\$eval(x^2 + 3x, 2) = 10

Should *eval* be an operation of poly? Discuss.

4.15   Categorize the operations of lists (figure 4.12) into the four categories of
section 4.9.2. Is this type adequate? (Hint: Consider implementing a procedure
to concatenate two lists, such as *concat*((1 2), (3 4)) = (1 2 3 4).)

4.16   Categorize the operations of olists (figure 4.16). Is this type adequate?

4.17   A student proposes a type *matrix* with operations to add and multiply
matrices and to invert a matrix. These matrices are mutable; for example, the
invert operation modifies its argument to contain the inverse of the original matrix. A second student claims that a matrix abstraction ought not to be mutable.
Discuss.

4.18   A student says that as long as programs outside a type's implementation cannot modify the rep, we have achieved as much as is possible from data
abstraction. Discuss.

5

Exceptions

A procedural abstraction is a mapping from arguments to results, with possible modification of some of the arguments. The arguments are members of the *domain* of the procedure, and the results are members of its *range*. If a procedure is implemented in a strongly typed language like CLU, then compile-time type checking ensures that the actual arguments really are members of the domain and the results are members of the range.

Often a procedure makes sense only for arguments in a subset of its domain. For example, the *choose* operation of intsets makes sense only if its intset argument is not empty. So far we have dealt with this situation by using a procedure whose behavior is defined over only part of its domain, such as

> choose = **proc** (s: intset) **returns** (int)
>   **requires** *s* is not empty.
>   **effects** Returns an arbitrary member of *s*.

The caller of a partial procedure must ensure that the arguments are in the permitted subset of the domain, and the implementer can ignore arguments outside this subset. Thus in implementing *choose*, we ignore the case of the empty intset.

There are times when partial procedures are a bad idea. The compiler cannot guarantee (in general) that the arguments to a procedure such as *choose* are actually in the permitted subset, and the procedure may thus be called with arguments outside the subset. When this happens, the result is likely to be an obscure error that is difficult to track down. For example, the program may continue to execute for some time after the error occurs and damage important data bases.

Partial procedures lead to programs that are not robust. A *robust* program is one that continues to behave reasonably even in the presence of errors. If an error occurs, the program may not be able to provide exactly the same behavior as if there were no error, but it should behave in a well-defined way. Ideally, it should continue after the error by providing some approximation to its behavior in the absence of an error; a program like this is said to provide *graceful degradation*. At worst, it should halt with a meaningful error message and without causing damage to permanent data.

A method that enhances robustness is to use procedures whose behavior is defined for all inputs in the domain. If the procedure is unable to perform its "intended" function for some of these inputs, at least it can inform its caller of the problem. In this way, the situation is brought to the attention of the caller, which may be able to do something about it, or at least avoid harmful consequences of the error.

Of course, checking whether inputs are in the permitted subset of the domain takes time, and it is tempting not to bother with the checks, or to use them only while debugging, and suppress them during production. This is generally an unwise practice. It is better to develop the habit of *defensive programming*, that is, writing each procedure to defend itself against errors. Errors can be introduced by other procedures, by the hardware, or by the user entering data; these latter errors will continue to exist even if the software is error-free.

Defensive programming makes it easier to debug programs. During production it is even more valuable because it can prevent a small error from causing a large problem, such as a damaged data base. Removing checks during production is analogous to disconnecting warning lights in an airplane; a pilot would never do this because the results could be catastrophic. Checks for illegal arguments should be removed only if we have proved that the errors can never occur or if the checks are prohibitively expensive.

How should the caller be notified if a problem arises? One possibility is to define a specific error result. For example, a factorial procedure might return zero if its argument is not positive:

fact = **proc** (n: int) **returns** (int)
    effects If $n > 0$ returns $n!$; otherwise returns 0.

This solution is not very satisfactory. Since the call with illegal arguments is probably an error, it is more constructive to treat this case in a specific way, so that a programmer who uses the procedure is less likely to ignore the error by mistake. Also, if every value of the return type is a possible result of the procedure, as is the case with *choose*, this solution is impossible, since there is no value available to indicate the error. A method that works uniformly, whether a distinguishable error value is available or not, is preferable.

Such a method can be achieved by generalizing procedures. Instead of having a single input domain, we divide the domain D into a number of subsets whose union is D:

$$D = D_0 \cup D_1 \cup \ldots \cup D_n$$

Each subset is "interesting" in the sense that the procedure behaves differently in each case. For example, the input of *choose* would be divided into two subsets:

$D_0 = \{$all nonempty intsets$\}$
$D_1 = \{$the empty intset$\}$

For an input in D_0, *choose* returns the chosen integer as before. For an input in D_1, it still returns, but in a way that brings the problem to the attention of its caller. As noted, it is not possible to do this notification by returning an integer, since any integer is a potential outcome for inputs in D_0. Therefore, some other method of notification is needed.

We provide notification by having different output ranges for each domain subset and by naming the different cases. We arbitrarily name the case of input from D_0 the "normal" case; the names of the other cases can be chosen by the definer of the procedure. Thus we have

p: $D_0 \rightarrow$ normal (R_0)
$\quad D_1 \rightarrow$ ename$_1$ (R_1)
$\quad \ldots$
$\quad D_n \rightarrow$ ename$_n$ (R_n)

where *ename$_1$*, $\ldots$, *ename$_n$* are the names of the "exceptional" situations corresponding to input in D_1, $\ldots$, D_n, respectively, and R_0, $\ldots$, R_n describe the number, order, and types of results returned in each case. *Ename$_1$*, $\ldots$, *ename$_n$* are called *exceptions*. For example, we have

choose: $D_0 \rightarrow$ normal (int)
$\quad\quad\quad D_1 \rightarrow$ empty ( )

where R_1 is empty because no result is returned to the caller in this case.

In this chapter we discuss how to specify, implement, and use procedures with exceptions. We also discuss a number of related design issues.

5.1 Specifications

To specify procedures that raise exceptions we extend our notation by adding a *signals clause* to the specification:

signals   % list names and results of exceptions here

This clause is part of the header, following the returns clause. It can be omitted if the procedure has no exceptions. The exceptions should be

choose = **proc** (s: intset) **returns** (int) **signals** (empty)
      effects If $size(s)$ = 0, signals *empty*; otherwise returns an arbitrary
          element of *s*.

search = **proc** (a: array[int], x: int) **returns** (ind: int)
             signals (not_in, duplicate (ind1: int))
      requires *a* is sorted in ascending order.
      effects If $x \in a$ exactly once, returns *ind* such that $a[ind] = x$; if $x \in a$
          more than once, signals *duplicate(ind1)*, where *ind1* is the index of
          one of the occurrences of *x*; otherwise signals *not_in*.

Figure 5.1 Some specifications with exceptions.

separated by commas, with their results (if any) enclosed by parentheses.
For example,

    choose = **proc** (i: intset) **returns** (int) **signals** (empty)

states that *choose* can raise the exception *empty*, and that no results are
returned in this case, while

    search = **proc** (a: array[int], x: int) **returns** (ind: int)
              signals (not_in, duplicate(ind1: int))

states that *search* can raise two exceptions: *not_in*, in which case no result
is returned, and *duplicate*, in which case a single integer result is returned.
The name *ind1* is introduced to refer to this result in the rest of the specification.

    As before, the effects section should define the behavior of the procedure
for all inputs not ruled out by the requires clause. Since this behavior includes exceptions, the effects section must define what causes the procedure
to terminate with each exception, and what its behavior is in each case. If
a procedure signals an exception for arguments in some D_i, inputs in D_i
should not be excluded in the requires clause. Termination by signaling an
exception is part of the ordinary behavior of the procedure.

    Figure 5.1 shows specifications of *choose* and *search*. Note that the spec-
ification of *search* contains a requires clause and that, as usual, its effects
section assumes that the requires clause is satisfied. The domains of *search*
are

    D_0 = { $\langle a, x \rangle$ | x ∈ a exactly once }
    D_1 = { $\langle a, x \rangle$ | x ∉ a }
    D_2 = { $\langle a, x \rangle$ | x ∈ a more than once }

Here the domain subsets are pairs consisting of an array and an integer.

rep = array[int]

choose = **proc** (s: **cvt**) **returns** (int) **signals** (empty)
    if rep$size(s) = 0 **then signal** empty **end**
    return (rep$bottom(s))
    end choose

Figure 5.2 Implementation of *choose.*

5.2 The CLU Exception Mechanism

A programming language can make exceptions easy to use by providing an exception mechanism. The mechanism must provide a way for a procedure to *signal* exceptions, that is, bring them to the attention of the caller. In addition, the calling procedure must be able to receive this information in order to *handle* the exceptions. However, the calling procedure should not have to handle exceptions that ought not to occur. For example, suppose we perform the following statement:

   f intset$size(s) > 0 **then** x:= intset$choose(s) **end**

It would certainly be surprising if the call of *choose* signaled the *empty* exception. It should not be necessary for the programmer to handle exceptions in cases like this.

5.2.1 Signaling Exceptions

A CLU proc can terminate in one of a number of conditions. Its header names the exceptions and defines the number and types of results returned in each case. For example, we have the headers

   choose = **proc** (s: intset) **returns** (int) **signals** (empty)
   search = **proc** (a: array[int], x: int) **returns** (int)
                signals (not_in, duplicate (int))

To terminate its execution in the "normal" way, a proc either falls off the end of its body or uses the **return** statement. To terminate with exception *ename*, a proc uses the **signal** statement

   signal *ename* (*exprs*)

The **signal** statement is legal only if the exception name is listed in the procedure header (or is *failure*, as discussed in section 5.2.4). Figure 5.2 contains an implementation of *choose* that illustrates signaling. In this figure we are assuming the intset implementation of figure 4.5.

5.2.2 Handling Exceptions

When a proc terminates with an exception, the calling program must be able to transfer control to some code that handles the exception. CLU provides the **except** statement for this purpose. This statement has the form

statement **except** *handler list* **end**

The *handler list* includes some subset (possibly all) of the exceptions signaled by calls in the statement. Each *handler* in the list names one or more exceptions, followed by a list of statements (called the *handler body*) describing what to do if that exception occurs. Permitting several exceptions to be named in the same handler avoids code duplication when more than one exception is handled in the same way.

When an exception occurs, the program no longer continues its normal flow. Instead control goes to the nearest containing **except** statement with a handler that names the exception. For example, if the call of intset$*choose* in

i := intset$choose(s)

terminates with the *empty* exception, the assignment to *i* is not made, and control transfers to the nearest **except** statement with a handler for *empty*.

Several different forms are available for handlers depending on whether the named exceptions have associated result objects and whether those objects are used in the handler body. To handle one or more exceptions with no associated objects, we simply list the exception names. For example,

when overflow, zero_divide: *body*

will handle exceptions named *overflow* and *zero_divide*, neither of which has any associated result objects, by executing the code in *body*; an example is

z := x/y
    except when overflow, zero_divide: z := 0 **end**

To handle exceptions with result objects that are to be used in the handler body, names must be associated with the objects. Again, a list of exception names is given, but it is followed by declarations of local variables to name the result objects; an example is

when duplicate (j: int): *body*

The scope of the declarations is the handler body. All the named exceptions must return objects of the declared types in the stated order. When the handler is executed, these objects are assigned to the local variables and the body is executed. For example, if $search(a, x)$ signals *duplicate*, then after execution of

x := search(a, x)
    except when duplicate (j: int): i := j **end**

i contains an index at which x occurs in a.

To handle exceptions with result objects when the objects are not used in the handler body, the list of exception names is followed by (*); an example is

  when not_in, duplicate (*): *body*

The number and types of result objects associated with the exceptions in this form need not agree; for example, *duplicate* has a single result, while *not_in* has none. This form encourages a programming style in which a procedure returns all possibly useful information when signaling; if this information is not needed in the calling procedure, it can be ignored.

If the programmer wishes to handle all remaining exceptions without listing their names, one of the following two forms can be used as the last handler in an **except** statement. The form

  others: *body*

is used when information about exception names and result objects is not important. If information about the exception name is desired, the form

  others (name: string): *body*

can be used. Here the name of the exception is given to the handler body as a string.

Note that any exception results are lost in the **others** handler. The **others** handler is used when cleaning up in a uniform way after several exceptions whose results do not matter. An example is a program that must close a stream before returning, no matter what problem arose.

Except statements can be nested. An **except** statement as a whole may raise any of the exceptions of its *statement* that are not included in the *handler list* plus any exceptions raised by the *handler list*. Figure 5.3 illustrates various uses of the **except** statement. The procedure *index_sum* attempts to produce the sum of the indexes in array a in which elements of intset s are stored; if it encounters difficulties, it signals *problem*. In the

```
index_sum = proc (s: intset, a: array[int]) returns (int) signals (problem)
    sum: int := 0
    while true do
        x: int := intset$choose(s)      % can signal empty
        sum := sum + search(a, x)        % overflow, not_in & duplicate
            except when duplicate (j: int): sum := sum + j end    % overflow
        intset$delete(s, x)
    end except when empty: return (sum)
            others: signal problem
            end
    end index_sum
```

Figure 5.3 Examples of **except** statements.

figure, possible signals are shown as comments on each statement. The *empty* exception from the call of *choose* is used to terminate the loop. The *duplicate* exception from *search* is handled in the inner **except** statement, while the *not_in* exception is handled in the outer **except** statement, in the **others** arm. This arm also handles *overflow*.

5.2.3 The Resignal Statement

Sometimes an exception is handled simply by signaling the same exception with the same results. CLU provides the **resignal** statement for this case:

   resignal *enames*

This is just a short form for an **except** statement in which every *ename* is signaled explicitly with exactly the same results. For example,

   i := search(a, x) **resignal** duplicate, not_in

is short for

```
i := search(a, x)
    except when not_in: signal not_in
        when duplicate (j: int): signal duplicate (j)
    end
```

5.2.4 Unhandled Exceptions

CLU programs need not contain handlers for all exceptions. Instead every procedure in CLU has one additional exception beyond those named in the signals list. This exception is named *failure*, and it has one argument, a string. Every exception raised by a called procedure and not handled explicitly by the caller is turned automatically into the *failure* exception. This transformation is the same as what would be accomplished by attaching the following **except** statement to the procedure body:

except when failure (s: string): **signal** failure (s)
    others (s: string): **signal** failure ("unhandled exception:" ‖ s)
  end

If the unhandled exception is not *failure*, its name becomes part of the string argument; if the unhandled exception is *failure*, its string argument is retained. Thus the string argument contains the name of the first unhandled exception.

The implementation of *choose* in figure 5.2 contains an unhandled exception. Since we know that *s* is not empty when the **return** statement is executed, we do not bother to catch the *bounds* exception. In the unlikely event that this exception is raised, it will be propagated automatically as *failure* ("unhandled_exception: bounds"). (Note that the *bounds* exception can occur only if the implementation of arrays is broken.)

The **others** clause of the **except** statement treats *failure* differently from other exceptions. If the exception caught by the arm

  others (s: string): *body*

is *failure*, then *s* is assigned the string argument of the *failure* exception rather than the string "failure." The effect is the same as if the **others** arm were broken into two parts:

when failure (s: string): *body*
others (s: string): *body*

Failure can also be signaled explicitly; if this is done, it must be given a string argument.

All exceptions should be handled except for those that ought not to occur. Therefore failures generally mean that a program or system error has occurred. At this point programmer analysis is usually needed. The information in the failure string is intended to be used primarily by programmers, not by programs.

5.2.5 The Exit Statement

In chapter 2 we introduced the **break** and **continue** statements; **break** is used to exit from a loop, while **continue** is used to continue with the next loop iteration. Sometimes more general transfers of control are needed. The **exit** and **except** statements together provide the needed ability.

The form of the **exit** statement is like that of the **signal** statement:

exit *ename* (*exprs*)

Ename names an *exit*; this name must occur in a containing **except** statement, and control is transferred to the nearest containing **except** statement with a handler for the name. The (*exprs*) part is omitted if no values are to be passed with the exit.

The following example uses the **exit** statement to terminate a search loop:

```
ai = array[int]
x: int
i: int := ai$low(a)
begin
    while i <= ai$high(a) do
        x := a[i]
        if special(x) then exit found end
        i := i + 1
    end
    x := make_new_one( )      % didn't find one, so make one
    end except when found: end
%  At this point we have a suitable value for x
```

5.3 Using Exceptions in Programs

When implementing a procedure with exceptions, the programmer's job, as always, is to provide the behavior defined by the specification. If this behavior includes exceptions, the program must signal the proper exceptions at the proper times with the meaning as described in the specification. To accomplish this task, the program will probably need to handle exceptions that are raised by procedures it calls.

Exceptions can be handled in two different ways. Sometimes an exception is *propagated* up another level; that is, the caller also terminates by signaling an exception with either the same or a different name. Before propagating the exception, the caller may do some local processing. Such processing is sometimes needed to satisfy the caller's specification. For example, an operation of a data type must ensure that objects satisfy the rep invariant before it terminates.

A second possibility is that the caller *masks* the exception, that is, handles the exception itself. For example, a procedure might read all characters

```
intset = cluster is create, insert, delete, member, size, choose

    rep = array[int]

    create = proc ( ) returns (cvt)
        return (rep$new( ))
        end create

    insert = proc (s: intset, x: int)
        if ~member(s, x) then rep$addh(down(s), x) end
        end insert

    delete = proc (s: cvt, x: int)
        j: int := getind(s, x) except when not_in: return end
        s[j] := rep$top(s)
        rep$remh(s)
        end delete

    member = proc (s: cvt, x: int) returns (bool)
        getind(s, x) except when not_in: return (false) end
        return (true)
        end member

    size = proc (s: cvt) returns (int)
        return (rep$size(s))
        end size

    choose = proc (s: cvt) returns (int) signals (empty)
        return (rep$bottom(s))
            except when bounds: signal empty end
        end choose

    getind = proc (s: rep, x: int) returns (int) signals (not_in)
        i: int := rep$low(s)
        while true do
            if x = s[i] then return (i) end
                except when bounds: signal not_in end
            i := i + 1
            end
        end getind

end intset
```

Figure 5.4 Implementation of intset.

in a stream by using stream$*getc*. When *getc* signals *end_of_file*, this simply
means that all characters have been read and it is time to finish up.

This section gives examples of using exceptions in specifying and implementing procedures. Our first example is intsets. These are the same as
before (figure 4.5), except that *choose* signals when its argument is empty.
An implementation is given in figure 5.4. Note that we have changed *getind*

to signal *not_in* when x is not in s. This signal is caught in *delete*, which simply masks the exception and returns normally. The implementation of *member* also masks the exception.

The implementation of *getind* uses the *bounds* exception signaled by the array *fetch* operation to indicate that it has checked all the elements of the array. When it catches this signal, it signals *not_in*—an example of exception propagation. *Choose* provides another example of propagation.

The implementation of *delete* contains an example of unhandled exceptions. If *getind* returns normally, we know that the index is within bounds and that the array is not empty. Therefore we do not bother to catch the *bounds* exception from the calls of array operations *top*, *fetch*, and *remh*.

The next example is ordered lists. By deleting the requires clauses and adding exceptions to operations *addel*, *remel*, and *least*, we get a more robust abstraction. The resulting specification is shown in figure 5.5.

The implementation of olist is shown in figure 5.6. Several of the operations make use of the **resignals** statement. For example, *addel* calls itself recursively on the right and left subtrees; if either of these call signals *dupl*, *addel* resignals it. **Resignal** is often useful in recursive implementations like this.

As a final example, consider the *summation* procedure specified and implemented in figure 5.7. Note that both the requires and **where** clauses list the exceptions associated with the *add* operation. *Summation* makes use of lists as defined in figure 4.12, except that operations *first* and *rest* signal *empty* if the argument list is empty. The *empty* exception arising from the call of list$*first* is masked; when it occurs, all elements of the list have been examined. The *underflow* and *overflow* exceptions are propagated.

5.4 Design Issues

Exceptions should be used to eliminate most constraints listed in requires clauses. The requires clause should remain only for efficiency reasons or if the context of use is so limited that we can be sure the constraint is satisfied. For example, *search* should probably still require that the array be sorted, since it can then be implemented much more efficiently. Also, the *merge* procedure used in merge sort (figure 3.5) should require its arguments to be sorted, for efficiency and also because the context of use is so limited.

Exceptions should also be used to avoid encoding information in ordinary results. For example, the *getind* procedure of the intset cluster (figure 5.4) signals if the element is not in the array instead of returning one greater than the high bound. It is better to convey this information with

olist = **data type** [t: **type**] **is** create, addel, remel, is_in, empty, least

Requires *t* has operations
  lt, equal: **proctype** (t, t) **returns** (t)
  that define a total order on *t*.

Overview

 Olists are mutable lists of elements. Operations *addel* and *remel* modify the
olist. Operation *least* returns the smallest element of the list.

Operations

 create = **proc** ( ) **returns** (olist[t])
  effects Returns a new, empty olist.

 addel = **proc** (s: olist[t], x: t) **signals** (dupl)
  modifies *s*
  effects If *x* is already in *s*, signals *dupl*; otherwise inserts *x* in *s*.

 remel = **proc** (s: olist[t], x: t) **signals** (not_in)
  modifies *s*
  effects If *x* is not in *s*, signals *not_in*; otherwise removes *x* from *s*.

 is_in = **proc** (s: olist[t], x: t) **returns** (bool)
  effects Returns true if *s* contains some element equal to *x*; otherwise
  returns false.

 empty = **proc** (s: olist[t]) **returns** (bool)
  effects Returns true if *s* contains no elements; otherwise returns
  false.

 least = **proc** (s: olist[t]) **returns** (t) **signals** (empty)
  effects If *s* is empty, signals *empty*; otherwise returns element *e* of *s*
  such that no element of *s* is < *e* (as determined by *t$lt*).

end olist

Figure 5.5 Specification of ordered lists.

an exception, since the result returned in this case cannot be used like a
regular result. By using an exception, we make it easy to distinguish this
result from a regular one, thus avoiding a potential error.

 When a procedure's behavior is defined only for arguments in a subset of
its domain, its implementation is permitted to do anything for arguments
outside the subset. Of course, not all implementations are equally desirable.
By far the best approach is to signal *failure*. Often this happens naturally,
either by not handling an exception that would not occur for arguments
in the permitted subset or by falling off the end of a procedure that must
return results. (The CLU compiler inserts code to signal *failure* in this
case.) In other cases, it may be worthwhile to spend a little effort. For

olist = **cluster** [t: **type**] **is** create, addel, remel, is_ in, least, empty
   where t **has** equal, lt: **proctype** (t, t) **returns** (bool)

node = record[val: t, left, right: olist[t]]
rep = variant[some: node, empty: null]

% A typical olist is [e1, ..., en]
% The abstraction function is
% A(r) = [ ]   if r is empty
%   = A(n.left) ‖ [n.val] ‖ A(n.right)   if r has some elements
%   where n = value_some(r)

% The rep invariant is
% If is_some(r) then
%   all components of n.left are < n.val
%   & n.val < all components of n.right
%   where n = value_some(r)

create = **proc** ( ) **returns** (**cvt**)
   return (**rep**$make_empty(**nil**))
   end create

addel = **proc** (s: **cvt**, v: t) **signals** (dupl)
   tagcase s
       tag some (n: node):
           if v = n.val **then** **signal** dupl
               elseif v < n.val **then** addel(n.left, v)
               else addel(n.right, v)
               end **resignal** dupl
       tag empty:
           rep$change_some(s, node${val: v, left: create( ), right: create( )})
       end
   end addel

Figure 5.6 Implementation of ordered lists (continues on next page).

example, in *merge* (figure 3.6) we could compare array elements as we look at them to see whether they are ordered properly and signal *failure* if they are not. Such a test costs little, but improves the robustness of the program considerably.

   Not all errors give rise to exceptions. Consider an erroneous record in a large input file, where it is possible to continue processing the file by skipping that record. In such a case, it may be appropriate to inform a person (not a program) about the error. Exceptions are a mechanism for communication among programs, not for communication from programs to people. To communicate with people, an error message might be written on some output device. Note, by the way, that what is done with the error is defined in the abstraction's specification.

```
remel = proc (s: cvt, v: t) signals (not_in)
    tagcase s
        tag empty: signal not_in
        tag some (n: node):
            if v = n.val
                then if empty(n.right)
                    then    % replace this node with left subtree
                        rep$v_gets_v(s, down(n.left))
                    else    % make n.val be value from right subtree
                        n.val := least(n.right)
                        remel(n.right, n.val)
                    end
                elseif v < n.val then remel(n.left, v)
                else remel(n.right, v)
                end resignal not_in
        end
    end remel

is_in = proc (s: cvt, v: t) returns (bool)
    tagcase s
        tag empty: return (false)
        tag some (n: node):
            if v = n.val then return (true)
                elseif v < n.val then return (is_in(n.left, v))
                else return (is_in(n.right, v))
                end
        end
    end is_in

least = proc (s: cvt) returns (t) signals (empty)
    tagcase s
        tag empty: signal empty
        tag some (n: node): return (least(n.left))
        end except when empty: return (n.val) end
    end least

empty = proc (s: cvt) returns (bool)
    return (rep$is_empty(s))
    end empty

end olist
```

Figure 5.6 (continued)

summation = **proc** [t: **type**] (x: list[t], zero: t) **returns** (t)
            signals (underflow, overflow)
      requires *t* has an addition operation
            add: **proctype** (t, t) **returns** (t) **signals** (underflow, overflow)
      effects Produces the sum of the elements in *x* starting from *zero*;
            signals *underflow* or *overflow* if an underflow or overflow occurs
            while computing the sum.
summation = **proc** [t: **type**] (x: list[t], zero: t) **returns** (t)
            signals (underflow, overflow)
      where t **has** add: **proctype**(t, t) **returns** (t) **signals** (underflow, overflow)
      sum: t := zero
      while true do
          sum := sum + list[t]$first(x)   % use of t$add
              except when empty: **return** (sum) **end**
          x := list[t]$rest(x)
          end resignal underflow, overflow
      end summation

Figure 5.7 The *summation* procedure.

Exceptions are not always associated with errors, though. For some
abstractions there may be more than one kind of normal behavior, and
here exceptions are a convenient tool. They provide a means for allowing
several kinds of behavior and enabling the caller to distinguish among the
different cases.

For example, the *lookup* operation on a symbol table has a twofold purpose. Given an identifier, it determines whether a declaration for this
identifier has already been processed and, if it has, returns information
about the identifier. A header for this operation might be

lookup = **proc** (s: symbol_table, id: string) **returns** (info)
            signals (not_ in)

Here we have chosen to treat the case in which the declaration exists
as the normal case, but we could easily have made the opposite choice.
One rationale for choice is efficiency. It is likely that exceptions will be
more expensive than normal returns. (This is true in CLU, although the
difference is small.) Therefore the case that is expected to occur most
frequently should be considered the normal case. When a distinction is
made between cases based on performance considerations, it is obvious
that no notion of "error" can be attached to the exception case.

The relationship of modification of arguments to termination in an exception condition is worth further discussion. The modifies section of a
specification indicates that an argument may be modified, but does not

say when this will happen. If there are exceptions, it is likely that the modification will happen only for some of them. Exactly what happens must be described in the effects section. Modifications must be described explicitly in each case where they occur; if no modifications are described, this means none happens. For example, consider

addel = **proc** (s: olist[t], x: t) **signals** (dupl)
    modifies s
    effects If x is already in s, signals *dupl*; otherwise inserts
      x in s.

Since no modification is described when *addel* signals *dupl*, s is modified only when *addel* returns normally.

The failure exception is an implicit exception of every procedure, yet no mention of failure was made in any of the specifications. This omission is appropriate. A specification describes the behavior of a procedure when it is working and when its arguments satisfy the requires clause. Failure occurs when the procedure is broken and no longer meets its specification or when the arguments are improper.

When failure occurs, it usually cannot be handled by the program. A failure means either a software error or a hardware fault. A very robust program, with redundant ways of doing things, may be able to shut down the failed part and continue running with the rest. In this case it is important to log the error so that it can be corrected later. A less robust program would terminate gracefully and log the error. (If the program is running in debugging mode, the person at the console can be notified immediately.)

The exceptions signaled by operations of data types are associated with specific operations. However, the names of exceptions are related to the type as a whole; they refer either to the status of the type's objects (for example, a list is either empty or not) or to the fact that no such object exists (for example, there is a largest and a smallest integer). Therefore the type's operations should use the same exception name for similar situations. For example, list operations *first* and *rest* both signal *empty* when the argument list is empty, and integer operations that attempt to produce too big an integer all signal *overflow*.

Exceptions provide information that can usually be obtained by calling operations directly. For example, the fact that a list is empty can be learned either by calling *first* and having it signal or by calling *empty*. Such redundancy is not required. For example, there might be a type in which the only way to find out whether an object is empty is through an exception.

In our sample implementations, we did not try to avoid raising exceptions; instead we used them to control program flow. Exceptions can improve performance by reducing the number of calls. For example, the *choose* implementation in figure 5.4 uses the *bounds* exception to avoid calling the array *size* operation.

5.5 Summary

In this chapter, we have generalized procedures to include exceptions. Exceptions are needed in robust programs because they provide a way to respond to errors. If an argument is not what is expected, a procedure can notify the caller of this fact rather than simply fail. Since this notification is distinct from the normal case, the caller cannot confuse the two.

Exceptions are introduced when procedures are designed. Most procedures should be defined over the entire domain; exceptions are used to take care of situations in which the "normal" behavior cannot happen. Partial procedures are suitable only for efficiency reasons or when the procedure is used in a limited context in which it can be proved that all calls have proper arguments. In either case, when the procedures are implemented, it is a good idea to practice defensive programming, causing failure to be signaled in as many cases as possible for inputs outside the accepted domain subset.

In implementing a procedure, the programmer must ensure that it terminates as specified in all situations. Only exceptions permitted by the specification should be signaled, and each should be signaled for the right reason. In doing the implementation, it is reasonable to use exceptions of called procedures and to mask or propagate as appropriate.

Further Reading

Goodenough, John B., 1975. Exception handling: issues and a proposed notation. *Communications of the ACM* 18(12): 683–696.

Liskov, Barbara, H., and Alan Snyder, 1979. Exception handling in CLU. *IEEE Transactions on Software Engineering* SE-5(6): 546–558.

Exercises

5.1 Modify the poly data abstraction defined in chapter 4 (figure 4.3) to take advantage of exceptions. Specify the new abstraction and then implement it.

5.2 Modify the list data abstraction defined in chapter 4 (figure 4.12) to take advantage of exceptions. Specify the new abstraction and then implement it.

5.3 Implement *remove_dupls* in terms of ordered lists, taking advantage of the exceptions signaled by the olist operations of figure 5.5.

5.4 Exercise 9 in chapter 4 concerned a bounded queue abstraction. Redefine this abstraction using exceptions and provide an implementation of the modified abstraction.

5.5 A map is a table that associates elements (of some arbitrary type) with strings. Each string is mapped to at most one associated element. Map operations include *create*, to create an empty map; *insert*, to add a string and its associated element; *change*, to change the element associated with a string; *delete*, to delete a string and its associated element; and *eval*, to look up the element associated with a string. Specify maps, being careful to include appropriate exceptions and to provide extra operations as needed for adequacy. Then implement your specification and provide the rep invariant and abstraction function. Your implementation must be efficient: The running time of *eval* must be much less than Order(n), where n is the number of entries in the map.

5.6 Discuss the adequacy of the map abstraction defined in exercise 5. Is it possible always to avoid having calls of map operations raise exceptions, for example, by calling other appropriate map operations? Discuss whether the decisions made in this regard were right or wrong.

5.7 A procedure to compute the minimum value of an array might require a nonempty array, return the smallest integer if the array is empty, or signal an exception if the array is empty. Discuss which alternative is best.

5.8 Instead of computing the minimum value, suppose that the procedure of exercise 7 were adding the elements of the array. Would this change your idea of which alternative is best?

6

Iteration Abstraction

This chapter discusses our final abstraction mechanism, the *iteration abstraction*, or *iterator* for short. Iterators are a generalization of the iteration methods available in most programming languages. They permit users to iterate over arbitrary types of data in a convenient and efficient way.

For example, an obvious use of a set is to perform some action for each of its elements:

> **for all** elements of the set
> > **do** action

Such a loop might go through the set completely—for example, to print all elements of a set. Or we might search for an element that satisfies some criterion, in which case the loop can stop as soon as the desired element has been found.

Intsets as we have defined them so far provide no convenient way to perform such loops. For example, suppose we want to compute the sum of the elements in an intset:

> setsum = **proc** (s: intset) **returns** (int)
> > **effects** Returns the sum of the elements of *s*.

The implementation of *setsum* shown in figure 6.1 illustrates the two main defects of our intset abstraction. First, to loop through all elements, we delete each element returned by *choose* so that it will not be chosen again. Thus two operations, *choose* and *delete*, must be called on each iteration. This inefficiency could be avoided by having *choose* remove the chosen element, but we still have the second problem, which is that iterating over an intset destroys it by removing all its elements. Such destruction may be acceptable at times, but cannot be satisfactory in general. Although we can collect the removed elements and reinsert them later, as is done in figure 6.1, such a method is clumsy and inefficient.

If *setsum* were an intset operation, we could implement it efficiently using array operations. However, *setsum* does not really make sense as an intset operation; it seems peripheral to the concept of a set. Furthermore, even if we could justify making it an operation, what about other similar

ai = array[int]

```
setsum = proc (s: intset) returns (int)
    a: ai := ai$new( )
    %  Compute the sum
    sum: int := 0
    while true do
        x: int := intset$choose(s)
        sum := sum + x
        intset$delete(s, x)
        ai$addh(a, x)
        end except when empty: end
    %  Restore the elements of s
    i: int := 1
    while true do
        intset$insert(s, a[i]) except when bounds: return (sum) end
        i := i + 1
        end
    end setsum
```

Figure 6.1 An implementation of *setsum*.

procedures we might want, such as printing all the elements? There must be a way to implement such procedures outside the type.

To support iteration adequately, we need to access all elements in a collection efficiently and without destroying the collection. How might we do this for intsets? We might provide an *el_seq* operation:

```
el_seq = proc (s: intset) returns (seq[int])
            effects Returns a sequence containing all the elements of s, each
                    exactly once, in some arbitrary order.
```

Given this operation, we can implement *setsum* as shown in figure 6.2. Since *el_seq* does not modify its argument, we no longer need to rebuild the intset after iterating.

Although *el_seq* makes it easier to use intsets, it is inefficient, especially if the intset is large. First, we have two data structures—the intset itself and the sequence. Second, in the case of a search loop we have probably done too much work; on average such a loop need not examine all elements of the collection being searched. For example, if we were searching an intset for a negative element, we could stop as soon as we encountered the first negative element. However, we must process the entire collection to build the sequence.

An alternative to *el_seq* is an operation that simply returns the representing array. However, this solution is very bad because it destroys abstraction

```
setsum = proc (s: intset) returns (int)
    items: sequence[int] := intset$el_seq(s)
    i: int := 1
    sum: int := 0
    while true do
        sum := sum + items[i]
            except when bounds: return (sum) end
        i := i + 1
        end
    end setsum
```

Figure 6.2 Implementation of *setsum* using the *el_seq* operation.

by specification. We have, in effect, exported the **down** operation, and no
longer have local control over the representation.

What is needed is a general method of iteration that is convenient and
efficient and that preserves abstraction by specification. The iterator provides the needed support. It is called like a procedure, but instead of
terminating with a result, it has many results, which it produces, or *yields*,
one at a time. The produced items can be used in other modules that specify actions to be performed for each item. The using module will contain
some sort of looping structure, such as

> **for each** result item i produced by iterator A
>   **do** perform some action on i

Each time the iterator yields an item, that item is acted on by the body
of the loop. Then control continues in the iterator so that it can yield the
next item.

Note the separation of concerns in such a form. The iterator is responsible for producing the items, while the module containing the loop defines
the action to be performed on them. The iterator can be used in different modules that perform different actions on the items, and it can be
implemented in different ways without affecting these modules.

Since the iterator yields items one at a time, it avoids the space and time
problems discussed earlier. We need not construct a potentially large data
structure to contain the items. Moreover, if the using module is performing
a search loop, the iterator can be stopped as soon as the item of interest
is found.

As mentioned earlier, iterators are a generalization of the iteration methods available in most programming languages. In addition to some form

of "while" loop, programming languages typically provide a "for" loop for iterating over integers. Such iteration is useful in conjunction with arrays, which are indexed, but does not mesh well with nonindexed abstractions like intset.

6.1 Specification

Iterators, like other abstractions, must be defined by specifications. The form of an iterator specification is similar to that of a procedure. The header has the form

iname = **iter** (...) **yields** (...) **signals** (...)

Here we use **iter** to identify the abstraction as an iterator. An iterator can yield zero or more objects in each iteration; the number and types of these objects are described in the *yields clause*. (If no objects are produced at each yield, the yields clause can be omitted.) The iterator may not return any results when it terminates normally, but it can terminate with an exception, with name and results as indicated in the signals clause. For example,

elements = **iter** (s: intset) **yields** (int)
  requires s is not modified by the loop body.
  effects Yields the elements of s, each exactly once, in some
        arbitrary order.

is a plausible operation for intset. Note that *elements* has no exceptions; if it is given the empty intset as an argument, it simply terminates without ever yielding any items. It is typical that the use of iterators eliminates problems associated with certain arguments (like the empty intset) for related procedures (like *choose*).

We require that s not be modified by the using loop. Such constraints are typical for iterators that work on mutable objects like intsets, since it is then clear what is meant by "the elements of s." These constraints will be discussed further in section 6.4.

6.2 CLU Iterators

Iterators can be implemented in CLU by an *iter* module. CLU also provides a **for** statement for using iterators.

rep = array [int]

elements = **iter** (s: **cvt**) **yields** (int)
  i: int := **rep**$low(s)
  while true do
    yield (s[i])
      except when bounds: **return end**
    i := i + 1
    end
  end elements

Figure 6.3 The *elements* operation of intset.

6.2.1 Implementing Iterators

A CLU iter is similar to a proc. It has a header like the header of the iterator specification. Within its body it uses the **yield** statement to produce the next item. However, the iter does not terminate at this point. Instead, it remains ready to yield more items if its caller is interested. In this case, it continues execution at the statement following the **yield** statement, and its local variables have the same values as before the **yield**.

Figure 6.3 shows an implementation of the *elements* iterator for intsets. We are assuming here the same array representation as in figures 4.5 and 5.5. When *elements* continues execution after a **yield**, its variable *i* is the index of the previously yielded element. *Elements* then goes on to produce the next element and terminates when all elements have been produced.

6.2.2 Using Iterators

CLU provides a **for** statement for using iterators. This statement defines some loop variables. If the variables are local to the loop, the form of the statement is

  for *decl_list* **in** *iter_call* **do** *body* **end**

If the variables are nonlocal, the form is

  for *idn_list* **in** *iter_call* **do** *body* **end**

When the **for** statement begins execution, loop variables are created if necessary and the *iter_call* is performed. When the iterator yields, the yielded objects are assigned to the loop variables, and then the body of the loop is executed. When the body finishes, control resumes in the iterator. When the iterator terminates, the **for** loop also terminates, and control in the using program continues at the statement following the **for** statement. Also, if the loop body executes a statement that terminates the **for**

```
setsum = proc (s: intset) returns (int)
    sum: int := 0
    for e: int in intset$elements(s) do
        sum := sum + e
        end
    return (sum)
    end setsum
```

Figure 6.4 Using an iterator.

	sum	e	i	pc in setsum	pc in elements
After first yield	0	4	1	start of body	after **yield**
After second yield	4	1	2	start of body	after **yield**
After third yield	5	7	3	start of body	after **yield**
Termination	12	—	—	**return** statement	—

Figure 6.5 A trace of an execution of *setsum*.

statement, the iterator is terminated automatically.

Figure 6.4 shows a third implementation of *setsum*. Each time the loop body is executed, another element of the set has been assigned to the loop variable *e*. This element is added to *sum*, and then the loop body terminates, causing execution of the iterator to be resumed. When all items have been yielded, *elements* terminates, causing the **for** statement to terminate. Then *setsum* returns the computed sum.

An iterator and the for loop that uses it pass control back and forth between them. First the iterator is called. It produces a result item for its user, but it does not terminate. Instead it yields control to its user. When the user is ready for the next item, it resumes execution of the iterator.

Figure 6.5 traces the execution of *setsum* when called with the intset $\{1, 4, 7\}$, assuming that the representing array contains these elements in the order 4, 1, 7. The figure shows the values of local variables *sum* and *e* in *setsum* and *i* in *elements* immediately after each yield. It also shows the value of *sum* after termination of the loop in *setsum*; no values are shown for *i* and *e* in this case because the call of *elements* and the loop body have terminated. After each yield, the program counter (*pc*) for *setsum* is at the beginning of the body of the **for** statement, and the program counter for *elements* is at the statement following the **yield**, that is, at the assignment to *i*.

Figure 6.6 shows a second use of *elements*, a search to find a negative element. If the yielded integer is negative, *find_neg* returns immediately. This return terminates the **for** statement and thus terminates *elements* before other integers in the intset have been produced.

find_neg = **proc** (s: intset) **returns** (int) **signals** (not_found)
   for x: int **in** intset$elements(s) **do**
      if x < 0 **then return** (x) **end**
     end
   signal not_found
   end find_neg

Figure 6.6 A searching use of an iterator.

Iterators can be called only in **for** statements, and each **for** statement can call only one iterator. However, **for** statements can be nested. Also, iterators can be implemented using **for** statements, and they can have recursive implementations. Some examples are given in section 6.3.

6.2.3 Built-In Iterators

CLU provides a number of iterator operations for the built-in types. For integers, there is int$*from_to*:

   from_to = **iter** (x, y: int) **yields** (int)
      effects Yields the integers between x and y, inclusive, in order.

There is also int$*from_to_by*, whose third argument specifies the amount by which x is incremented on each iteration. For arrays (and sequences), there are two iterators: *elements*, which yields the elements of the array (or sequence) from the low bound to the high bound; and *indexes*, which yields the indexes of the array (or sequence) from the low bound to the high bound. Finally, strings have an iterator *chars*, which yields the characters of the string from first to last.

6.3 Examples

This section contains several examples of iterators. The first example is the simple "filter" in figure 6.7; it is called a filter because it removes some elements produced by another iterator, yielding only the others to its caller:

   filter = **iter** (s: intset, els: **itertype** (intset) **yields** (int),
              pred: **proctype** (int) **returns** (bool)) **yields** (int)
      requires s is not modified by loop body.
      effects Yields, in arbitrary order, all elements produced by $els(s)$
         that satisfy *pred*.

filter = **iter** (s: intset, els: **itertype** (intset) **yields** (int),
                    pred: **proctype** (int) **returns** (bool)) **yields** (int)
          for e: int **in** els(s) **do**
              if pred(e) **then yield** (e) **end**
              end
          end filter

Figure 6.7 A filter.

node = record[val: t, left, right: olist[t]]
rep = variant[some: node, empty: null]

small_to_big = **iter** (s: **cvt**) **yields** (t)
    tagcase s
        tag empty: **return**
        tag some (n: node):
            for e: t **in** small_to_big(n.left) **do** % produce elements of left subtree
                yield (e)
                end
            yield (n.val)
            for e: t **in** small_to_big(n.right) **do** % produce elements of right subtree
                yield (e)
                end
        end
    end small_to_big

Figure 6.8 The *small_to_big* operation of olist.

For example, if *odd* returns true only if its argument integer is odd, then

filter (s, intset$elements, odd)

yields the odd integers in *s*.
    As a second example, we add a *small_to_big* operation to olist:

small_to_big = **iter** (s: olist[t]) **yields** (t)
        requires *s* is not modified by loop body.
        effects Yields elements of *s*, each exactly once, in ascending
            order as determined by *t$lt*.

This operation might replace the *least* operation provided previously. The
implementation is shown in figure 6.8. It is recursive; first *small_to_big*
calls itself recursively on the left subtree, next it yields the value at the
node, and then it calls itself recursively on the right subtree. Note, by the
way, that *small_to_big* is a parameterized abstraction. It inherits the type
parameter *t* from olist.

si = sequence[int]

permutations = **iter** (seq: si) **yields** (si)
   % if sequence is single integer, yield it and return
   if si\$size(seq) = 1 **then yield** (seq)
     else for index: int **in** si\$indexes(seq) **do**
        % create a subsequence omitting the value at index
        val: int := seq[index]
        prefix: si := si\$subseq(seq, 1, index − 1)
        suffix: si := si\$subseq(seq, index + 1, si\$size(seq))
        % append subsequence permutations to value, and yield
        for perm: si **in** permutations(prefix ‖ suffix) **do** % ‖ is concat
          yield (si\$addl(perm, val))
          end % for perm
       end % for index
     end % if
   end permutations

Figure 6.9 The *permutations* iterator.

Our next example is the *permutations* iterator in figure 6.9:

permutations = **iter** (seq: sequence[int]) **yields** (sequence[int])
     effects Yields, in an arbitrary order, all sequences constructed
        by permuting the elements of *seq*.

The implementation uses the *indexes* iterator of sequences. It works by producing all permutations that have the first element of the original sequence first, then all permutations with the original second element first, and so on. To produce permutations with the *n*th element first, it creates a subsequence containing all the other elements and produces permutations of that subsequence by calling itself recursively. It adds the *n*th element to the front of each subsequence permutation and yields the resulting sequence.

Our final example is the *allprimes* iterator in figure 6.10:

allprimes = **iter** ( ) **yields** (int)
     effects Yields the prime numbers in increasing order.

Our implementation keeps all primes found so far in the *primes* array. The **while** loop produces odd numbers; those that are multiples of previously found primes are eliminated in the **for** loop. We stop the **for** loop as soon as we reach a prime whose square is greater than the candidate, since that candidate cannot possibly be divisible by a larger prime.

ai = array[int]

```
allprimes = iter ( ) yields (int)
    yield (2)   % yield the first prime
    primes: ai := ai$new( )
    x: int := 1
    while true do
        x := x + 2   % the next candidate
        for p: int in ai$elements(primes) do
            if x // p = 0 then exit not_found end   % x is not a prime
            if p * p > x then break end   % x is a prime
            end except when not_found: continue end
        % have a prime
        yield (x)
        ai$addh(primes, x)
        end
    end allprimes
```

Figure 6.10 The *allprimes* iterator.

6.4 Design Issues

Most data types will include iterators among their operations, especially types like intset and olist whose objects are collections of other objects. Iterators are frequently needed for adequacy; they make elements of a collection accessible in a way that is both efficient and convenient.

A type might have several iterators. For example, olist might have the operation

```
big_to_small = iter (s: olist[t]) yields (t)
        requires s is not modified by the loop body.
        effects Yields elements of s, each exactly once, in descending
            order as determined by t$lt.
```

in addition to the *small_to_big* operation discussed earlier.

For mutable collections we have consistently required that the loop body not modify the collection being iterated over. If we omit this requirement, the iterator must behave in a well-defined way even when modifications occur. For example, suppose integer n is deleted from an intset while *elements* is running; should n be yielded by *elements* or not?

One approach is to require that an iterator yield the elements contained in its collection argument at the time of the call, even if modifications occur later. The behavior of an iterator specified in this way is well defined, but the implementation is likely to be inefficient. For example, if the *elements* iterator were specified like this, its implementation would have to copy the intset when it was first called—just what we objected to in the *el_seq* operation. Because the approach of constraining the loop body avoids such inefficiencies, it will be preferred most of the time. A related issue is whether the iterator itself can modify the collection. As a general convention, such modifications should be avoided.

Modifications by the loop body or the iterator can sometimes be useful. For example, consider a program that performs tasks waiting on a task queue:

```
for t: task in task_queue$all_tasks(queue) do
    % perform t.
    % if t generates a new task, nt, then enqueue it by performing
    %     taskqueue$enq(queue, nt)
end
```

When the task being performed generates another task, we simply enqueue it to be performed later; the *all_tasks* iterator will present it for execution at the appropriate time. However, examples like this are rare; usually neither iterator nor loop body will modify the collection.

6.5 Summary

This chapter identified a problem in the adequacy of data types that are collections of objects. Since a common use of a collection is to perform some action for its elements, we need a way to access all elements. This method should be efficient in space and time, convenient to use, and nondestructive of the collection. In addition, it should support abstraction by specification.

Iterators are a mechanism that solves this problem. Since they produce the objects one at a time, extra space to store the objects is not needed, and production can be stopped as soon as the desired object has been found. Iterators support abstraction by specification for the containing type by encapsulating the method of producing the objects; this method depends on knowledge of the rep, but using programs are shielded from this knowledge.

Iterators are efficient to execute. Yielding from an iterator is like a call: The body of the **for** loop is "called" from the iterator. Resuming an iterator is similar to a return: The loop body "returns" to the iterator. Therefore the cost of using iterators is at most one procedure call per execution of the loop body; the cost may be less because of compiler optimizations.

Iterators are useful in their own right, as was indicated by the *permutations* and *primes* examples. However, their main use is as operations of data types. We shall see other examples of such use in the rest of the book.

Further Reading

Atkinson, Russell R., Barbara H. Liskov, and Robert W. Scheifler, 1978. Aspects of implementing CLU. In *Proceedings of the ACM 1978 Annual Conference*, pp. 123–129.

Exercises

6.1 Specify a procedure *is_prime*, which determines whether an integer is prime, and then implement it using *allprimes* (figure 6.10).

6.2 Implement the *big_to_small* iterator for olists that was specified in section 6.4.

6.3 It would be useful to have a *terms* iterator for poly (figure 4.3) that yielded all the nonzero terms. Give a specification for this iterator. Should it be a poly operation?

6.4 Implement the *terms* iterator specified in exercise 3, either as a poly operation or not as you prefer. Then use *terms* to implement

diff = **proc** (p: poly) **returns** (poly)
  effects Returns the poly that is the result of differentiating *p*.

6.5 Give a specification of a *stack* abstraction that provides access to its elements in last-in/first-out order. Stacks are mutable. Note that for stacks to be useful in some applications (for example, to implement the type in exercise 6), there must be some way to access all the stack elements. We could provide an iterator that yields the stack elements from the most recently pushed to the least recently pushed, or we could provide a *fetch* operation that returns the *i*th element for *i* between 1 and the number of elements on the stack. Which is the better choice?

6.6 Figure 6.11 gives a specification of a block-structured symbol table abstraction that provides operations well-suited to compiling programs written in a block-structured language. Implement this abstraction and give the rep invariant and abstraction function. (Hint: Use stacks (see exercise 5) and maps (see exercise 5 in chapter 5); add extra operations to stack and map as needed, and give specifications for any such added operations.)

symtab = **data type** [etype: **type**] **is** create, enter_scope, leave_scope,
        add_id, lookup, depth

Overview

Symtab is a symbol table designed for use in a compiler for a block-
structured programming language. It provides operations to enter and leave
scopes as well as operations to add and lookup information about
identifiers. Symtabs are mutable.

Operations

create = **proc** ( ) **returns** (symtab)
        effects Returns a new, empty symtab, which is ready to accept
        information about the outermost scope.

enter_scope = **proc** (s: symtab)
        modifies s
        effects Modifies s so that it is prepared to accept information about
        a new scope.

leave_scope = **proc** (s: symtab) **signals** (outermost_scope)
        modifies s
        effects If s is positioned at the outermost scope, signals
        outermost_scope. Otherwise removes information about the
        innermost scope from s; s is now positioned to accept information
        about the scope that immediately contains the removed scope.

add_id = **proc** (s: symtab, id: string, info: etype) **signals** (duplicate)
        modifies s
        effects If *id* is already defined in the innermost scope in s, signals
        duplicate. Otherwise adds *id* with the associated *info* to the
        innermost scope of s.

lookup = **proc** (s: symtab, id: string) **returns** (etype)
                signals (not_defined)
        effects If *id* is not defined in any scope in s, signals *not_defined*.
        Otherwise returns the information associated with s. If s is
        defined in several scopes, the information in the innermost of
        these scopes is returned.

depth = **proc** (s: symtab) **returns** (int)
        effects Returns the number of scopes in s.

end symtab

Figure 6.11 Specification of a block-structured symbol table.

7

Using Pascal

This book presents a programming method based on abstractions. So far we have discussed three kinds of abstractions—procedures, data types, and iterators—and used CLU to provide concrete examples. It is important, however, not to confuse the approach to programming with the programming language used to illustrate it. In this chapter we therefore discuss how one might use our approach in conjunction with Pascal.

CLU was designed specifically to support our method and contains syntactic entities corresponding to each of the three kinds of abstractions. Pascal contains no direct support for either data or iteration abstractions, nor does it provide for parameterized abstractions. Nevertheless, the design of programs to be implemented in Pascal can still make productive use of these abstractions. To be sure, a design should take advantage of those design methods that Pascal supports well and avoid methods that will be inordinately difficult to implement, but the programming language should not be allowed to influence the program design unduly. Long ago good assembly language programmers learned to use kinds of abstractions that were not provided by their programming language. Many Pascal programmers have learned the same lesson.

In this chapter we discuss procedure, data, and iteration abstractions in turn, and then conclude with an example. There are a number of plausible ways to provide these abstractions in Pascal. They can be broadly partitioned into methods that store abstract objects on the stack and methods that store abstract objects on the heap. The heap-based methods tend to provide more flexibility, at the cost of some efficiency. We shall confine our present discussion to one heap-based method. In chapter 15, we discuss other alternatives.

Our discussion presumes only a limited familiarity with Pascal and explains the more unusual features of Pascal when they are introduced. There are many dialects of Pascal in use today. For the most part we have tried to stick to a subset of Pascal that is common to most dialects. However, some of the examples appearing in this chapter are not compilable in some dialects of Pascal. All of our programs have been compiled and run using Turbo Pascal®.

7.1 Procedure and Function Abstractions

In Pascal, operations that return values are called functions, and those that do not are called procedures. We shall use the word *operation* when our remarks are pertinent to both functions and procedures.

There are several important differences between CLU procedures and Pascal operations. The most significant are the following:

1. Pascal operations can read and modify variables that are global to the called operation.

2. Pascal formals are declared to be either **var** or value. Neither of these argument-passing mechanisms is the same as that used in CLU.

3. Pascal functions can return only one argument, and this argument must belong to a rather limited set of types.

4. Pascal has no exception mechanism.

7.1.1 Global Variables

There are situations in which global variables are useful. The most prominent occur when it is useful to save some state between invocations of a procedure. (In many other programming languages, this can be accomplished using own variables; for a description of own variables in CLU, see section A.8.5 of the appendix.)

In general, it is unwise to design abstractions that depend upon global variables, since the meaning of the abstraction will then depend upon the context in which it is placed. This can severely limit its reusability. We shall not use global variables in this chapter.

7.1.2 Parameter Passing and Returning Values

Pascal supports two parameter-passing mechanisms: **var** parameters are passed by reference, and non-**var** parameters by value. With *call by value*, the actual argument is an expression (for example, $x + y$). This expression is evaluated, and its value is copied into a formal variable of the operation. There are two important points to observe about call by value. The first point is that modifying the formal parameter has no effect on the actual. Failure to keep this in mind is a frequent source of bugs in Pascal programs. The second point is that call by value is expensive for large values, such as arrays, since a copy must be made.

In *call by reference* the actual must be a variable or a "pseudovariable," such as an element of an array. It cannot be an expression. The formal is thought of as providing another name for the actual; that is, using the formal in the called operation is equivalent to using the actual in the caller.

Pascal functions are like CLU procedures in that they can modify their arguments and return a result. They may, however, return only one result, and there are limitations on its type. Only scalars and pointers can be returned in most versions of Pascal, and this will have an impact on the conventions we use to implement abstract types.

7.1.3 Handling Exceptions

In designing CLU procedures we make heavy use of CLU's exception mechanism. Sometimes we use the special exception *failure* to indicate that an error has occurred. More often, however, we use exceptions simply as another way for the called procedure to communicate information to the caller. In Pascal we use different conventions to replace these two distinct uses of exceptions.

To deal with failures in Pascal programs we provide a procedure called *failure*. This procedure is placed at the front of the program, where it can be called by every other operation in the program. Its specification is

> **procedure** failure (s: error_msg)
> > **modifies** output stream on primary output device.
> > **effects** Prints the following string on the primary output
> > device:
> > > 'Failure. Program terminated because:' + s
> > and then halts program execution.

(We use + to denote a concatenation of strings. This is common in versions of Pascal that support strings.) The *failure* procedure is called whenever *failure* would be signaled, either implicitly or explicitly, in CLU. The argument of the call should, of course, identify the problem and the operation in which it occurred. Ordinarily type *error_msg* would be declared to be some type of string. If the Pascal being used does not support strings, the argument can be represented as a *packed array* of characters.

The specification

> search = **proc** (a: int_array, x: int) **returns** (i: int) **signals** (not_in)
> > **effects** If *x* occurs in *a*, returns *i* such that $a[i] = x$; otherwise
> > signals *not_in*.

illustrates a typical use of nonfailure exceptions in CLU. In Pascal we achieve a similar effect with

function search1 (**var** a: int_array; x: integer; **var** i: integer):
        search1_exceptions
    modifies i
    effects If x occurs in a, sets i such that $a[i] = x$ and returns ok;
        otherwise returns not_in.

where *search1_exceptions* is the enumerated type

type search1_exceptions = (ok, not_ in)

Note that *search1* takes a third argument, the **var** parameter i. This variable is used to hold the result in the normal case.

By convention, we embed all calls to this function in a **case** statement, such as

    case search1(a, x, i) **of**
        ok: a[i] := a[i] + 1;
        not_ in: **write**('Element not found.')
    end

The **case** statement should have an arm corresponding to each value that can be returned by the function. (If only two alternatives are possible, as in *search1*, the result could be a boolean instead of an enumerated type, and an **if** statement would be used instead of a **case** statement.)

This approach has some drawbacks. It is inconvenient that *search1* modifies a **var** parameter rather than returning an index, since we cannot use it in expressions such as $a[search(a, x)]$. Furthermore, *search1* is not as safe to use as *search*. For example, suppose we decide to add another value to *search1_exceptions*. According to our convention, an extra arm needs to be added to each **case** statement in which *search1* is called. Failure to do this, however, will not result in either a compile-time or a runtime error. (In Pascal, when the selector of a **case** statement does not match any of the arms, the program continues to execute at the statement following the **case**.)

7.2 Data Abstractions

In this section we outline an approach to specifying, using, and implementing data abstractions when programming in Pascal. The approach relies heavily upon programming conventions. This is in contrast to CLU, where the compiler provides direct support. The conventions are designed primarily to ensure that we can implement or reimplement a data abstraction

intqueue = **data type is** q_ new, q_ isempty, q_append, q_ remfirst

Overview

  Intqueues are used to store values of type integer. Elements can be
  retrieved and deleted from the queue in first-in/first-out order only.

Operations

  function q_ new: intqueue
    effects Returns a new queue with no elements in it.

  function q_ isempty (q: intqueue): boolean
    effects Returns true if q has no elements in it and false otherwise.

  procedure q_append (q: intqueue; e: integer)
    modifies q
    effects Adds e to the end of q.

  function q_ remfirst (q: intqueue): integer
    requires q is not empty.
    modifies q
    effects Returns the element at the front of q and removes that
      element from q.

end intqueue

Figure 7.1 Specification of a queue of integers.

without reimplementing any of the modules that use it and that we can
return abstract values from functions.

  Figure 7.1 contains a specification of a queue-of-integers data abstraction.
This abstraction supplies a type, *intqueue*, and the set of all operations that,
by convention, are allowed to deal with the representation of that type. All
of the operation names start with the prefix q. We rely on this convention
to avoid conflicts with operation names used in other data types.

  Note that despite the fact that *q_append* and *q_remfirst* both modify
the formal q, q is not passed as a **var**. We shall implement intqueue, and
all other abstract types, using a pointer into the heap. When we pass an
abstract object to a procedure, we are therefore always passing a reference.
To modify the abstract object, we modify the object to which the reference
points, not the reference itself. Consequently, abstract objects are never
passed as **var** parameters.

  We choose to store objects of abstract types on the heap to facilitate
implementing types whose objects change size at runtime. We insist that
the rep be a reference to ensure that abstract objects can be returned by

functions. If, for example, the rep were a record containing a pointer as a component, it could not be returned by a function. Finally, using a pointer ensures that it is efficient to pass abstract objects by value.

By convention, the declaration of the rep type will always be of the form

type
    typename = ↑typename_rep;

where *typename_rep* is a type that describes the "real" rep. For example, for type intqueue we use the rep type:

type
    intqueue = ↑intqueue_rep;
    intqueue_rep = ↑intqueue_elem;
    intqueue_elem = **record**
        val: **integer**;
        next: intqueue_rep;
    end;

Given our conventions, it is simple to declare variables of an abstract type, for example,

var q: intqueue;

The meaning of this declaration is that space for a new uninitialized variable, *q*, is allocated on the stack. The amount of space allocated is simply enough to hold a pointer, just as in CLU.

When *q* is declared, only a pointer is allocated, and therefore only a pointer is freed when the procedure in which *q* is declared is exited. Storage for the elements of *q* is allocated in the operations of type intqueue by explicit calls to the Pascal primitive **new**. Since garbage is not collected automatically in Pascal, storage for elements must be freed explicitly by the program in which the queue is used. Pascal provides a **dispose** primitive for this purpose. **Dispose** takes a pointer as an argument, and it follows the pointer and deallocates the storage pointed to by that pointer. The amount of storage deallocated depends upon the pointer's type. **Dispose**(*q*) will deallocate the space occupied by a single pointer of type *intqueue_rep*. This means that all the storage occupied by elements will remain allocated, which can lead to a particularly pernicious kind of bug called a *storage leak*. A storage leak occurs when we fail to deallocate storage that should be freed. The only symptom of the bug is that over time the heap is gradually exhausted. By the time the program runs out of space, it may be very far from the place where the leak occurs.

To ensure proper storage deallocation, each type should provide an additional operation called *destroy*. *Destroy* takes a single argument, an object of the type, and deallocates all heap storage associated with that object. The value of the object on return from *destroy* is undefined. Here is a specification of *q_destroy*:

> **procedure** q_destroy (q: intqueue)
> **modifies** q
> **effects** All heap storage occupied by q is deallocated.

By convention, users never free abstract objects directly; instead they call the *destroy* operation for the object's type. This convention controls the amount of space freed. It does not, however, ensure that there are no references to the deallocated object. If such a reference exists, it becomes a *dangling reference* after the object is freed. If an attempt is made to access storage via a dangling reference, meaningless information can be read. Even worse are modifications made through dangling references; they can destroy consistency constraints on other objects. Although it is possible for a Pascal implementation to detect dangling references, such detection is expensive and therefore generally not provided. Figure 7.2 contains an implementation of intqueue, including a *destroy* operation.

Following the convention that each abstract type be represented by a pointer ensures that assignment of abstract values has the same meaning as assignment in CLU. This is not the same as the meaning of assignment for built-in types in Pascal. In Pascal, for example, the array assignment *a1* := *a2* involves making a copy of *a2*. A subsequent change to the value of *a2* will not affect *a1*. If, on the other hand, we write *q1* := *q2*, where *q1* and *q2* are intqueues, only the pointer will be copied, and *q1* and *q2* will share the intqueue.

It is important that sharing work properly; subsequent changes to the queue *q2* must be reflected in *q1* and vice versa. We accomplish this by having an extra level of indirection in the rep: Instead of pointing directly to the first item on the queue, *q1* and *q2* point to a pointer to that first item (see figure 7.3). This allows us to remove the first element of the queue and have that change be visible to both *q1* and *q2*. The indirection could also occur through a record instead of just a pointer. For example, suppose we decided to store the queue size in the rep (this would be useful if intqueues had a *size* operation). Then we might have

{ Declaration of the rep type for intqueue }

type
   intqueue = ↑intqueue_rep;
   intqueue_rep = ↑intqueue_elem;
   intqueue_elem = **record**
     val: **integer**;
     next: intqueue_rep;
     end;

{ Start of implementations of operations of type intqueue }

function q_new: intqueue;
   var q: intqueue;
   begin new(q); q↑ := **nil**; q_new := q **end** { q_new };

function q_isempty (q: intqueue): **boolean**;
   begin q_isempty := (q↑ = **nil**) **end** { q_isempty };

procedure q_append (q: intqueue; e: **integer**);
   var last_elem, elem: intqueue_rep;
   begin
     new(elem);
     elem↑.val := e;
     elem↑.next := **nil**;
     if q_isempty(q)
       then q↑ := elem
       else begin
         last_elem := q↑;
         while last_elem↑.next <> **nil do**
           last_elem := last_elem↑.next;
         last_elem↑.next := elem
         end;
   end { q_append };

function q_remfirst (q: intqueue): **integer**;
   var oldq: intqueue_rep;
   begin
     if q_isempty(q) {Requires not met}
       then failure('q_remfirst called with empty queue')
       else begin
         q_remfirst := q↑↑.val;
         oldq := q↑;
         q↑ := q↑↑.next;
         dispose(oldq) {deallocate space for removed element}
        end
   end { q_remfirst };

Figure 7.2 Implementations of intqueue operations (continues on next page).

```
procedure q_destroy (q: intqueue);
    var next_elem, old_elem: intqueue_rep;
    begin
       next_elem := q↑;
       while next_elem <> nil do
          begin
          old_elem := next_elem;
          next_elem := next_elem↑.next;
          dispose(old_elem);
          end;
       dispose(q)
    end { q_destroy };
```

{ End of implementations of operations of type intqueue }

Figure 7.2 (continued)

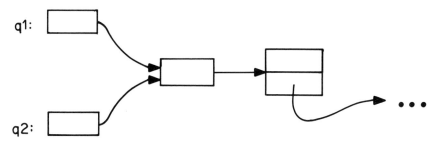

Figure 7.3 $q1$ and $q2$ share the same queue.

```
intqueue_rep = record
    size: integer;
    first: ↑intqueue_elem;
    end
```

but sharing would still work properly.

It is frequently useful to include in a type a copy function, for example, *q_copy*, along the lines discussed in chapter 4. Copy functions should provide a complete, nonsharing copy of their argument. They can be used whenever we want an assignment with the normal Pascal semantics. For example, we could write $q1 := q\_copy(q2)$ rather than $q1 := q2$. Figure 7.4 illustrates the absence of sharing when $q\_copy(q2)$ is assigned to $q1$. Keep in mind, however, that these copy functions will not be called by the compiler as part of implicit assignments—for example, in call by value.

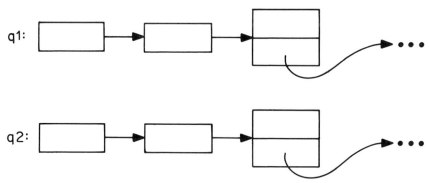

Figure 7.4 $q1$ and $q2$ do not share the same queue.

Finally, note that our method does not prevent programs outside a type's implementation from accessing the rep. We rule out such access by convention.

7.3 Polymorphic Abstractions

The type parameters of CLU allow us to build polymorphic abstractions, that is, abstractions that can be used for objects of different types. Pascal has some built-in polymorphism. The array type generator, for example, can be used to declare arrays with different bounds and different types of elements. However, Pascal provides almost no support for user-defined polymorphism. Of course, this need not prevent us from designing polymorphic abstractions. Specifying such abstractions is relatively straightforward. We use a syntax similar to that used for specifying parameterized abstractions to be implemented in CLU. The example of a parameterized queue is given in figure 7.5.

Since Pascal provides no support for implementing parameterized abstractions, we again rely on programming conventions. An implementation of the parameterized queue can look much like the implementation of intqueue. The declaration of the rep type is given in figure 7.6.

These declarations are not legal Pascal since they include illegal identifiers, such as *queue[etype]*. We also introduce syntax errors into the implementations of the operations. These syntax errors prevent programmers from accidentally using uninstantiated parameterized abstractions. The implementations are like those in figure 7.2 except that we replace the prefix *q* by the prefix *queue[etype]* and the identifiers *intqueue* and *integer* by *queue[etype]* and *etype*, respectively. An implementation of the operation *q[etype]_append* is given in figure 7.7.

queue = **data type** [etype: **type**] **is** q[etype]_new, q[etype]_isempty,
          q[etype]_append, q[etype]_remfirst, q[etype]_destroy

Overview

Queues are used to store values of type etype. Elements can be retrieved
and deleted from the queue in first-in/first-out order only.

Operations

. . .

procedure q[etype]_append (q: queue[etype]; e: etype)
    modifies q
    effects Adds e to the end of q.

. . .

end queue

Figure 7.5 Partial specification of a parameterized queue.

type
    queue[etype] = ↑queue[etype]_rep;
    queue[etype]_rep = ↑queue[etype]_elem;
    queue[etype]_elem = **record**
        val: etype;
        next: queue[etype]_rep;
        end;

Figure 7.6 Rep type for implementation of *queue[etype]*.

procedure q[etype]_append (q: queue[etype]; e: etype);
    var last_elem, elem: queue[etype]_rep;
    begin
        new(elem);
        elem↑.val := e;
        elem↑.next := **nil**;
        if q[etype]_isempty(q)
            then q↑ := elem
            else begin
                last_elem := q↑;
                while last_elem↑.next <> **nil do**
                    last_elem := last_elem↑.next;
                last_elem↑.next := elem
                end
    end { q[etype]_append };

Figure 7.7 Implementation of *q[etype]_append*.

```
type
    intq = ↑intq_rep;
    intq_rep = ↑intq_elem;
    intq_elem = record
        val: integer;
        next: intq_rep;
    end;
procedure intq_append (q: intq; e: integer);
    var last_elem, elem: intq_rep;
    begin
        new(elem);
        elem↑.val := e;
        elem↑.next := nil;
        if intq_isempty(q)
            then q↑ := elem
            else begin
                last_elem := q↑;
                while last_elem↑.next <> nil do
                    last_elem := last_elem↑.next;
                last_elem↑.next := elem
            end
    end { intq_append };
```

Figure 7.8 Instantiation of *queue[etype]*.

Before using the implementation of *queue[etype]* in a program, we must use a text editor to replace every occurrence of *queue[etype]* by a convenient, legal identifier, for example, *intq*; to replace all remaining occurrences of the identifier *etype* by some already known type, for example, **integer**; and to change the names of the operations by replacing the prefix, for example, renaming *q[etype]_append* to *intq_append*. Figure 7.8 shows an instantiation of the type declaration of figure 7.6 and the implementation of figure 7.7.

7.4 Generators

Pascal has a **for** statement that permits iteration over subranges of scalar types (except real), but it provides no mechanism for user-defined iterators. As noted in chapter 6, this can be inconvenient. For example, to sum the elements in an intqueue, we might write code like that in figure 7.9. This implementation would be a bit less awkward if we added a *q_copy* function to intqueue, but it would still be unsatisfactory. It wastes both time and space. At the very least it involves making a copy of the entire queue.

```
sum := 0;
t := q_new;
while not(q_isempty(q)) do
   begin
   i := q_remfirst(q);
   sum := sum + i;
   q_append(t, i)
   end;
while not(isempty(t)) do
   q_append(q, q_remfirst(t));
```

Figure 7.9 Summing the elements of a queue without an iterator.

To avoid this problem we create a special kind of type called a *generator*. Generators simulate iterators. Each generator has three functions and a procedure. The three functions create the generator object, produce the next item, and test whether all items have been yielded; the procedure destroys the generator object.

For example, for intqueue we might provide the generator *qelems*. Function *qelems_create* does a job that corresponds to the initialization that is done before the first **yield** in an iterator in CLU. It returns an object of type qelems. The second function, *qelems_next*, returns one element each time it is called and modifies the qelems generator object to indicate that that element has already been yielded. The last function, *qelems_done*, is used to determine whether or not all elements have been yielded. The procedure *qelems_destroy* frees up any space that has been allocated by *qelems_create* or *qelems_next*.

A specification of qelems is contained in figure 7.10. Both *qelems_done* and *qelems_next* require that "the queue used to create *qi* has not been modified since *qi* was created." This is analogous to a common requires clause in CLU iterators, which requires that the object being iterated over not be modified in the body of the loop.

Using qelems, we can reimplement summation as

```
sum := 0;
qi := qelems_create(q);
while not(qelems_done(qi)) do
    sum := sum + qelems_next(qi);
qelems_destroy(qi)
```

Other loops are implemented similarly.

Figure 7.11 contains an implementation of the operations of qelems. The implementation is compatible with the implementation of intqueue in

qelems = **generator type is** qelems_create, qelems_done, qelems_ next,
qelems_destroy

Overview

Qelems are used to iterate over the elements of intqueues. The implementation may depend on the rep of intqueue.

Operations

function qelems_create (q: intqueue): qelems
effects Returns a qelems object that can be used to "yield" all the elements of *q*.

function qelems_done (qi: qelems): boolean
requires The intqueue used to create *qi* has not been modified since *qi* was created.
effects Returns true if all elements of the intqueue used to create *qi* have been "yielded" since *qi* was created. Returns false otherwise.

function qelems_ next (qi: qelems): integer
requires The intqueue used to create *qi* has not been modified since *qi* was created and there exists an element of the intqueue that has not yet been yielded.
modifies *qi*
effects Returns an element of the intqueue used to create *qi*. This element has not been returned since *qi* was created. Modifies *qi* to record the fact that the element has been "yielded."

procedure qelems_destroy (qi: qelems)
modifies *qi*
effects All heap storage occupied by *qi* is deallocated.

end qelems

Figure 7.10 Specification of a generator type for queues.

figure 7.2; in fact, it accesses the rep of intqueue. Such access is typical of generators for collections like intqueue. They really are part of the collection type, just as the iterator would be in CLU. Note that the rep has an extra level of indirection to ensure that sharing works properly.

7.5 A Full Example

This section provides a concrete illustration of how to assemble a Pascal program using the approach we have discussed. Assume that we have available a library of abstractions that includes the *intbintree* type specified in figure 7.12 and that we wish to implement a function that searches an intbintree in breadth-first order.

type
    qelems = ↑qelems_rep;
    qelems_rep = ↑intqueue_elem;

function qelems_create (q: intqueue): qelems;
    var qi: qelems;
    begin
        new(qi);
        qi↑ := q↑;
        qelems_create := qi
    end { qelems_create };

function qelems_done (qi: qelems): **boolean**;
    begin
        qelems_done := (qi↑ = **nil**)
    end { qelems_done };

function qelems_next (qi: qelems): **integer**;
    begin
        if qi↑ = **nil**
            then failure('qelems_next called with empty queue generator')
            else begin
                qelems_next := qi↑↑.val;
                qi↑ := qi↑↑.next
            end
    end { qelems_next };

procedure qelems_destroy (qi: qelems);
    begin
        dispose(qi);
    end { qelems_destroy };

Figure 7.11 Implementation of a generator for type intqueue.

Breadth-first search is typically implemented using a queue of trees. We begin by placing the tree to be searched on an empty queue. We next check to see whether the root node of that tree contains the element we are looking for. If not, we remove the tree from the queue and enqueue its left and right subtrees. We then repeat the process for each of those trees. If we empty the queue, then the element was not in the tree.

Figure 7.13 contains an implementation of function *breadth_first_search*. It assumes that the abstraction *queue[etype]* has been instantiated by replacing *etype* by *intbintree* and naming the resulting type *btq*.

Note that *breadth_first_search* uses *btq_destroy* to deallocate the queue it creates. If it did not do this, the storage occupied by the elements of the local variable *q* would be lost. On the other hand, *bt_destroy(t)* is not called. Before leaving an operation, we must invoke destroy operations to dispose of objects that have been allocated dynamically by that operation.

intbintree = **data type is** bt_new, bt_isempty, bt_append_left,
bt_append_right, bt_left, bt_right, bt_rootval, bt_destroy

Overview

Intbintrees (binary trees) are used to store values of type integer.
An element may appear at more than one node of the tree. Both
interior nodes and leaves have values. Intbintrees are mutable.

Operations

function bt_new: intbintree
effects Returns an empty binary tree.

function bt_isempty (bt: intbintree): boolean
effects Returns true if *bt* is empty, and returns false otherwise.

procedure bt_append_left (bt: intbintree; e: integer)
modifies *bt*
effects Adds a node with value *e* as the leftmost node of *bt*.

procedure bt_append_right (bt: intbintree; e: integer)
modifies *bt*
effects Adds a node with value *e* as the rightmost node of *bt*.

function bt_left (bt: intbintree): intbintree
requires *bt* is non-empty.
effects Returns the subtree of *bt* whose root is the left immediate
descendent of the root of *bt*. This can be the empty tree.

function bt_right (bt: intbintree): intbintree
requires *bt* is non-empty.
effects Returns the subtree of *bt* whose root is the right immediate
descendent of the root of *bt*. This can be the empty tree.

function bt_rootval (bt: intbintree): integer
requires *bt* is non-empty.
effects Returns the value of the root node of *bt*.

procedure bt_destroy (bt: intbintree)
modifies *bt*
effects All heap storage occupied by *bt* is deallocated.

end intbintree

Figure 7.12 Specification of intbintree.

Although *breadth_first_search* declares a local variable of type intbintree, it
does not create any objects of that type.

We now have all the pieces necessary to provide a breadth-first search
function that could be embedded in a larger program. The way in which
we put the pieces together depends on the Pascal being used. In many

```
function breadth_first_search (bt: intbintree; e: integer): boolean;
    var
        q: btq;
        t: intbintree;
        found: boolean;
    begin
        found := false;
        if not(bt_isempty(bt))
            then begin
                q := btq_new;
                btq_append(q, bt);
                while (not(btq_isempty(q)) and not(found)) do
                    begin
                    t := btq_remfirst(q);
                    if bt_rootval(t) = e
                        then found := true
                        else begin
                            if not(bt_isempty(bt_left(t)))
                                then btq_append(q, bt_left(t));
                            if not(bt_isempty(bt_right(t)))
                                then btq_append(q, bt_right(t));
                            end
                    end;
                btq_destroy(q)
                end;
        breadth_first_search := found
    end { breadth_first_search }
```

Figure 7.13 Implementation of *breadth_first_search*.

dialects of Pascal (including Standard Pascal) the order of declarations within a block is restricted to all the **const** declarations, followed by all the **type** declarations, then all the **var** declarations, and finally all the operation declarations. This implies that the declaration of the rep of an abstract type will not be adjacent to the implementations of the type's operations.

Here we assume that we are using a Pascal in which this restriction has been relaxed to allow an interleaving of declarations. If a large amount of programming must be done in a more constrained Pascal, it probably pays to write a program that translates programs presented in the way outlined here into programs acceptable to the compiler. Such a program can be written fairly quickly in Pascal itself.

Pascal requires that all identifiers be declared before they are used. *Forward declarations* are used to write mutually recursive operations. A forward declaration gives the header of an operation without giving its body. Subject to the declaration-before-use constraint, the general format of a Pascal program will always be

1. An implementation of the *failure* procedure.

2. Implementations of all the abstract types to be used in the program, including generators.

3. Implementations of all the operations that are not part of an abstract type, and of constant, type, variable, and forward declarations.

4. The body of the program.

The structure of a program corresponding to our current example would be

1. An implementation of the *failure* procedure.

2. Implementation of intbintree.

3. Implementation of queue[intbintree].

4. Implementation of *breadth_first_search*.

5. Other declarations, and the body of the program.

7.6 Summary

The approach to programming advocated in this book is, to a large extent, independent of the programming language used. We have suggested that while the choice of implementation language should exert some influence on program design, it should not be a dominant concern. In particular, there is no need to limit the kinds of abstractions appearing in a design to those directly supported by the programming language.

  As a case study, we looked at Pascal—a language with relatively primitive abstraction mechanisms. We defined how to modify procedural, iteration, and data abstractions to fit Pascal and proposed programming conventions that can be used to implement data abstractions, generators, polymorphic abstractions, and exception handling in Pascal. Our approach to Pascal can be relatively easily adapted to a variety of other programming languages (we taught a course that used PL/I for several years). We discuss such adaptations in chapter 15.

Further Reading

Garland, Stephen J., 1986. *Introduction to Computer Science with Applications in Pascal.* Reading, Mass.: Addison-Wesley Publishing Co.

```
program test;
    var
        i, sum: integer;
        q: intqueue;
        qi, qi1: qelems;
    begin
        q := q_new;
        for i := 1 to 3 do q_append(q, i);
        sum := 0;
        qi := qelems_create(q);
        while not(qelems_done(qi)) do begin
            qi1 := qelems_create(q);
            while not(qelems_done(qi1)) do
                sum := sum + qelems_next(qi1);
            qelems_destroy(qi1);
            sum := sum + qelems_next(qi);
            end;
        qelems_destroy(qi);
        q_destroy(q);
        writeLn(sum)
    end { test }
```

Figure 7.14 A Pascal program.

Exercises

7.1 Implement the intbintree type specified in figure 7.12.

7.2 Specify a *bintree[etype]* abstraction based upon the intbintree abstraction specified in figure 7.12.

7.3 Implement intqueue (figure 7.1) so that accessing and removing the first and the last element can be done in constant time. Must one change the implementation of *breadth_first_search* to take advantage of this improved implementation of intqueue?

7.4 Specify and implement a general tree abstraction, that is, one in which there can be any finite number of branches at each a node. (Hint: In your implementation, you might consider using a queue of children at each node.)

7.5 Modify *breadth_first_search* (figure 7.13) so that it will work on general trees, as specified in your answer to exercise 4.

7.6 Specify and implement a stack abstraction and use it to implement a function that does depth-first search.

7.7 Specify and implement a generator that returns all the leaves of an intbintree (figure 7.12).

7.8 What does the program in figure 7.14 implement? (Assume the specifications of figures 7.1 and 7.10.)

8

More on Specifications

Throughout this book we emphasize the importance of specifications in all stages of program development. Our main premise is that the proper use of abstraction is the key to good programming. Without specifications, abstractions are too intangible to be helpful. In this chapter we discuss the meaning of specifications and some criteria to consider when writing them. We also discuss two primary uses of specifications.

8.1 Specifications and Specificand Sets

The purpose of a specification is to define the behavior of an abstraction. Users will rely on this behavior, while implementers must provide it. An implementation that provides the described behavior is said to *satisfy* the specification.

We define the meaning of a specification to be the set of all program modules that satisfy it. We call this the *specificand set* of the specification. As an example, consider the specification

> p = **proc** (y: int) **returns** (x: int)
>   **requires** $y > 0$
>   **effects** $x > y$

This is satisfied by any procedure named p that, when called with an argument greater than zero, returns a value greater than its argument. Members of the specificand set include

> p = **proc** (y: int) **returns** (int) **return** (y + 1) **end** p
> p = **proc** (y: int) **returns** (int) **return** (y * 2) **end** p
> p = **proc** (y: int) **returns** (int) **return** (y + 3) **end** p

Like every specification, this one is satisfied by an infinite number of programs.

It is important to remember that a specification, its specificand set, and a particular member of the specificand set are very different kinds of things, as different as a program, the set of all possible executions of that program, and an execution of that program on a single set of data.

elems = **iter** (b: bag[t]) **yields** (e: t)
    effects Yields every element in b, one at a time.

elems = **iter** (b: bag[t]) **yields** (e: t)
    requires b not changed in loop using *elems*.
    effects Yields every element in b, one at a time.

elems = **iter** (b: bag[t]) **yields** (e: t)
    requires b not changed in loop using *elems*.
    effects Yields exactly the elements of b, one at a time, in an arbitrary
        order. Every element in b is yielded exactly the number of times it
        occurs in b.

Figure 8.1 Three specifications of *elems*.

8.2 Some Criteria for Specifications

Good specifications take many forms, but all of them have certain attributes in common. Three important attributes—restrictiveness, generality, and clarity—are discussed in this section.

8.2.1 Restrictiveness

There is a vast difference between knowing that some members of a specification's specificand set are appropriate and knowing that all members are appropriate. This is similar to the difference between knowing that a program works on some inputs and knowing that it works on all inputs, a difference we shall emphasize when we talk about testing programs (see chapter 9). A good specification should be restrictive enough to rule out any implementation that is unacceptable to its abstraction's users. This requirement is the basis of almost all uses of specifications.

In general, discussing whether or not a specification is sufficiently restrictive involves discussing the uses to which members of the specificand set might be put. There are certain common mistakes, however, that almost always lead to inadequately restrictive specifications. One such mistake is failing to state needed requirements in the requires clause. For example, figure 8.1 gives three specifications for an *elems* iterator for a bag of integers. (A bag is like a set except that elements can occur in it more than once. For example, a bag[int] could contain 3 twice. Bags are sometimes called multisets.) The first specification fails to address the question of what happens if b is changed within a loop using *elems*. It therefore allows implementations exhibiting radically different behavior. For example, does changing b affect the values returned by *elems*?

One way to deal with this particular problem is to require that b not be changed within the loop using *elems*, as is done in the second specification.

indexs = **proc** (s1, s2: string) **returns** (i: int)
    effects If *s1* occurs as a substring in *s2*, then *i* is the least index at
    which *s1* occurs. *i* is 0 if *s1* does not occur in *s2*, and 1 if *s1* is the
    empty string. E.g.,
        indexs("bc", "abcbc") = 2
        indexs("", "a") = 1

Figure 8.2 Specification of string$*indexs* operation.

This specification may or may not be sufficiently restrictive, since it does
not constrain the order in which the elements are returned. It would be
better if it either defined an order or included the phrase "in arbitrary
order." In addition, the specification fails to make clear what is done when
an element is contained in *b* more than once. For that matter, it does not
even say explicitly that *elems* returns only elements that are in *b*. The
third specification corrects these deficiencies.

Other mistakes are failing to identify when exceptions should be signaled and failing to specify behavior at boundary cases. For example, the
string$*indexs* operation takes strings *s1* and *s2* and, if *s1* is a substring of
s2, returns the index at which *s1*'s first character occurs in *s2*; an example is

string$indexs("ab", "babc") = 2

A specification that contained only this information would not be restrictive
enough because it does not explain what would happen if *s1* were not a
substring of *s2* or if it occurred multiple times in *s2*, or if *s1* or *s2* were
empty. The specification in figure 8.2 is restrictive enough.

The moral is that it takes considerable care to write sufficiently restrictive
specifications. A formal specification language can play a helpful role here
(formal specifications will be discussed in chapter 10).

8.2.2 Generality

A good specification should be general enough to ensure that few, if any,
acceptable programs are precluded. The importance of the generality criterion may be less obvious than that of restrictiveness. It is not essential
to ensure that no acceptable implementation is precluded, but the more
desirable (that is, efficient or elegant) implementations should not be ruled
out. For example, the specification

sqrt = **proc** (sq: real, e: real) **returns** (root: real)
    requires $sq \geq 0$ & $e > .001$.
    effects $0 \leq (root * root - sq) \leq e$.

constrains the implementer to algorithms that find approximations that are greater than or equal to the actual square root. The constraint may well result in a needless loss of efficiency.

It is our desire to make specifications as general as possible that has led us to the *definitional* style of specification used in this book. A definitional specification explicitly lists properties that the members of the specificand set are to exhibit. The alternative to a definitional specification is an *operational* one. An operational specification, instead of describing the properties of the specificands, gives a recipe for constructing them. For example,

> search = **proc** (a: array[int], x: int) **returns** (i: int) **signals** (not_in)
>   **effects** Examines $a[low]$, $a[low + 1]$, ..., in turn and returns
>   the index of the first one that is equal to x. Signals *not_in* if
>   none equals x.

is an operational specification of *search*, while

> search = **proc** (a: array[int], x: int) **returns** (i: int) **signals** (not_in)
>   **effects** Returns i such that $a[i] = x$; signals *not_in* if there is no
>   such i.

is definitional. The first specification explains how to implement *search*, while the second merely describes a property that its inputs and outputs must satisfy. Not only is the definitional specification shorter, but it also allows greater freedom to the implementer, who may choose to examine the array elements in some order other than first to last.

Operational specifications have some advantages. Most significantly, they seem to be relatively easily constructed by trained programmers— chiefly because their construction so closely resembles programming. They are generally longer than definitional specifications, however, and they often lead to overspecification. The operational specification of *search*, for example, specifies which index is to be returned if x occurs more than once in a. As another example, consider trying to write an operational specification for a square root procedure.

A good check for generality is to examine every property required by a specification, in both the requires and effects clauses, and ask whether it is really needed. If it is not, then it should be eliminated or weakened. Also, any portion of a specification that is operational rather than definitional should be viewed with suspicion.

8.2.3 Clarity

When we talk about what makes a program "good," we consider not only the computations it describes but also properties of the program text itself—for example, whether it is well modularized and nicely commented. Similarly, when we evaluate a specification, we must consider not only properties of the specificand set but also properties of the specification itself—for example, whether it is easy to read.

A good specification should facilitate communication among people. A specification may be sufficiently restrictive and sufficiently general—that is, it may have exactly the right meaning—but this is not enough. If this meaning is hard for readers to discover, the specification's utility is severely limited.

There are two distinct ways in which people may fail to understand a specification. They may study it and come away knowing that they do not understand it. For example, a reader of the second specification of *elems* in figure 8.1 may be confused about what to do if an element occurs in the bag more than once. This is troublesome, but not as dangerous as when people come away thinking that they understand a specification when in fact they do not. In such a case, the user and the implementer of an abstraction may each interpret its specification differently, leading to modules that cannot work together. For example, the implementer of *elems* may decide to yield each element the number of times it occurs in the bag, while the user expects each element to be yielded only once.

Clarity is an important but amorphous criterion. It is easy enough to say that a good specification should be easy to understand, but much harder to say how to achieve this. There are many factors contributing to clarity, of which conciseness, redundancy, and structure are perhaps the most important.

The most concise presentation may not always be the best specification, but it is frequently the best starting point. There are good reasons to increase the size of a specification by adding redundant information or levels of structure, as we shall discuss, but it is important to avoid pointless verbosity. Generally, as a specification grows longer it is more likely to contain errors, less likely to be completely and carefully read, and more likely to be misunderstood.

It is important not to confuse length with completeness. A stream-of-consciousness technique will easily lead to specifications that are both long and incomplete. Like programs, specifications that grow by accretion are often longer than they need to be. Instead of just adding to a specification, it is important to step back from that local change and see whether there is a way to consolidate the information. It takes more time to write a complete short description than to write a complete long one, but the author of a specification owes it to the readers to make this investment.

Any specification containing redundant text is less concise than it could be. Redundancy should not be introduced without a good reason, but it can be justified in two ways: to reduce the likelihood that a specification will be misunderstood by its readers and to catch errors.

In many respects, the role of a specification is like that of a textbook. It should be designed not merely to contain information, but to communicate that information effectively. Redundancy can be used to reduce the likelihood that an important point will be missed. The old dictum, "Tell 'em what you're gonna tell 'em, tell 'em, then tell 'em what you told 'em," has some pedagogical validity. The key is to present the same information in more than one way, to be redundant without being repetitious. Consider, for example,

p = **proc** (s1, s2: set) **returns** (b: bool)
    effects b is true if $s1$ is a subset of $s2$, and false otherwise.

p = **proc** (s1, s2: set) **returns** (b: bool)
    effects $b = \forall x[x \in s1 \text{ implies } x \in s2]$.

p = **proc** (s1, s2: set) **returns** (b: bool)
    effects b is true if $s1$ is a subset of $s2$, and false otherwise, i.e.,
        $b = \forall x[x \in s1 \text{ implies } x \in s2]$.

The first specification is concise and for most readers quite clear. However, some readers might be left with a nagging doubt: Was the word "subset" carefully chosen, or might the the author have meant *proper* subset? The second specification, while a bit harder to read than the first, leaves no doubt on this point. The question it raises is why, if the specifier intended that p be a subset test, was this not stated explicitly? The third specification, of course, answers both of these questions.

By stating the same thing in more than one way, a specification provides readers with a benchmark against which they can check their understanding. This helps to prevent misunderstandings and thus allows readers to spend less time studying a specification. A particularly useful kind of redundancy in this regard is one or more well-chosen examples, such as those given in the specification of *index* in figure 8.2.

A specification that states the same thing in more than one way also allows for the fact that different readers will find different presentations of the same information easier to understand. Frequently, some critical part of a specification is a concept with a name that will be meaningful to some readers, but not to others. For example, consider

> pv = **proc** (inc, r: real, n: int) **returns** (value: real)
>   **requires** $inc > 0 \mathbin{\&} r > 0 \mathbin{\&} \text{n} > 0$.
>   **effects** Returns the present value of an annual income of inc for
> n years at a risk-free interest rate of r.
>   I.e., $value = inc + (inc/(1 + r)) + \ldots + (inc/(1 + r)^{n-1})$.
>   E.g., $pv(100, .10, 3) = 100 + 100/1.1 + 100/1.21$

For readers well versed in financial matters, a specification that did not use the phrase "present value" would not be as easy to understand as one that did. For readers lacking that background, the part of the specification following "I.e." is invaluable. The part following "E.g." can be used by either group of readers to confirm their understanding.

If readers are to benefit from redundancy, it is critical that all redundant information be clearly marked as such. Otherwise a reader can waste a lot of time trying to understand what new information is being presented when, in fact, none is. A good way to indicate that information is redundant is to preface it with "i.e." or "e.g."

Redundancy does not reduce the number of mistakes in a specification. Instead it makes them more evident and provides the reader with the opportunity to notice them. For example, consider

> too_cold = **proc** (temp: int) **returns** (b: bool)
>   **effects** $b =$ true if $temp$ is ≤ 0 degrees Fahrenheit;
>   otherwise $b =$ false.
> too_cold = **proc** (temp: int) **returns** (b: bool)
>   **effects** $b =$ true if $temp$ is ≤ 0 degrees Fahrenheit; otherwise
> $b =$ false. I.e., $b =$ true exactly when $temp$ is not greater
>   than the freezing point of water at standard temperature
>   and pressure.

The first specification offers a reader no reason for suspicion, but the second should ring a useful warning bell for most readers. This warning should eventually lead to a revised specification.

One of the primary problems with informal specifications is that each reader brings a somewhat different knowledge and perspective to the task of reading a specification. Introducing redundancy can go a long way toward coping with this. For example, consider

> billion = **proc** ( ) **returns** (b: int)
>   **effects** Returns the integer one billion.

billion = **proc** ( ) **returns** (b: int)

    effects Returns the integer one billion, i.e., 10^9.

Both American and British readers are likely to find the first specification perfectly unambiguous. Unfortunately, they are also likely to interpret it in completely different ways, for in the United States a billion is 10^9, whereas in Britain it is 10^{12}. The insertion of what an American author might consider redundant information in the second specification precludes any confusion.

8.3 Why Specifications?

Specifications are essential for achieving program modularity. Abstraction is used to decompose a program into modules. However, an abstraction is intangible. Without some description, we have no way to know what it is or to distinguish it from one of its implementations. The specification is this description.

A specification describes an agreement between providers and users of a service. The provider agrees to write a module that belongs to the specificand set. The user agrees not to rely on knowing which member of this set is provided, that is, not to assume anything except what is stated by the specification. This agreement makes it possible to separate consideration of the implementation from the use of a program unit. Specifications provide the logical fire walls that permit divide-and-conquer to succeed.

Specifications are obviously useful for program documentation. The very act of writing a specification is also beneficial because it sheds light on the abstraction being specified. Our experience is that we often profit as much from this activity as from our use of the result. Writing a specification almost always teaches us something important about the specificand set being described. It does this by encouraging prompt attention to inconsistencies, incompleteness, and ambiguities. In some cases, such improved understanding is the most important result of a specification effort.

The goal is to write specifications that are both restrictive enough and general enough. Thus we pay special attention to requirements, exceptions, and boundary conditions. Doing this involves posing questions about the behavior of the abstractions, questions like those posed about *indexs*—for example, what to do if either string is empty. The point is that posing and answering such questions forces us to think carefully about an abstraction and its intended use.

The construction of a specification focuses attention on what is required of a program. It serves as a mechanism for generating questions that should be answered in consultation with users of a program or a module, rather than by implementers. By encouraging the asking of these questions in the early stages of system development, specification helps us debug our understanding of a system's requirements and design before we start implementation.

As we shall discuss in chapters 12 and 13, specifications should be written as soon as the decisions they record have been made. Since specifications become irrelevant only when their abstraction is obsolete, they should continue to evolve as long as the program evolves. It is a serious mistake to treat the process of writing specifications as a separate phase of a software project.

Once written, specifications can serve many different purposes. They are helpful to designers, implementers, and maintainers alike. During the implementation phase of the software life cycle, the presence of a good specification helps both those implementing the specified module and those implementing modules that use it. As discussed previously, a good specification strikes a careful balance between restrictiveness and generality. It tells the implementer what service to provide, but does not place any unnecessary constraints on how that service is provided. In this way it allows the implementer as much flexibility as is consistent with the needs of users. Of course, specifications are crucial for users, who otherwise would have no way to know what they can rely on in implementing their modules. Without specifications, all that exists is the code, and it is unclear how much of that code will remain unchanged over time. During testing, specifications provide information that can be used in generating test data and in building stubs that simulate the specified module. (We discuss this use in chapter 9.) During the system-integration phase, the existence of good specifications can reduce the number and severity of interfacing problems by reducing the number of implicit assumptions about module interfaces. When an error does appear, specifications can be used to pinpoint where the fault lies. Moreover, they define the constraints that must be observed in correcting the error, which helps us to avoid introducing new errors while correcting old ones.

Finally, a specification can be a helpful maintenance tool. The existence of clear and accurate documentation is a prerequisite for efficient and effective maintenance. We need to know what each module does and, if it is at all complex, how it does it. All too often, these two aspects of documentation are intimately intertwined. The use of specifications as documentation, however, helps to keep them separate and makes it easier to see the ramifications of proposed modifications. For example, a proposed modification that requires us only to reimplement a single abstraction without changing its specification has a much smaller impact than one that changes the specification as well.

8.4 Summary

This chapter has discussed specifications and offered criteria to follow in writing them. We defined the meaning of a specification to be the set of all program modules that satisfy it. This definition captures the intuitive purpose of a specification, namely, to state what all legal implementations of an abstraction have in common. Such a specification tells users what they can rely on and tells implementers what they must provide.

Good specifications should be restrictive, general, and clear. Restrictiveness and generality involve the set of modules that satisfy the specification: No implementations that would be unacceptable to users of an abstraction should be permitted, and desirable implementations (ones that are efficient or elegant, for example) should not be ruled out. Generality is made easier when specifications are written using a definitional approach, which just states properties of the specificand set. An operational approach, which explains a way to implement the abstraction, tends to yield specifications that are too restrictive.

Clarity refers to the ease with which users understand the specification. The main way to enhance clarity is to start with a concise statement and then add some redundancy, often in the form of an example. Redundancy allows readers to check their understanding of the specification. It also makes errors more evident, since these often show up as inconsistencies in the redundant descriptions. To make the reader's job as simple as possible, all redundant information should be clearly marked as such.

Specifications have two main uses. First, the act of writing a specification sheds light on the abstraction being specified by focusing attention on the properties of that abstraction. This use can be enhanced by careful attention to properties that might be overlooked, including what should be stated in the requires clause, exactly when exceptions should be signaled, and the treatment of boundary cases. Sometimes this use is the main benefit of a specification because it points out a problem with the abstraction that requires further study.

The second use is as documentation. Specifications are valuable during every phase of software development, from design to maintenance. Of course, they are not the only program documentation required. A specification describes what a module does, but any module whose implementation is clever should also have documentation that explains how it works. Program modification and maintenance are eased if these two forms of documentation are clearly distinguished.

A specification is the only tangible record of an abstraction. Specifications are a crucial part of our methodology, since without them abstractions would be too imprecise to be useful. We shall continue to emphasize them in the chapters that follow.

Further Reading

Parnas, David L., 1977. The use of precise specifications in the development of software. In *Proceedings of IFIP Congress 77*, pp. 861–868.

Exercises

8.1 Provide a concise but readable specification of a bag of integers abstraction, with operations to create an empty bag, insert and delete an element, test an element for membership, give the size of the bag, give the number of times an element occurs in a bag, and yield the elements of the bag.

8.2 Take a specification you have given for a problem in an earlier chapter and discuss its restrictiveness, generality, and clarity.

8.3 Is it meaningful to ask whether a specification is correct? Explain.

8.4 Discuss how specifications can be used during system integration.

8.5 Discuss the relationship between an abstraction, its specification, and its implementation.

9

Testing and Debugging

So far we have talked a bit about program design and quite a lot about program specification and implementation. We now turn to the related issues of ascertaining whether or not a program works as we hope it will and discovering why not when it does not.

We use the word *validation* to refer to a process designed to increase our confidence that a program will function as we intend it to. Most commonly we do validation through a combination of testing and some form of reasoning about why we believe the program to be correct. We shall use the term *debugging* to refer to the process of ascertaining why a program is not functioning properly, and *defensive programming* to refer to the practice of writing programs in a way designed specifically to ease the process of validation and debugging.

Before we can say much about how to validate a program, we need to discuss what we hope to accomplish by that process. The most desirable outcome would be an ironclad guarantee that all users of the program will be happy at all times with all aspects of its behavior. This is not an attainable goal. Such a guarantee presumes an ability to know exactly what it would mean to make all users happy. The best result we can hope for is a guarantee that a program satisfies its specification. Experience indicates that even this modest goal can be difficult to attain. Most of the time we settle for doing things to increase our confidence that a program meets its specification.

There are two ways to go about validation. We can argue that the program will work on all possible inputs. This must involve careful reasoning about the text of the program, and is generally referred to as *verification*. We shall discuss techniques that allow us to reason rigorously about programs in chapter 11. As we shall see then, formal program verification is generally too tedious to do successfully without machine aids. Unfortunately, because only relatively primitive aids exist today, most program verification is still rather informal. Even informal verification, however, can be a difficult process.

The alternative to verification is testing. We can easily be convinced that a program works on some set of inputs merely by running it on each member of the set and checking the results. If the set of possible inputs is small, exhaustive testing (checking every input) is possible. For most programs, however, the set of possible inputs is so large (indeed, it is often infinite) that exhaustive testing is impossible. Nevertheless, a carefully chosen set of test cases can greatly increase our confidence that the program works as specified. If well done, testing can detect most of the errors in programs.

In this chapter we focus on testing as a method of validating programs. We discuss how to select test cases and how to organize the testing process. We also discuss debugging and defensive programming.

9.1 Testing

Testing is the process of executing a program on a set of test cases and comparing the actual results with the expected results. Its purpose is to reveal the existence of errors. Testing does not pinpoint the location of errors, however; this is done through debugging. When we test a program, we examine the relationship between its inputs and outputs. When we debug a program, we worry about this relationship but also pay close attention to the intermediate states of the computation.

The key to successful testing is choosing the proper test data. As mentioned earlier, exhaustive testing is impossible for almost all programs. For example, if a program has three integer inputs, each of which ranges over the values 1 to 1,000, exhaustive testing would require running the program 1 billion times. If each run took 1 second, this would take slightly more than 31 years.

Faced with the impossibility of exhausting the input space, what do we do? Our goal must be to find a reasonably small set of tests that will allow us to approximate the information we would have obtained through exhaustive testing. For example, suppose a program accepts a single integer as its argument and happens to work in one way on all odd integers and in a second way on all even ones; in this case testing it on any even integer, any odd integer, and zero is a pretty good approximation to exhaustive testing.

9.1.1 Black-Box Testing

Test cases are generated by considering both the specification and the implementation. In black-box testing we generate test data from the specification alone, without regard for the internal structure of the module being tested. This approach, which is common across many engineering disciplines, has several significant advantages. The most important advantage is that the testing procedure is not adversely influenced by the component being tested. For example, suppose the author of a program made the implicit invalid assumption that the program would never be called with a certain class of inputs. Acting upon this assumption, the author might fail to include any code dealing with that class. If test data were generated by examining the program, one might easily be misled into generating data based upon the invalid assumption. A second advantage of black-box testing is that it is robust with respect to changes in the implementation. Good black-box test data need not be changed even when major changes are made to the program being tested. A final advantage is that the results of a test can be interpreted by people unfamiliar with the internals of the program being tested.

Testing Paths through the Specification

A good way to generate black-box test data is to explore alternate paths through the specification. These paths can be through both the requires and effects clauses. As an example of a path through the requires clause, consider the specification

sqrt = **proc** (x: real, epsilon: real) **returns** (ans: real)
      requires $x \geq 0$ & $(.00001 < epsilon < .001)$
      effects $(x - epsilon \leq ans * ans \leq x + epsilon)$

The requires clause of this specification is the conjunction of two terms:

1. $x \geq 0$.

2. $(.00001 < epsilon < .001)$.

To explore the distinct ways in which the requires clause might be satisfied, we must explore the pairwise combinations of the ways each conjunct might be satisfied. Since the first conjunct is a disjunct of two primitive terms ($x \geq 0$ is just a shorthand for $x = 0 \mid x > 0$), it can be satisfied in one of two ways. This leaves us with two interesting ways to satisfy the requires clause:

1. $x = 0$ & $.00001 < epsilon < .001$.

2. $x > 0$ & $.00001 < epsilon < .001$.

Any set of test data for *sqrt* should certainly test each of these cases.

It can be difficult to formulate test data that explore many different paths through the effects clause of the specification. It may be difficult even to know which paths can be explored. For example, given the above specification of *sqrt*, we might expect the program sometimes to return an exact result, sometimes a result a little less than the square root, and sometimes a result a little greater. However, a program that always returned a result greater than or equal to the actual square root would be a perfectly acceptable implementation. We would not be able to find test data that forced this program to return a result less than the square root, but we could not know this without examining the code. In fact, without examining the code, we would have no idea which classes of inputs would lead to results in the three categories.

Nevertheless, we should always examine the effects clause carefully and try to find test data that exercise different ways to satisfy it. For example, consider the intset$*member* operation:

> member = **proc** (s: intset, x: int) **returns** (bool)
> **effects** Returns true if x is in s; otherwise returns false.

The effects clause of this specification is a conjunction: Either x is in s or it is not. Both conjuncts should be tested by the test cases.

Often paths through the effects clause pertain to error handling. Failing to signal an exception when called with exceptional input is just as serious as failing to do the proper thing with normal input. Therefore the test data should cause every possible signal to be raised. For example, consider the specification

> search = **proc** (a: array[int], x: int) **returns** (i: int) **signals** (not_in)
> **effects** If x is in a then returns i such that $a[i]$ = x; otherwise
> signals *not_in*.

Here we must include test cases for both the case in which x is in a and the case in which it is not. Similarly, if *sqrt* signaled exceptions rather than having a requires clause, we would want to include test data that should cause the exceptions.

Testing Boundary Conditions
A program should always be tested on "typical" input values—for example, an array or a set containing several elements, or an integer between the smallest and largest values expected by a program. It is also important to test atypical inputs, which tend to show up as boundary conditions.

Considering all paths through the requires clause tests certain kinds of boundary conditions—for example, the case in which *sqrt* is asked to find the square root of zero. A lot of boundary conditions, however, do not emerge from such analysis. It is important to check as many boundary conditions as possible. Such checks catch two very common kind of errors:

1. logical errors, in which a path to handle a special case presented by a boundary condition is omitted, and

2. failure to check for conditions that may cause the underlying language or hardware system to raise an exception (for example, arithmetic overflow).

To generate tests designed to detect the latter kind of error, it is a good idea to use test data that cover all combinations of the largest and smallest allowable values for all bounded numerical arguments. For example, tests for *sqrt* should include cases for *epsilon* very close to .001 and .00001. For strings, tests should include the empty string and a one-character string; for arrays, we should test the empty array and a one-element array.

Aliasing Errors
Another kind of boundary condition occurs when a single mutable object is bound to two different formals. For example, suppose procedure

```
append_array = proc (a1, a2: array[int])
        modifies a1 and a2.
        effects Removes the elements of a2 and appends them to the end
            of a1.
```

were implemented by

```
append_array = proc (a1, a2: array[int])
    ai = array[int]
    while ai$size(a2) > 0 do
        ai$addh(a1, a2[ai$low(a2)])
        ai$reml(a2)
        end
    end append_array
```

Any test data that did not include an input in which *a1* and *a2* are bound to the same nonempty array would fail to turn up a very serious error in *append_array*.

9.1.2 Testing Based on the Program Text

While black-box testing is generally the best place to start when attempting to test a program thoroughly, it is rarely sufficient. Without looking at the internal structure of a program it is impossible to know which test cases are likely to give new information. It is therefore impossible to tell how much coverage we get from a set of black-box test data. For example, suppose a program relies on table lookup for some inputs and computation for others. If the black-box test data happened to include only those values for which table lookup is used, the tests would give no information about the part of the program that computed values.

A good way to supplement black-box testing is with inputs that exercise the different paths through the program. The goal here is to have a test set such that each path is exercised by at least one member of the set. We say that such a test set is *path-complete*. In chapter 11 we discuss a rigorous technique for enumerating paths through a program. Here we rely on informal arguments.

Consider the program

```
max_of_three = proc (x, y, z: int) returns (int)
    if x > y
        then if x > z then return (x) else return (z) end
        else if y > z then return (y) else return (z) end
    end
end max_of_three
```

Despite the fact that there are n^3 inputs, where n is the range of integers allowed by the programming language, there are are only four paths through the program. Therefore the path-complete property leads us to partition the test data into four classes. In one class x is greater than y and z. In another, x is greater than z but smaller than y, and so forth. Representatives of the four classes are

3, 2, 1   3, 2, 4   1, 2, 1   1, 2, 3

It is easy to show that path-completeness is not sufficient to catch all errors. Consider the program

```
max_of_three = proc (x, y, z: int) returns (int)
    return (x)
end max_of_three
```

The test set containing just the input

2, 1, 1

is path-complete for this program. Using this test might mislead us into believing that our program was correct, since the test would certainly fail to uncover any error. The problem is that a testing strategy based upon exercising all paths through a program is not likely to reveal the existence of missing paths, and omitting a path is a fairly common programming error. This problem is a specific instance of the general fact mentioned earlier: No set of test data based solely upon analysis of the program text is going to be sufficient. One must always take the specification into account.

```
j := k
for i: int in int$from_to(1, 100) do
   if pred(i * j) then j := j + 1 end
   end
```

Figure 9.1 A program with many paths.

Another potential problem with a testing strategy based upon selecting path-complete test data is that there are often too many different paths through a program to make that practical. Consider the program fragment in figure 9.1. There are 2^{100} different paths through this program, as can be seen from the following analysis. The **if** statement causes either the true or the false branch to be taken, and both of these paths go on to the next iteration of the loop. Thus for each path entering the ith iteration, there are two paths entering the $(i + 1)$st iteration. Since there is one path entering the first iteration, the number of paths leaving the ith iteration is 2^i. Therefore there are 2^{100} paths leaving the 100th iteration.

Testing each of 2^{100} paths is not likely to be practical. In such cases we generally settle for an approximation to path-complete test data. The most common approximation is based upon considering two or more iterations through a loop as equivalent and two or more recursive calls to a procedure as equivalent. To derive a set of test data for the above program, for example, we find a path-complete set of test data for the program

```
j := k
for i: int in int$from_to(1, 2) do
   if pred(i * j) then j := j + 1 end
   end
```

There are only four paths through this program. A path-complete set of test data would have representatives that made each of the following true:

1. $pred(k)$ & $pred(2k + 2)$

2. $pred(k)$ & $\sim pred(2k + 2)$

3. $\sim pred(k)$ & $pred(2k)$

4. $\sim pred(k)$ & $\sim pred(2k)$

To sum up, we always include test cases for each branch of a conditional. However, we approximate path-complete testing for loops and recursion as follows:

1. For loops with a fixed amount of iteration, as in the example just shown, we use two iterations. We choose to go through the loop twice rather than once because failing to reinitialize after the first time through a loop is a

common programming error. We also make certain to include among our tests all possible ways to terminate the loop.

2. For loops with a variable amount of iteration, we include zero, one, and two iterations, and in addition we include test cases for all possible ways to terminate the loop. For example, consider

```
while x > 0 do
    % do something
end
```

With a loop like this, it is possible that no iterations will be performed. This situation should always be handled by the test cases because not executing the loop is another situation that is likely to be a source of program error.

3. For recursive procedures we include test cases that cause the procedure to return without any recursive calls and test cases that cause exactly one recursive call.

This approximation to path-complete testing is, of course, far from fail-safe. Like engineer's induction, "One, two, three—that's good enough for me," it frequently uncovers errors, but offers no guarantees.

9.1.3 An Example

To illustrate these points, we consider a simple procedure for determining whether a string is a palindrome. A palindrome is a string the reads the same backward and forward (an example is "deed"). Figure 9.2 gives a specification and implementation of this procedure. By looking at the specification we can see that we need tests that cause both true and false to be returned. In addition, we must include the empty string as a boundary condition. This might lead to the strings " ", "deed", and "ceed". Examination of the code indicates that we should test the following cases:

1. not executing the loop;

2. returning false in the first iteration;

3. returning true after the first iteration;

4. returning false in the second iteration;

5. returning true after the second iteration.

Cases 1, 4, and 5 are covered already. For cases 2 and 3 we might add "ab" and "aa". At this point we should ask ourselves whether there is any case we have missed, and we might notice that all our test strings have an even number of characters. Therefore we should add a number of odd-length test strings. These strings should include the boundary case of a string

palindrome = **proc** (s: string) **returns** (bool)
        effects Returns true if *s* reads the same forward and backward; other-
            wise returns false. E.g., "deed" and " " are both palindromes.

palindrome = **proc** (s: string) **returns** (bool)
   low: int := 1
   high: int := string$size(s)
   while high > low **do**
     if s[low] ~= s[high] **then return** (**false**) **end**
     low := low + 1
     high := high − 1
     end
   return (**true**)
   end palindrome

Figure 9.2 The *palindrome* procedure.

of length 1. Finally, we should arrange the test cases in a sensible order, with the shortest first. Such an arrangement helps in finding errors (see section 9.4).

9.1.4 Testing Iterators

Generating test cases for iterators is similar to generating them for procedures. The only point of interest is that iterators have paths in their specifications that are similar to those for loops. In other words, we must be sure to include cases in which the iterator yields once and twice, and if it is possible for the iterator to yield zero times, we should include a case for that. For example, consider the *primes* iterator:

primes = **iter** (n: int) **yields** (int)
        effects If $n \geq 2$, yields all primes less than or equal to n;
           otherwise yields nothing.

Test cases here could include calls with n equal to 1, 2, and 3. Whether more tests are needed or not can be determined by looking at the iterator's implementation.

9.1.5 Testing Data Types

In testing data types, as usual we generate test cases by considering the specifications and implementations of each of the operations. We must now test the operations as a group rather than individually, however, because some operations (the constructors and primitive constructors) produce the objects that are used in testing others. In the intset operations, for example,

intset = **data type is** create, insert, delete. member, size, elements

Overview

Intsets are unbounded mathematical sets of integers. Intsets are mutable: *insert* and *delete* add and remove integers from the set.

Operations

create = **proc** ( ) **returns** (intset)
effects Returns a new, empty intset.

insert = **proc** (s: intset, x: int)
modifies s
effects Adds x to the elements of s; after insert returns, $s_{post} = s \cup \{x\}$, where s_{post} is the set of values in s when *insert* returns.

delete = **proc** (s: intset, x: int)
modifies s
effects Removes x from s (i.e., $s_{post} = s - \{x\}$).

member = **proc** (s: intset, x: int) **returns** (bool)
effects Returns ($x \in s$).

size = **proc** (s: intset) **returns** (int)
effects Returns the number of elements in s.

elements = **iter** (s: intset) **yields** (int)
requires s is not modified by loop body.
effects Yields the elements of s, each exactly once, in arbitrary order.

end intset

Figure 9.3 Specification of intset.

the constructors *create*, *insert*, and *delete* must be used to generate the arguments for the other operations and for each other. (The specification for intset is repeated in figure 9.3.)

We begin by looking at paths in specifications. The specifications of *member* and *elements* have obvious paths to explore. For *member* we must generate cases that produce both true and false as results. Because *elements* is an iterator, we must look at least at paths of lengths zero, one, and two. Therefore we shall need intsets containing zero, one, and two elements. The empty intset and the one-element intset also test boundary conditions. Thus to test the observers, we might start with the following intsets:

1. the empty intset produced by calling intset$*create*( );

2. the one-element intset produced by inserting 3 into the empty set; and

3. the two-element intset produced by inserting 3 and 4 into the empty set.

For each, we would do calls on *member*, *size*, and *elements* and check the results. In the case of *member*, we would do calls in which the element is in the set, and others in which it is not.

We obviously do not yet have enough cases. For example, *delete* is not tested at all, and there are also paths in other specifications that have not yet been discussed. These paths are somewhat hidden in our informal specifications; they would be more evident if we had used formal specifications.

The size of an intset remains the same when we insert an element that is already in the set, and we must therefore look at a case in which we insert the same element twice. Similarly, the size decreases when we delete an element only if it is in the set, so that we must look at one case in which we delete an element after inserting it and another in which we delete an element that is not in the set. We might use these additional intsets:

4. the set obtained by inserting 3 twice into the empty set;

5. the set obtained by inserting and then deleting 3; and

6. the set obtained by inserting 3 and deleting 4.

To find these hidden paths we must look explicitly for paths in the constructors. Thus *insert* must work properly whether the element being inserted is already in the set or not, and similarly for *delete*. This simple approach will produce the three cases just given.

In addition, of course, we must look for paths in the implementations of the operations. The cases identified so far provide quite good coverage for the implementation using the array with no duplicates (see figure 9.4). One possible problem is in *member*, which contains a loop. To cover all paths in this loop, we must test the case of a two-element array with either no match or a match with the first or the second element. (It is not possible to find such tests cases by considering only the specification. At the level of the specification we are concerned only with whether or not the element is in the set; its position in the array is not of interest.) Similarly, in *delete* we must be sure to delete both the first and second elements of the array. In doing so we cover the potential error of not handling the last element of the array properly. The implementation of *delete* moves the last element of the array into the location of the deleted element before shrinking the array. If we reversed the order of these two actions, the program would not work when the element being deleted was stored at the high bound of the array.

intset = **cluster is** create, insert, delete, member, size, elements

  rep = array[int]

  create = **proc** ( ) **returns** (**cvt**)
    return (rep$new( ))
    end create

  insert = **proc** (s: intset, x: int)
    if ~member(s, x) **then** rep$addh(**down**(s), x) **end**
    end insert

  delete = **proc** (s: **cvt**, x: int)
    for j: int **in** rep$indexes(s) **do**
      if s[j] = x
        then s[j] := **rep**$top(s)
          rep$remh(s)
        end
      end
    end delete

  member = **proc** (s: **cvt**, x: int) **returns** (bool)
    for y: int **in** rep$elements(s) **do**
      if y = x **then return** (**true**) **end**
      end
    return (**false**)
    end member

  size = **proc** (s: **cvt**) **returns** (int)
    return (rep$size(s))
    end size

  elements = **iter** (s: **cvt**) **yields** (int)
    for y: int **in** rep$elements(s) **do**
      yield (y)
      end
    end elements

end intset

Figure 9.4 Implementation of intset.

9.2 Unit and Integration Testing

Testing typically occurs in two phases. During *unit testing* we attempt to convince ourselves that each individual module functions properly in isolation. During *integration testing* we attempt to convince ourselves that when all the modules are put together, the entire program functions properly.

    Integration testing is generally more difficult than unit testing. First, the intended behavior of an entire program is often much harder to characterize than the intended behavior of its parts. Second, problems of scale tend to arise in integration testing that do not arise in unit testing; for example, it may take much longer to run a test. Finally, specifications play rather different roles in the two kinds of validation.

The acceptance of the specification as a given is a key factor that distinguishes unit testing from integration testing. During unit testing, when a module fails to meet its specification, we generally conclude that it is incorrect. When validating a whole program, we must accept the fact that the most serious errors are often errors of specification. In these cases, each unit does what it is supposed to, but the program as a whole does not. A prime cause of this kind of problem is ambiguous specifications. When this occurs, a module may perform as expected by those doing its unit testing while failing to meet the expectations of those writing some of the modules that call it. This makes errors detected during integration testing particularly difficult to isolate.

Consider a program implemented by module P, which calls module Q. During unit testing, P and Q are tested individually. (To test either individually, it is necessary to simulate the behavior of the other, as will be discussed.) When each of them has run correctly on its own test cases, we test them together to see whether they jointly conform to P's specification. In doing this joint test, we use P's test cases. Now suppose that an error is discovered. These are the possibilities:

1. Q is being tested on an input that was not covered in its test cases.

2. Q does not behave as was assumed in testing P.

It is tempting when dealing with multiple modules like P and Q to test them jointly rather than to do unit tests for each first. Sometimes such joint tests are a reasonable approach, but unit testing is usually better. For example, to test the program shown in figure 9.1, we only care that each of the four paths be covered; the various ways in which *pred* produces its results are not of concern. However, testing *pred* thoroughly probably involves many test cases. Combining all these test cases has many disadvantages: There are more tests to run, tests may take longer to run, and if either of these modules is reimplemented, we shall have to rethink the whole set of test cases. Testing each module individually is more efficient.

9.3 Tools for Testing

It is useful to automate as much of the testing process as possible. Usually we cannot automate the generation of test data; generating appropriate inputs for testing a program is a nonalgorithmic process requiring serious

```
% accept as inputs the files:
%     file_of_tests, bad_tests, correct_results, and incorrect_results

for  % each test in file_of_tests
    do if test.square < 0 | test.epsilon < .00001 | test.epsilon > .001
       then  % add test to bad_tests
       else result := sqrt(test.square, test.epsilon)
          if real$abs(square − result ∗ result) <= epsilon
             then  % add ⟨test, result⟩ to correct_results
             else  % add ⟨test, result⟩ to incorrect_results
             end
       end
    end
```

Figure 9.5 Driver for *sqrt*.

thought. Furthermore, to automate the process of deciding what outputs are appropriate for any set of inputs is often as difficult and error prone as writing the program being tested.

What we can automate are the processes of invoking a program with a predefined sequence of inputs and checking the results with a predefined sequence of tests for the acceptability of outputs. A mechanism that does this is called a *driver*. A driver should call the unit being tested and keep track of how it performs. More specifically, it should

1. Set up the environment needed to call the unit being tested. In some languages (but not CLU) this may involve creating and initializing certain global variables. In most languages it may involve setting up and perhaps opening some files.

2. Make a series of calls. The arguments for these calls could be read from a file or embedded in the code of the driver. If arguments are read from a file, they should be checked for appropriateness, if possible.

3. Save the results and check their appropriateness.

The most common way to check the appropriateness of results is to compare them to a sequence of expected results that has been stored in a file. Sometimes, however, it is better to write a program that compares the results directly to the input. For example, if a program is supposed to find the roots of a polynomial, it is easy enough to write a driver that checks whether or not the values returned are indeed roots. Similarly, it is easy to check the results of *sqrt* by a computation. A driver for testing an implementation of the *sqrt* specification given previously is shown in figure 9.5.

In addition to drivers, testing often involves the use of *stubs*. A driver simulates the parts of the program that call the unit being tested. Stubs simulate the parts of the program called by the unit being tested. A stub must

1. Check the reasonableness of the environment provided by the caller.

2. Check the reasonableness of the arguments passed by the caller.

3. Modify arguments and the environment and return values in such a way that the caller can proceed. It is best if these effects match the specification of the unit the stub is simulating. Unfortunately, this is not always possible. Sometimes the "right" value can only be found by writing the program the stub is supposed to replace. In such cases we must settle for a "reasonable" value.

(If all communication is only via arguments and results, then it is not necessary to check or modify the environment.)

Drivers are clearly necessary when testing modules before the modules that invoke them have been written. Stubs are necessary when testing modules before the modules that they invoke have been written. Both are needed for unit testing, in which we want to isolate the unit being tested as much as possible from the other parts of the program.

In practice, it is common to implement drivers and stubs that rely on interaction with a person. A very simple implementation of a stub might merely print out the arguments it was called with and ask the person doing the testing to supply the values that should be returned. Similarly, a simple driver might rely on the person doing the testing to verify the correctness of the results returned by the unit being tested. Although drivers and stubs of this nature are easy to implement, they should be avoided whenever possible. They are far more prone to error than automated drivers and stubs, they make it hard to build up a good database of test data, and they make it hard to reproduce tests.

The reproducibility of tests is particularly important. The following testing scenario is all too typical:

1. The program is tested on inputs 1 through n without uncovering an error.

2. Testing the program on input $n + 1$ reveals the existence of an error.

3. Debugging leads to a fix that makes the program work on input $n + 1$.

4. Testing continues at input $n + 2$.

This is an unwise practice, for there is a nonnegligible probability that the change that made the program work on input $n+1$ will cause it to fail on some input between 1 and n. Whenever any change is made, no matter how small, it is important to make sure that the program still passes all the tests it used to pass. This is called *regression testing*. Regression testing is practical only when there are tools available that make it relatively easy to rerun old tests.

It is important to implement drivers and stubs with care. When an error is detected, we want it to be in the code being tested. If drivers and stubs are implemented carelessly, however, they are at least as likely to contain errors as the program being tested. In this case the programmer wastes lots of time testing and debugging the testing environment.

9.4 Debugging

Testing tells us that something is wrong with a program, but knowing the symptom is a far cry from knowing its cause. Once we know that a problem exists, the tactics to be used in locating and fixing the problem—in debugging—are extremely important. The variance in the efficiency with which people debug is quite high, and we can offer no magic nostrums to make debugging easy. Most of what we have to say on the subject is simple common sense.

Debugging is the process of understanding and correcting errors. In debugging we try to narrow the scope of the problem by looking for simple test cases that manifest the bug and by looking at intermediate values to locate the responsible region in the code. As we collect evidence about the bug, we formulate hypotheses and attempt to refute them by running further tests. When we think we understand the cause of the bug, we study the appropriate region of code to find and correct the error. In chapter 11 we discuss ways to annotate and reason about code that facilitate such study.

The word "bug" is in many ways misleading. Bugs do not crawl unbidden into programs. We put them there. *Do not think of your program as "having bugs"; think of yourself as having made a mistake.* Bugs do not breed in programs. If there are many bugs in a program, it is because the programmer has made many mistakes.

Always keep in mind that debugging consumes more time than programming. It is worth trying very hard to get your program right the first time. Read your code very carefully and *understand exactly why you expect it to work* before you begin to test it. No matter how hard you try and no matter how clever you are, though, the odds against your program working properly the first time are very long. Consequently, you should *design, write, and document your programs in ways that will make them easier to test and debug.* The key is making sure that you have relatively small modules that can be tested independently of the rest of your program. To a large extent, this can be achieved by following the design paradigms outlined earlier in this book. Introduce data abstractions and associate with each the most restrictive possible rep invariant. Write careful specifications for each procedure, so that when it comes time to test it, you know both what input values it should be prepared to deal with and what it should do in response to each of the possible inputs.

Just as you need an overall testing strategy, you also need a careful plan for every debugging session. *Before beginning, decide exactly what you want to accomplish and how you plan to accomplish it.* Know what input you are going to give your program, and exactly what you expect it to do with that input. If you have not thought carefully about your inputs, you will probably waste a lot of time doing things that are not likely to help isolate the problem.

The so-called scientific method provides a good paradigm for systematic debugging. The crux of the scientific method is to

1. begin by studying already available data,

2. form a hypothesis that is consistent with those data, and

3. design and run a repeatable experiment that has the potential to refute the hypothesis.

Consider a program that accepts a positive integer as input and is supposed to return true if the number is prime and false otherwise. As our first test case we try 2, and our program returns the correct answer, true. We next try 3, and the program returns the incorrect result, false. We now have two pieces of data on which to form a hypothesis. One plausible hypothesis is that somehow we have failed to reinitialize something after the first input and that the program will always work on the first input and fail on the second. (This error is unlikely in CLU.) To check this hypothesis we can test the program on the same two arguments but reverse the order of the tests; that is, we can try 3 and then 2. Before running the tests, we decide which results would support our hypothesis and which would refute it:

1. results supporting hypothesis: (true, false);

2. results refuting hypothesis: (false, true), (false, false), (true, true).

When we try the experiment, the program returns false and then true. We immediately reject our first hypothesis and look for another, for example, that the program will fail on all odd primes.

When debugging, a good starting goal is to *find a simple input that causes the problem to occur*. This input may not be the test data that first revealed the existence of the bug. It is often possible to find simpler input that is sufficient to provoke a manifestation of the bug. Thus a good way to start is to pare down the test data and then run the program on variants of that subset.

For example, suppose we are testing the *palindrome* procedure of figure 9.2, and when we run it on the famous (allegedly Napoleonic) palindrome "able was I ere I saw elba", it returns false. This is a rather long palindrome, so we should try to find a shorter one on which the program fails. We might begin by taking the middle character of this palindrome and seeing whether the program succeeds in recognizing that the single character "r" is itself a palindrome. If it fails on that, we might hypothesize that the program does not work on palindromes containing an odd number of characters, and we should examine our other tests to see whether they support this hypothesis. If it succeeds in recognizing that "r" is a palindrome, we might try "ere" on the hypothesis that it will fail on odd palindromes containing more than one character. If "ere" fails to provoke the error, we should probably try "I ere I". Suppose the program fails on this input. Two hypotheses come to mind: Perhaps the blanks are the root problem, or perhaps it is the uppercase letters. We should now test our program on the shortest inputs that might confirm or refute each hypothesis, for example, " " and "I".

Once we have found a small input that causes the error to occur, we use this information to locate where in the program the bug is likely to be. Finding the kind of input necessary to provoke a symptom is often tantamount to locating a bug. If not, however, the next step is to narrow the scope of the problem by examining intermediate results.

The goal is to rule out parts of the program that cannot be causing the problem and then look in more detail at what is left. We do this by *tracing* the program, that is, running it and looking at the values of variables at specific points in its control flow. If the program consists of several modules, our first goal is to discover which module is the source of the bug. We do this by tracing all calls and returns of procedures. (For iterators we trace yields and resumes.) For each call, we ask whether the arguments are what they should be; the arguments should satisfy the requires clause of the called procedure and should also follow from what we have learned in the trace so far. If the arguments are not right, then the error is in the calling module. Otherwise we ask whether the results of the call follow from the arguments. If not, the error is in the called procedure.

Localizing the problem to a single procedure or iterator is often enough, since we can then discover the error by examining the code of the faulty module. Sometimes, however, it is useful to narrow the bug to a subpart of the faulty module. To do this, we continue the trace and examine the values of local variable of the module. The goal is to detect the first manifestation of incorrect behavior. It is particularly important to *check the appropriateness of intermediate results* against values computed prior to beginning the trace. If you wait until you see the intermediate results before thinking about what they should be, you run the risk of being unduly influenced by your (erroneous) program.

Consider the following incorrect implementation of *palindrome*:

```
palindrome = proc (s: string) returns (bool)
    low: int := 1
    high: int := string$size(s)
    while high > low do
        if s[low] ~= s[high] then return (false) end
        low := low + 1
        if high > low + 1 then high := high - 1 end
    end
    return(true)
    end palindrome
```

This implementation does not work properly on odd-length palindromes that have more than one character. Therefore we might trace it on the string "ere". Suppose we expect that *low* and *high* equal the low and high bounds of the array at the end of initialization and that *low* increases and *high* decreases in each iteration. At the end of initialization, $low = 1$ and $high = 3$, as expected. However, at the end of the first iteration, we notice that $low = 2$ and $high = 3$, which is not expected. We should be able recognize the error at this point.

If you have access to an excellent debugger, it may be possible to examine intermediate results relatively conveniently, for example, merely by telling the debugger which variables you wish to trace. If not, it is worth your while to *write a considerable amount of code whose only purpose is to help you examine intermediate results*. One piece of code that should be written in any case is an operation for each type that displays the objects in abstract form. When debugging polynomials, for example, it is much easier to understand what is happening if the poly is displayed as the string

"3 + x**5"

instead of the array

   [1: 3, 0, 0, 0, 1]

Such a routine can be called either by a user of an interactive debugger or by a print statement. Also, it is often useful to have the reverse routine that accepts an abstract form for an object and produces an appropriate object of the rep type. (In CLU we follow the convention of naming the display operation *unparse* and the other operation *parse*.)

In planning debugging sessions, keep in mind that *the bug is probably not where you think it is*. If it were, you would have found it by now. It is all too easy to develop and cling to a fixed idea about the location of a bug. The first obvious manifestation of a bug can occur far from the code where the error lies. If you think that you know which procedure contains the error, and you have spent a significant amount of time examining that procedure, you are probably wrong. Keep an open mind. Examine the reasoning that led you to that procedure, and ask yourself whether there is any possibility that it is flawed.

One way to get a slightly different perspective on the problem is to *ask yourself where the bug is not*. It can be much easier to understand where a bug could not possibly be than where it is. Often, trying to demonstrate that a bug could not possibly be in a particular place will lead to the discovery that that is exactly where it is. In any event, the systematic elimination of possibilities is frequently the best way to home in on a bug.

While trying to eliminate possible locations for a bug, *take a careful look at your input* as well as at your code. Every programmer has spent time hunting for a bug in a program when in fact the problem is in the input. As mentioned, you should write your drivers and stubs carefully so as to avoid as many such errors as possible.

Looking carefully at the input is a good example of the more general principle: *Try the simple things first*. Most errors are not particularly subtle. Simple errors that occur frequently include

1. reversing the order of input arguments,

2. failing to initialize a variable,

3. failing to reinitialize a variable the second time through a program segment,

4. copying only the top level of a data structure when you intended to copy all levels, and

5. failing to parenthesize an expression correctly.

Remember that in designing your program you went through a reasoning process not unlike that involved in debugging it. The existence of a bug is evidence that your initial reasoning was flawed. It is easy to convince yourself that a procedure does not contain a bug by using the same reasoning that led you to introduce the bug in the first place. It can therefore be invaluable to *get somebody else to help you.* Asking for help is not an admission of failure; it is merely good practice. Try to explain your problem to somebody else. Often, the mere attempt to articulate your reasoning will lead you to discover the source of the problem. Failing that, a fresh viewpoint is almost certain to prevent you from getting stuck in too deep a rut.

One of the hardest problems in debugging is *deciding what to take for granted.* The naive thing to say is that one should take nothing for granted, but this is generally counterproductive. When a test yields a faulty result, the problem might lie in one of the modules the program calls, the compiler used to compile the program, the operating system, the hardware, the electrical system of the building housing the hardware, and so on. The most likely location, however, is your program. Begin by taking everything else for granted and looking for a bug in the program. If after a reasonable amount of effort you fail to find any problem with your program, start worrying about the modules your program calls. If after a reasonable amount of effort you can find no problem there, find out whether or not the compiler or operating system has been changed recently. If not, you should be very reluctant to attribute your problems to either of them.

When you encounter a bug that you just cannot track down, *make sure that you have the right source code.* In putting together large systems, you will almost always have to rely on separate compilation facilities. This can easily lead to a situation in which the object code exhibiting the bug does not match the source in which you are trying to find the bug. A particularly vexing variant of this problem can occur when either the compiler or the operating system is changed. When your program's behavior has changed and you are absolutely sure that you have not changed anything, make sure that you are not using code that has been compiled to run on a different version of the operating system.

When you have tried everything you can think of and still have not found the bug, *go away.* The goal of any programming project is to complete the program (including its documentation and testing) expeditiously. The goal is not to find a particular bug as soon as possible. The obsessive pursuit of a particular bug is almost always counterproductive. If you spend too long looking for the same bug, there is a high probability that you will become stuck in a rut. If you try to debug when you are overly tired, you will at best work inefficiently. At worst, you will make mistakes, such as making ill-considered changes to the program or accidentally deleting a crucial file.

When you do find a bug, *try to understand why you put it there.* Was it a clerical error, does it reflect a lack of understanding of the programming language, or is it indicative of some logical problem? Knowing why you inserted a bug may help you understand how to fix your program. It may also help you to discover other bugs and even to avoid bugs in the future.

Finally, when you think that you have found a bug and that you know how it got there, *do not be in too much of a rush to fix "the bug."* Make sure that the bug you found could indeed have caused the symptoms that you observed. If you have already spent a lot of time observing the behavior of your program, it may be counterproductive to change that behavior before you have completed your detective work. Not only is it often easier to repair many bugs at once than to repair many bugs one at a time, but it almost always leads to a cleaner and more efficient program.

When you do decide to make a change, *think through all of its ramifications.* Convince yourself that the change really will both cure the problem and not introduce new problems. The hardest bugs to find are often those we insert while fixing other bugs. This is because we are often not as systematic in designing these "patches" as in our original designs. We try to make local changes, when a more global approach might well be called for. It is often more efficient to reimplement a small procedure than to patch an old one.

9.5 Defensive Programming

In preparing to cope with mistakes, it pays to program defensively. In every good programmer there is a streak of suspicion. Assume that your program will be called with incorrect inputs, that files that are supposed to be open may be closed, that files that are supposed to be closed may be open, and so forth. Write your program in a way designed to call these mistakes to your attention as soon as possible. CLU has been designed in such a way that many conventional defensive programming methods are built into the language. Most notably, CLU provides complete compile-time type checking and complete runtime checking for out-of-bounds subscripts and arithmetic overflow and underflow. Two standard defensive programming methods not built into CLU are checking requirements and rep invariants, and exhaustive testing of all conditionals.

The violation of a rep invariant or a procedure's requirements is often the first manifestation of a bug. If code to check these explicitly is not included, the first observable symptom of the bug may occur quite far from the place where the mistake actually occurred. Consider, for example, a procedure with the specification

> in_range = **proc** (x, y: int, a: elem_array, e: elem) **returns** (b: bool)
>   **requires** $y \geq x$.
>   **effects** Returns true if e is an element of $a[x], \ldots, a[y]$, i.e.,
> $b = \exists z[x \leq z \leq y \ \& \ a[z] = e]$.

Suppose that a caller of this procedure reverses the order of the first two arguments. *In_range* will probably return false whether or not e is in a. The first observable symptom of this incorrect call might appear arbitrarily far from the call. In the worst case, the error would never be detected, and the program in which *in_range* occurs would simply return an incorrect answer. However, if *in_range* signals *failure* when $x > y$, the error can usually be found immediately.

Failure to perform exhaustive testing in conditionals can have a similar effect. For example, suppose the *receive* procedure delivers a string that has been sent over a communications network in a message and that for this particular call only the values "deliver" and "examine" are meaningful. The implementation

```
s := receive( )
if s = "deliver" then  % carry out the deliver request
    elseif s = "examine" then  % carry out the examine request
    else signal failure ("unexpected request:" || s)
    end
```

is far superior to the marginally more efficient implementation

```
s := receive( )
if s = "deliver"
    then  % carry out deliver request
    else  % carry out examine request
    end
```

The rep invariant should be tested at the beginning of each operation, and before returning if the rep is new or has been modified. To make this easy, a cluster should include an internal routine to do the test. This routine can return if the invariant is satisfied and signal *failure* otherwise.

Rep invariants and requirements are two examples of assertions about intermediate states of a computation. In chapter 11 we shall discuss several other kinds of assertions. It is frequently useful to write code to check these.

Defensive programming generally involves a certain amount of extra overhead—both for the programmer and at runtime. Most of the time, the programming overhead is not an issue, since defensive programming almost always reduces the total amount of programmer time over the course of a programming project. The runtime overhead cannot be dismissed so easily. For programs in which performance is an issue, some defensive programming methods can be prohibitively expensive. If, for example, the hardware does not detect arithmetic overflow, detecting it in software can more than double the cost of doing arithmetic. On a more abstract level, a binary search procedure can hardly afford to check that every array given to it is indeed sorted.

When it seems that defensive programming will be prohibitively expensive in a production version of a program, we should still give serious thought to putting the checks in while the program is under development. Removing error detection code just before a program is put into production use is much easier than inserting it during debugging. Removing a program's defenses, however, should not be done as a matter of course. If at all possible, these should be left in the production version. It is almost certain that when the program first goes into production use, it will still contain some bugs and that other bugs will be introduced as it is modified. It is important that these bugs be detected and repaired as expeditiously as possible. The actual economic cost of an undiscovered error in a program may exceed the cost of keeping the checks in during production runs. It is usually worthwhile to retain at least the inexpensive checks.

9.6 Summary

This chapter has discussed the related issues of testing and debugging. Testing is a method of validating a program's correctness. We have described a way to develop test cases methodically by examining both a module's specification and its implementation. The test cases should then be run by a driver that checks the results of each case; the driver either produces the inputs or reads them from a file, and either checks the results by computations or compares them to outputs in a file. If the test being run is a unit test, then lower-level modules are replaced by stubs that simulate their effects. Later, during integration testing, the stubs are replaced by the implementation.

Testing can exhibit the presence of a bug. Debugging is the process of understanding and correcting the cause of the bug. In debugging we try to narrow the scope of the problem by searching for simple test cases that manifest the bug and by looking at intermediate values to locate the responsible region in the code. As we collect evidence about the bug, we formulate hypotheses and attempt to refute them by running further tests. When we think we understand the cause of the bug, we study the responsible region of code to find and correct the error.

Debugging can be made easier if we practice defensive programming. This consists of inserting checks in the program to detect errors that are likely to occur. In particular, we should check that the requires clause is satisfied. It is also a good idea to check the rep invariant. These checks should be retained in the production code if possible.

The outcome of being methodical about testing, debugging, and defensive programming is a reduction of programmer effort. This work pays off not only when the program is written, but also later when it is modified.

Further Reading

Beizer, B., 1983. *Software Testing Techniques*. New York: Van Nostrand Reinhold.

Goodenough, John B., and Susan L. Gerhart, 1975. Toward a theory of test data selection. *IEEE Transactions on Software Engineering* SE-1(2): 156–173.

Howden, William E., 1981. A survey of dynamic analysis methods. In *Tutorial: Software Testing and Validation Techniques* (New York: IEEE Computer Society Press), pp. 209–231.

Weinberg, Gerald, M., 1971. *The Psychology of Computer Programming*. New York: Van Nostrand Reinhold.

Exercises

9.1 Develop a set of test cases for *merge* using the specification given in figure 3.5 and the implementation given in figure 3.6. Do the same thing for *merge_sort* and *sort*. Write a driver for *merge*.

9.2 Develop a set of test cases for poly (figures 4.3 and 4.7). Write a driver for poly.

9.3 Develop a set of test cases and write a driver for *permutations* (figure 6.9).

9.4 In exercise 5 of chapter 4 intsets were implemented using olists. Discuss what kind of stub you would use for olists in testing your implementation of intset.

9.5   The block-structured symbol table in exercise 6 of chapter 6 can be implemented using stacks and maps. Sketch such an implementation. Then discuss how you would use stubs for these two abstractions in testing symtab.

9.6   Implement an iterator that yields all Fibonacci numbers. (A Fibonacci number is the sum of the preceding two Fibonacci numbers, and the first Fibonacci number is 0. For example, the first seven Fibonacci numbers are 0, 1, 1, 2, 3, 5, and 8.) Define test cases in advance of debugging. Then debug your program and report on how successful your tests were.

9.7   Develop an error profile for yourself. Keep a log in which you record errors in your programs. For each error, record the reason for it and look for patterns.

10

Writing Formal Specifications

Most of the specifications in this book are written in a quasiformal way, using an undefined notation. In effect, they are highly stylized comments. We have chosen to emphasize informal specifications for several reasons:

1. While formal specifications are a promising area of research in programming methods, their utility across a wide range of software development projects has yet to be demonstrated.

2. Becoming adept at writing formal specifications takes some time. We do not think that in a one-semester course students can master both formal specifications and the other material contained in this book.

3. Writing formal specifications is difficult unless there is some sort of machine support available. At the very least, syntax and type checkers are needed. Most readers of this book will not have access to such support.

In this chapter we present a language for writing formal specifications. We include this material because we believe that all students of programming should acquire at least a reading knowledge of formal specifications and that such knowledge can be a great help in writing informal specifications.

Recall that our informal specifications of data abstractions include an overview section whose purpose is to provide an intuitive description of the type being defined. Usually this description refers to some domain with which readers are assumed to be familiar. For example, we defined intset in terms of mathematical sets and poly in terms of polynomials over the integers. The problem with informal specifications is that these auxiliary domains are never defined precisely. If readers have an intuitive understanding of the domain, and this intuition matches that of the specifier, they will be able to understand specifications based on the domain; otherwise they will not. Worst of all, as mentioned in chapter 8, there is no way to know whether a reader and the specifier are interpreting things in the same way. Undetected differences are a prime source of difficulties between implementers and users of abstractions.

Formal specification provides a means of defining these auxiliary domains precisely. It uses two kinds of specifications. An *interface specification* describes the interface that an abstraction presents to its users, that is, to implementers of other modules that use the abstraction. Each interface specification is built on top of an *auxiliary specification*, which provides a vocabulary that allows us to express the interface specification succinctly and clearly. An interface specification is always dependent on the programming language. Auxiliary specifications, on the other hand, are independent of the programming language.

, The role of auxiliary specifications in formal specifications is similar to that of the overview section in informal specifications. The difference is that auxiliary specifications define objects that have a precise meaning, so that the abstractions defined in terms of the auxiliary specifications also have a precise meaning that will be the same for users and implementers alike.

The language presented here is a subset of a more complete specification language, Larch/CLU. This subset contains only enough mechanism to define nonparameterized procedural and data abstractions and provides little support for the modular construction of specifications. We do not give a precise semantics for Larch/CLU in this chapter. That would not be difficult, but it would involve a considerable amount of moderately tedious detail. It would also require slightly more mathematical background than we are assuming for readers of this book. We shall use natural language and examples to explain the relationship between specifications and the programs intended to satisfy them. In chapter 11 we shall be more precise both about this satisfies relation and about the relation between a specification and the programs that use the module being specified.

10.1 An Introduction to Auxiliary Specifications

Auxiliary specifications provide two kinds of identifiers, *function identifiers* and *sort identifiers*. The notions of function and sort are closely related to the notions of procedure and type. It is important, however, not to confuse these concepts. When talking about auxiliary specifications, we shall consistently use the words function and sort. When talking about interface specifications, we shall use the words procedure and type.

We mean by a function what mathematicians mean—a mapping from a cross product of values (the *domain* of the function) to a single value (the *range* of the function). We mean by a sort an identifier used to characterize the domains and ranges of functions. We mean by procedures and types what (CLU) programmers mean.

In our specifications, we shall sometimes overload an identifier and use it to refer to both a function and a procedure or both a sort and a type. As we shall see, there is no syntactic context in which both functions and

STRING_TABLE_SPEC: **trait**

> **introduces**
>> NEW: → STRING_TABLE
>> ADD: STRING_TABLE, STRING, INT → STRING_TABLE
>> DELETE: STRING_TABLE, STRING → STRING_TABLE
>> IS_IN: STRING, STRING_TABLE → BOOL
>> EVAL: STRING_TABLE, STRING → INT
>> IS_EMPTY: STRING_TABLE → BOOL
>> SIZE: STRING_TABLE → INT

Figure 10.1 A partial trait.

procedures or sorts and types can appear, so there can be no ambiguity. However, to make the distinction absolutely clear, we shall adopt the convention of capitalizing the names of functions and sorts. Also, sometimes our specifications will use infix operators, such as +, for functions, but never for procedures, again maintaining the distinction.

Auxiliary specifications are built from pieces of text called *traits*. In the examples of this chapter, the purpose of the trait is to define a new sort with its functions. We begin by defining the form and meaning of traits. We then discuss the theory of a trait, that is, the theorems that can be proved about the introduced sort. These theorems form the basis of the interface specifications that use the trait. We next describe how traits can be combined and discuss two important properties of traits, consistency and sufficient-completeness. Finally, we discuss how to construct traits that are well-defined.

10.1.1 Traits

A trait has three basic parts: a *name*, an *introduces* section, and a *constrains* section. The *name* is the name of the trait, and we shall adopt the convention of having the trait name consist of the name of the introduced sort followed by "_SPEC." The *introduces* section lists the function identifiers introduced by the trait and gives a *signature* (the sorts of its domain and range) for each. Part of a trait is shown in figure 10.1. This trait defines a sort called STRING_TABLE, which is a mapping from strings to integers. This sort has functions, for example, to create an empty table and to look up the integer associated with a string.

The function identifiers in a trait can be used to form *terms* in much the same way that procedure calls can be used to form expressions in a programming language. Examples of terms are

NEW
ADD(NEW, s, v)
ADD(NEW, s_1, EVAL(ADD(NEW, s, v), s))
SIZE(ADD(NEW, s, v))
IS_EMPTY(NEW)

A term is called *variable-free* if it contains no variables. The first and last
terms in the list are variable-free. Note that when a nullary function, such
as NEW, occurs in a term, we do not follow it with a pair of parentheses.

Signatures are used to *sort-check* terms; sort checking is similar to type
checking in programming languages. We can deduce from the specification in figure 10.1 that if s and s_1 are of sort STRING and v is of sort
INT, then the first three terms in the list are sort-correct terms of sort
STRING_TABLE, the fourth is a sort-correct term of sort INT, and the
fifth is a sort-correct term of sort BOOL. An example of a sort-incorrect
term is

SIZE(SIZE(NEW))

The rest of the specification deals only with sort-correct terms.

Simply knowing all the sort-correct terms and their sorts is not very informative; we need to know something about the meaning of these terms.
This is the purpose of the *constrains section* of the specification, which
gives information about the meaning of the function identifiers. It does
this primarily by presenting a set of axioms relating sort-correct terms.
The list of function identifiers at the start of the constrains section contains all the function identifiers that the immediately following axioms
are intended to constrain. Figure 10.2 adds the constrains section to
STRING_TABLE_ SPEC. In this case all the introduced functions are being constrained.

The equations tell us when two terms mean the same thing. For example,
axiom 2 in figure 10.2 states that any term of the form

IS_IN(s, NEW)

where s is a variable of sort STRING, has the same meaning as the term

FALSE

The intuitive meaning of this axiom is that no element is contained in the
empty table, NEW. Axiom 7 states that deleting anything from the empty

STRING_TABLE_SPEC: **trait**

 introduces
  NEW: → STRING_TABLE
  ADD: STRING_TABLE, STRING, INT → STRING_TABLE
  DELETE: STRING_TABLE, STRING → STRING_TABLE
  IS_IN: STRING, STRING_TABLE → BOOL
  EVAL: STRING_TABLE, STRING → INT
  IS_EMPTY: STRING_TABLE → BOOL
  SIZE: STRING_TABLE → INT

 constrains NEW, ADD, DELETE, IS_IN, EVAL, IS_EMPTY, SIZE
 so that for all [s, s_1: STRING, v: INT, t: STRING_TABLE]

  1) EVAL(ADD(t, s, v), s_1) =
   if $s = s_1$ **then** v **else** EVAL(t, s_1)

  2) IS_IN(s, NEW) = FALSE
  3) IS_IN(s, ADD(t, s_1, v)) = (($s = s_1$) | IS_IN(s, t))

  4) SIZE(NEW) = 0
  5) SIZE(ADD(t, s, v)) =
   if IS_IN(s, t) **then** SIZE(t) **else** $1 +$ SIZE(t)

  6) IS_EMPTY(t) = (SIZE(t) = 0)

  7) DELETE(NEW, s) = NEW
  8) DELETE(ADD(t, s, v), s_1) =
   if $s = s_1$ **then** DELETE(t, s_1) **else** ADD(DELETE(t, s_1), s, v)

Figure 10.2 The STRING_TABLE_SPEC trait.

table, NEW, simply results in the empty table. Axiom 5 states that the size of the table,

 ADD(t, s, val)

is equal to the size of t if s is already in t, and otherwise is equal to 1 plus the size of t.

 The axioms always relate terms of the same sort. Some of them relate terms of the sort being defined, while others relate terms of the other sorts. The axioms relating terms of the sort being defined can be thought of as simplifications; they state that terms of certain forms can be reduced to terms of some other form. For example, axioms 7 and 8 in figure 10.2 imply that for any variable-free term containing DELETE there exists another term with the same meaning that consists of only ADDs and NEW. An example of such a variable-free term is

DELETE(ADD(NEW, "a", 3), "b")

By axiom 8 and the fact that "a" $\neq$ "b", this reduces to

ADD(DELETE(NEW, "b"), "a", 3)

By axiom 7, this reduces to

ADD(NEW, "a", 3)

thus eliminating DELETE from the term.

The axioms relating terms of sort STRING_TABLE do not tell us much about what STRING_TABLE means. For example, knowing that deleting anything from NEW results in NEW does not tell us much about what NEW means. Axioms 2 and 4, on the other hand, tell us quite a lot about the meaning of NEW. The point is that the only way to understand the meaning of a new sort is to relate it to other, known, sorts.

As was the case with data abstraction operations, the functions introduced by a trait include both *observers* and *constructors*. Constructors return values of the sort being specified; observers map values of that sort onto other sorts. In STRING_TABLE_ SPEC, NEW, DELETE, and ADD are constructors, while the other functions are observers.

In principle, there could be a sort with no observers, but such a sort would be singularly uninteresting. With no way to relate its values to those of other sorts, its values would be indistinguishable. The ability to distinguish among values of a sort rests solely upon the effects these values have when they appear as arguments to observers. This is similar in spirit to the notion that the only way to obtain any information about an abstract object in a program is to perform an observer operation on it.

10.1.2 The Theory of a Trait

The set of theorems, Th, that we can prove about the terms defined in a trait is called the *theory* of the trait. This theory is infinitely large and is built up deductively, from the axioms appearing in the trait, using a *logical system*. A logical system consists of a set of syntactically well-formed assertions, called *wffs* or *well-formed formulas*, and a method for proving the *validity* of a wff. A *proof* is an application of that method. It is useful to think of a proof as a sequence of wffs, the last of which is the wff whose validity we are proving. Each wff in the proof is either an axiom or can be derived from wffs preceding it in the proof and a rule of inference. An *axiom* is a wff that is assumed to be valid. *Rules of inference* allow us to deduce the validity of one wff, the *conclusion*, from the validity of others, the *hypotheses*.

The purpose of an auxiliary specification is to define a theory. Associated with a specification is an infinite set of wffs containing

1. all terms of sort BOOL, such as

TRUE
IS_IN(NEW, "a")

2. all equations of the form $t_1 = t_2$, where t_1 and t_2 are of the same sort, such as

$3 = (3 - 3)$
SIZE(NEW) = 0

3. all wffs that can be built from other wffs and the quantifiers of first-order predicate calculus, such as

∀ s: STRING [~IS_IN(NEW, s)]

The theory of a specification is a subset of this set. This subset is closed under a set of rules of inference; that is, it contains all wffs that can be derived from a set of axioms and a set of rules of inference. This section discusses the built-in rules of inference, the way in which a specification introduces new axioms, and two mechanisms through which a specification can introduce new rules of inference.

We use the axioms of a trait to reason about equality of the values denoted by terms, that is, to prove theorems of the form $t_1 = t_2$, where t_1 and t_2 are terms of the same sort. One example of such reasoning was given when we reduced a term containing DELETE to a term not containing DELETE. Formalizing such reasoning is straightforward. The equations appearing explicitly in the trait are the axioms. The five basic ways to make deductions are presented in figure 10.3. An example of an instantiation is

(SIZE(t) = 0)[NEW **for** t] = (SIZE(NEW) = 0)

In addition to the rules in figure 10.3, our specification language includes a set of axioms about BOOLs and a set of rules of inference. The BOOL axioms define the meaning of functions such as & (and) and | (or). We shall use these as we need them, but shall not bother to state them formally. We also include the axioms and rules of inference of first-order predicate

Reflexivity:
  For all terms t, $t = t$.

Symmetry:
  For all terms t_1 and t_2, $t_1 = t_2 \Rightarrow t_2 = t_1$.

Transitivity:
  For all terms t_1, t_2, and t_3, $t_1 = t_2 \& t_2 = t_3 \Rightarrow t_1 = t_3$.

Substitutivity:
  For all functions f of n arguments, if $t = t'$, then
    $f(t_1, \ldots, t, \ldots, t_n) = f(t_1, \ldots, t', \ldots, t_n)$.

Instantiation:
  If $t_1 = t_2$ and x is a variable, then $t_1[t$ **for** $x] = t_2[t$ **for** $x]$, where
  $t'[t$ **for** $x]$ stands for the term t' with all free occurrences of the
  variable x replaced by the term t.

Figure 10.3 Rules for equational reasoning.

calculus with equality. Among other things, these allow us to reason about
quantifiers.

This simple construction implies that the equations and inequations in
the theory of a trait are directly derivable from the axioms in the trait.
There is no metarule stating that if two terms are not provably equal, then
they may be considered to be unequal. Nor is there a metarule stating the
converse.

We now use the rules of figure 10.3 and the specification of figure 10.2
to prove that

SIZE(ADD(ADD(NEW, x, y), x, z)) = 1

(Informally, this equation is valid because adding a binding for an already
bound STRING replaces the old binding.) The proof proceeds by reducing
the left-hand side of the equation to the right-hand side; it is shown in
figure 10.4. (The leap of faith in the last step is required because we have
not yet given axioms about integers. In section 10.1.3 we shall discuss how
such outside information can be introduced into a trait. Until then, we
shall just assume it.)

Note that the proof in figure 10.4 holds for all strings x and integers y
and z; it shows that

∀ x: STRING, y, z: INT [SIZE(ADD(NEW, x, y), x, z) = 1]

All our formulas containing variables are universally quantified in this way.
Usually we shall omit the universal quantifiers; we shall include them only
when we want to emphasize the sorts of the variables.

SIZE(ADD(ADD(NEW, x, y), x, z)) =
   (by axiom 5, with the instantiations
      ADD(NEW, x, y) **for** t
      x **for** s
      z **for** v)

if IS_IN(x, ADD(NEW, x, y)) **then** SIZE(ADD(NEW, x, y))
      else 1 + SIZE(ADD(NEW, x, y))=
   (by axiom 3, with the instantiations
      x **for** s
      NEW **for** t
      x **for** s1
      y **for** v)

if (x = x) | IS_IN(x, NEW) **then** SIZE(ADD(NEW, x, y))
      else 1 + SIZE(ADD(NEW, x, y))=
   (by reflexivity of = and meaning of |)

if TRUE **then** SIZE(ADD(NEW, x, y)) **else** 1 + SIZE(ADD(NEW, x, y)) =
   (by meaning of if-then-else)

SIZE(ADD(NEW, x, y)) =
   (by axiom 5, with the instantiations
      NEW **for** t
      x **for** s
      y **for** v)

if IS_IN(x, NEW) **then** SIZE(NEW) **else** 1 + SIZE(NEW) =
   (by axiom 2, with the instantiation
      x **for** s)

if FALSE **then** SIZE(NEW) **else** 1 + SIZE(NEW) =
   (by meaning of if-then-else)

1 + SIZE(NEW) =
   (by axiom 4)

1 + 0 =
   (by leap of faith)

1

Figure 10.4 Example proof.

While the relatively small theory just developed is often useful, there are times when a larger theory is needed. For example, with what we have so far we cannot prove that

   ∀ t: STRING_TABLE [(t = NEW) | ∃ s: STRING [IS_IN(s, t)]]

Generated by and *partitioned by* give two ways to specify larger theories.

Saying that a sort S is *generated by* a set of functions F asserts that each term of sort S is equal to a term whose outermost function is in F. For example, we might say that the natural numbers are

generated by [0, successor]

and that the integers are

generated by [0, successor, predecessor]

Now consider adding to the specification of STRING_TABLE

STRING_TABLE **generated by** [NEW, ADD]

This states that NEW and ADD are sufficient to generate all terms of sort STRING_TABLE. In other words, every variable-free term of sort STRING_TABLE is equivalent to some term in which NEW and ADD are the only functions with range STRING_TABLE. Therefore, if we can show that something is true for all terms of this form, we know it is true for all STRING_TABLEs.

The addition of *generated by* allows us to prove theorems using complete induction. The induction is over the number of ADDs and NEWs that appear in the term. For example, to show that some theorem of the form

$\forall$ t: STRING_TABLE [P(t)]

holds, we need merely show

1. The basis step:

P(NEW)

This shows that the property P holds for the term with just one occurrence of ADDs and NEWs. (NEW is the only such term.)

2. The induction step:

$\forall$ s: STRING, v: INT, t: STRING_TABLE [P(t) $\Rightarrow$ P(ADD(t, s, v))]

Here we assume that P holds for table *t* and then show that it holds for "larger" tables obtained by ADDing to *t*.

For example, suppose we want to show that every nonempty STRING_TABLE contains a string; that is, we want to show the property

$\forall$ t: STRING_TABLE [(t = NEW) | ($\exists$ s: STRING [IS_IN(s, t)])]

To prove this theorem we have to show

1. The basis step: Here we do the substitution NEW **for** t to obtain

(NEW = NEW) | (∃ s: STRING [IS_IN(s, NEW)]))

2. The induction step: Here we assume our property for table t and prove it for larger tables:

∀ s_1: STRING, v: INT, t: STRING_TABLE
    [((t = NEW) | ∃ s: STRING [IS_IN(s, t)]) ⇒
        ((ADD(t, s_1, v) = NEW) |
            ∃ s: STRING [IS_IN(s, ADD(t, s_1, v))]))]

The validity of the basis step follows from the reflexivity of = and the meaning of |. The validity of the induction step can be shown by using axiom 3 of figure 10.2 and choosing s to be s_1.

Somewhat curiously, the above proof did not take advantage of the induction hypothesis. In showing the validity of the induction step, we were able to reduce the right-hand side of the implication to true without using the left-hand side of the implication. Nevertheless, the proof depended upon the induction rule afforded by the *generated by*. Without that rule, showing the validity of these formulas would not have told us that the theorem was true for all STRING_TABLEs. A *generated by* gives us a way of characterizing the notion "all STRING_TABLEs."

An example that uses the induction hypothesis is the proof of

∀ t: STRING_TABLE [SIZE(t) ≥ 0]

To prove this theorem we have to show

1. The basis step:

SIZE(NEW) ≥ 0

2. The induction step:

∀ s: STRING, v: INT, t: STRING_TABLE
    [SIZE(t) ≥ 0 ⇒ SIZE(ADD(t, s, v)) ≥ 0]

The basis step follows from axiom 4 and the fact that 0 ≥ 0. The induction step can be reduced, using axiom 5, to

SIZE(t) ≥ 0 ⇒ ((**if** IS_IN(s_1, t) **then** SIZE(t) **else** 1 + SIZE(t)) ≥ 0)

which simplifies by the meaning of if-then-else and $\Rightarrow$ to

[(SIZE(t) $\geq$ 0 & IS_IN(s_1, t)) $\Rightarrow$ SIZE(t) $\geq$ 0] &
[(SIZE(t) $\geq$ 0 & ~IS_IN(s_1, t)) $\Rightarrow$ (1 + SIZE(t)) $\geq$ 0]

which simplifies to TRUE.

Even with the addition of *generated by*, all proofs are ultimately based upon direct equational substitution. That is to say, nothing can be deduced from the absence of equations. Somewhat surprisingly, this is not always adequate. As mentioned earlier, the functions of a sort are either constructors or observers. Given the reasoning system just outlined, we can show that two terms of sort STRING_TABLE are different by showing that they can be distinguished using some observer—for example, that they are of different sizes. However, we can show that two terms of sort STRING_TABLE are equal only by direct equational substitution (perhaps combined with induction). For example, it is not possible to prove that

ADD(ADD(t, s, v), s_1, v) = ADD(ADD(t, s_1, v), s, v)

Suppose we were quite sure that we had a complete set of observers, so that we did not plan to add observers to gain new information about values of sort STRING_TABLE. We might then wish to consider equal all pairs of terms of sort STRING_TABLE that could not be distinguished using these observers. The *partitioned by* construct allows us to do this. Saying that sort S is *partitioned by* a set of functions F asserts that if two terms of sort S are unequal, a difference can always be observed using a function in F. In this case the terms are equal if they *cannot* be distinguished using any of the functions in F.

Consider adding

STRING_TABLE **partitioned by** [IS_IN, EVAL]

to STRING_TABLE_SPEC. Then to prove the theorem

ADD(ADD(t, s, v), s_1, v) = ADD(ADD(t, s_1, v), s, v)

we need to show we cannot distinguish these two terms using IS_IN or EVAL. Thus we must prove

1. $\forall$ s_2: STRING [IS_IN(s_2, ADD(ADD(t, s, v), s_1, v)) =
     IS_IN(s_2, ADD(ADD(t, s_1, v), s, v))]

2. $\forall$ s_2: STRING [EVAL(ADD(ADD(t, s, v), s_1, v), s_2) =
     EVAL(ADD(ADD(t, s_1, v), s, v), s_2)]

By applying axiom 3 twice to the terms on each side of the equal sign in the first formula we obtain

$$((s_2 = s_1) \mid ((s_2 = s) \mid IS\_IN(s_2, t))) =$$
$$((s_2 = s) \mid ((s_2 = s_1) \mid IS\_IN(s_2, t)))$$

which can be reduced, using the associativity and commutativity of $\mid$, to true. The reduction of the second formula can be done similarly, using axiom 1.

In this example we added to STRING_TABLE_SPEC a *partitioned by* that included only some of the observers introduced in the trait. We omitted those observers that provided no new information. Any two STRING_TABLEs that can be distinguished using the set of observers {IS_IN, EVAL, IS_EMPTY, SIZE} can also be distinguished using the smaller set {IS_IN, EVAL}. IS_EMPTY can be omitted because a STRING_TABLE is empty only if no string IS_IN it. Similarly, SIZE counts the number of strings in the STRING_TABLE; again the test that matters is whether a string IS_IN the table. Using the smaller set in the *partitioned by* has the advantage of halving the number of cases that need to be considered in a proof. (If we omit a needed function from *partitioned by*, the trait will be inconsistent, as discussed in section 10.1.4.)

10.1.3 Combining Traits

The specification given in STRING_TABLE_SPEC is incomplete because it depends on undefined sort and function identifiers. We mentioned earlier that there are several sorts in trait STRING_TABLE_SPEC—for example, INT and STRING as well as STRING_TABLE. Each of these sorts has functions associated with it, and some of these functions are used in STRING_TABLE_SPEC—for example, + for INT, = for STRING, and FALSE for BOOL. Sort BOOL and its functions have a predefined meaning in the specification language, and any use of BOOL functions thus has a meaning. However, no other definitions are assumed in a trait. In particular, no meaning is given in STRING_TABLE_SPEC for INT or STRING or their functions.

The most straightforward thing to do would be to augment the specification with additional clauses dealing with these functions. Thus we could add extra axioms defining the meaning of the INT functions used in STRING_TABLE_SPEC. However, adding extra axioms reduces modularity; it is like including an implementation of a procedure in a module that calls the procedure. What we would like to do is to specify a sort in one trait and then use it in other traits. This can be accomplished by trait *importation.* For example, we might add to trait STRING_TABLE_SPEC

imports INT_SPEC, STRING_SPEC

When a trait is imported into another trait, the importing trait acquires all of the function declarations, equations, *generated bys*, and *partitioned bys* of the imported trait.

Importation is used primarily to structure specifications to make them easier to read. It also introduces extra checking: Functions appearing in imported traits cannot be constrained in either the importing trait or any other imported trait. This means that imported traits do not "interfere" with one another. Each imported trait can be fully understood independently of the context into which it is imported. This tells us, for example, that within STRING_TABLE_SPEC the function + means exactly the same thing as it does in the simpler context of trait INT_SPEC.

10.1.4 Internal Consistency and Sufficient-Completeness

The ultimate criterion of appropriateness for an auxiliary specification is its utility in writing interface specifications. This criterion cannot be formalized. There are, however, two formal notions that can be very useful in evaluating and hence constructing auxiliary specifications: internal consistency and sufficient-completeness.

If the theory of a trait includes the equation TRUE = FALSE, we call the trait *internally inconsistent.* As soon as TRUE = FALSE is introduced into the theory associated with a trait, every well-formed formula becomes a member of that theory. In short, if we can prove TRUE = FALSE, we can prove anything. Clearly, internal inconsistency should always be avoided.

For example, the specification INTBAG_SPEC in figure 10.5 is internally inconsistent because axiom 5 implies that DELETE removes only one occurrence of an integer from the INTBAG, but axiom 3 states that DELETE removes all occurrences. We can prove TRUE = FALSE as follows: Consider the term

IS_IN(DELETE(INSERT(INSERT(NEW, 2), 2), 2), 2)

By axiom 3, this term equals FALSE, since $\sim(2 = 2)$ equals FALSE. However, using axiom 5, we can prove that this term equals

IS_IN(INSERT(NEW, 2), 2)

We can then use axiom 2 to prove that this term equals TRUE. Since equality is transitive, we may now infer that TRUE = FALSE.

INTBAG_SPEC: **trait**

   imports INT_SPEC

   introduces
      NEW: → INTBAG
      INSERT: INTBAG, INT → INTBAG
      DELETE: INTBAG, INT → INTBAG
      IS_IN: INTBAG, INT → BOOL

   constrains NEW, INSERT, DELETE, IS_IN **so that**
         for all [b: INTBAG, i, j: INT]

      1) IS_IN(NEW, i) = FALSE
      2) IS_IN(INSERT(b, i), j) = (i = j) | IS_IN(b, j)
      3) IS_IN(DELETE(b, i), j) = ~(i = j) & IS_IN(b, j)

      4) DELETE(NEW, i) = NEW
      5) DELETE(INSERT(b, i), j) =
            if i = j **then** b **else** INSERT(DELETE(b, j), i)

Figure 10.5 Specification of a bag abstraction.

   The internal inconsistency of INTBAG_SPEC can be eliminated by removing axiom 3. Removing this axiom, however, raises a question: How
do we know that terms of the form IS_IN(DELETE(b, i), j) have a well-
defined meaning? Internal inconsistency arises when we say too much in a
trait, that is, when we constrain two terms to be both equal and unequal
to each other. It is also possible to get in trouble by not saying enough in a
trait. It would be useful to have a way to decide whether sufficient axioms
have been given.

   The notion of a complete axiom set is familiar to mathematicians, although the precise definition varies. Sometimes an axiom set is said to be
complete if it is impossible to add an independent axiom—that is, if all
well-formed formulas either follow from or are inconsistent with the original set of axioms. Sometimes completeness is defined with respect to some
underlying domain that the axioms are intended to describe. If we have a
model of the underlying domain, we can define completeness in terms of
whether or not everything that is true in the model is provable from the
axioms.

   Unfortunately, none of the common definitions of completeness is appropriate for dealing with traits. We do not usually have an alternative
definition against which to measure the completeness of the trait. Requiring
that there be no independent axioms leads us to overconstrain things, as
discussed in chapters 3 and 8 (that is, it does not permit underdetermined
specifications).

Sufficient-completeness is a notion of completeness that is useful for traits. It is applicable to sorts for which there is a *generated by*. A *generated by* defines a set of functions that are sufficient to generate all values of a sort. For a specification to be sufficiently complete with respect to a *generated by* of the form

S **generated by** $[f_1, \ldots, f_n]$

it must be possible to show that

1. All variable-free terms of sort S that contain functions with range S are provably equal to terms whose only functions with range S are $f_1, \ldots, f_n$.

2. All variable-free terms that are not of sort S are provably equal to terms that contain no function with range S.

If these conditions hold, then we have enough axioms.

Consider the specification in figure 10.6. This is like the specification of figure 10.5, except that the axiom relating IS_IN to DELETE has been removed and a *generated by* has been added. To show that this specification is sufficiently complete we need to show that

1. Any term of the form DELETE(b, i), where b and i are variable-free terms, is equal to a term in which DELETE does not occur.

2. Any term of the form IS_IN(b, i), where b and i are variable-free terms, is equal to a term in which none of NEW, INSERT, and DELETE occurs.

Both of these can be proved using induction on the depth of nesting, in the term b, of functions in the generator set {NEW, INSERT}. The proof that DELETE can be eliminated is as follows:

1. The basis step: Let b be NEW. By axiom 3, DELETE(NEW, i) = NEW, and thus DELETE can be eliminated.

2. The induction step: Assume that if the depth of nesting in b is less than n, DELETE can be eliminated. Let INSERT(b_1, i) have depth of nesting n. By axiom 4, DELETE(INSERT(b_1, i), j) equals either b_1 or INSERT(DELETE(b_1, j), i). The depth of nesting of INSERT in both b_1 and DELETE(b_1, j) is less than n, and therefore DELETE can be eliminated in either case. Therefore DELETE can be eliminated from DELETE(INSERT(b_1, i), j).

10.1.5 Constructing Traits

A major problem in constructing traits is deciding exactly which axioms are needed. The notion of sufficient-completeness leads to a technique that is helpful here. Suppose we wish to complete the specification started in

INTBAG_SPEC1: **trait**

  imports INT_SPEC

  introduces
    NEW: $\rightarrow$ INTBAG
    INSERT: INTBAG, INT $\rightarrow$ INTBAG
    DELETE: INTBAG, INT $\rightarrow$ INTBAG
    IS_IN: INTBAG, INT $\rightarrow$ BOOL

  constrains NEW, INSERT, DELETE, IS_IN **so that**
    INTBAG **generated by** [NEW, INSERT]
    for all [b: INTBAG, i, j: INT]

    1) IS_IN(NEW, i) = FALSE
    2) IS_IN(INSERT(b, i), j) = (i = j) | IS_IN(b, j)

    3) DELETE(NEW, i) = NEW
    4) DELETE(INSERT(b, i), j) =
        if i = j **then** b **else** INSERT(DELETE(b, j), i)

Figure 10.6 Sufficiently complete specification of a bag abstraction.

figure 10.7 so that the functions will have the meanings connoted by their names. We begin by partitioning these functions into four classes:

1. *Basic constructors*: {NEW, PUSH} is a minimal generator set for INTSTACKs. All distinct values of sort INTSTACK can be defined in terms of the functions in this set. Having chosen a set of basic constructors, we add to the specification

  INTSTACK **generated by** [NEW, PUSH]

2. *Extra constructors*: {POP}.

3. *Basic observers*: {TOP, SIZE} is a minimal observer set for INTSTACKs All other functions that map INTSTACKs to a value of some other sort can be defined in terms of the functions in this set. This information will be used in determining the *partitioned by* as discussed further below.

4. *Extra observers*: {ISEMPTY}.

    The next step in our technique is to construct a set of candidate left-hand sides for axioms. We do this using our two sufficient-completeness criteria. The first criterion tells us that we must be able to show that every variable-free term of the form POP(s) is convertible to a term in which POP does not occur. The *generated by* tells us that it will be sufficient to consider terms of the form

INTSTACK_SPEC: **trait**

  imports INT_SPEC

  introduces
     NEW: → INTSTACK
     PUSH: INTSTACK, INT → INTSTACK
     POP: INTSTACK → INTSTACK
     TOP: INTSTACK → INT
     SIZE: INTSTACK → INT
     ISEMPTY: INTSTACK → BOOL

Figure 10.7 Start of a specification of INTSTACK.

  POP(NEW)
  POP(PUSH(s, i))

Similarly, the second criterion leads us to consider terms that capture the possible interactions between the basic observers and the basic constructors:

  TOP(NEW)
  TOP(PUSH(s, i))
  SIZE(NEW)
  SIZE(PUSH(s, i))

Finally, we generate left-hand sides in which the extra observers are applied to variables. Here this gives us one additional left-hand side:

  ISEMPTY(s)

Figure 10.8 shows right-hand sides for all but one of these left-hand sides. We have not supplied a right-hand side for TOP(NEW), since there does not seem to be a reasonable value to supply. This means that our specification will not be sufficiently complete. We should keep this in mind when using INTSTACK_SPEC in an interface specification and make sure that the interface specification's meaning does not depend upon the value of TOP(NEW). Frequently, the places where a trait is not sufficiently complete correspond to exceptions in interface specifications built from the trait.

The last step in constructing this trait is the addition of a *partitioned by*. At first glance, it may seem that the basic observers are the appropriate functions for a *partitioned by*. Suppose, however, that we added to INTSTACK_SPEC

  INTSTACK **partitioned by** [TOP, SIZE]

POP(NEW) = NEW
POP(PUSH(s, i)) = s
TOP(PUSH(s, i)) = i
SIZE(NEW) = 0
SIZE(PUSH(s, i)) = 1 + SIZE(s)
ISEMPTY(s) = (SIZE(s) = 0)

Figure 10.8 Axioms for INTSTACK.

Now consider the stacks denoted by

s_1 = PUSH(PUSH(NEW, 1), 2)
s_2 = PUSH(PUSH(NEW, 2), 2)

There is no way to distinguish s_1 from s_2 using only TOP and SIZE, yet they clearly represent different values. We need to use the extra constructor POP to observe values within the stacks. Thus we have

INTSTACK **partitioned by** [TOP, SIZE, POP]

Frequently, the functions that should appear in a *partitioned by* are those in the union of the basic observers and extra constructors. This is the case for INTSTACK_SPEC.

It is worth noting that very often the set of axioms in a trait is not sufficiently complete, as was the case with INTSTACK. Sufficient-completeness is still a useful property, though, because it allows us to generate a candidate set of axioms. We may then decide explicitly to omit some axiom, but starting with a list of candidates helps us avoid omitting an axiom by accident.

10.1.6 The Auxiliary Specification INT_ARRAY_SPEC

Figure 10.9 defines a trait INT_ARRAY_SPEC, which can be used to define a general CLU-like array of integers. This trait is used in the interface specifications in the remainder of this chapter and in chapter 11. It has been constructed using the technique outlined in the previous section.

There are several things to note in the INT_ARRAY_SPEC trait. The use of the sort identifier INT_ARRAY instead of a list of operations in the *constrains* list is an example of a commonly used shorthand. The sort identifier stands for the list of all functions in the trait that have the sort INT_ARRAY in their signature. In this example, all the introduced functions are being constrained.

The set of functions specified in this trait is not minimal. The *generated by* tells us that ASSIGN and REMOVEH cannot be used to generate any

INT_ARRAY_SPEC: **trait**

    imports INT_SPEC

    introduces
        NEW: INT $\rightarrow$ INT_ARRAY
        APPENDH: INT_ARRAY, VAL $\rightarrow$ INT_ARRAY
        REMOVEH: INT_ARRAY $\rightarrow$ INT_ARRAY
        ASSIGN: INT_ARRAY, INT, VAL $\rightarrow$ INT_ARRAY
        FETCH: INT_ARRAY, INT $\rightarrow$ VAL
        IS_IN: VAL, INT_ARRAY $\rightarrow$ BOOL
        LBOUND: INT_ARRAY $\rightarrow$ INT
        HBOUND: INT_ARRAY $\rightarrow$ INT
        LENGTH: INT_ARRAY $\rightarrow$ INT
        IS_DEFINED: INT_ARRAY, INT $\rightarrow$ BOOL
        SAME_BOUNDS: INT_ARRAY, INT_ARRAY $\rightarrow$ BOOL

    constrains INT_ARRAY **so that**
        INT_ARRAY **generated by** [NEW, APPENDH]
        INT_ARRAY **partitioned by** [FETCH, SAME_BOUNDS]
        for all [a, a1, a2: INT_ARRAY, v, v1: VAL, i, j: INT]

        LBOUND(NEW(i)) = i
        LBOUND(APPENDH(a, v)) = LBOUND(a)

        HBOUND(NEW(i)) = i $-$ 1
        HBOUND(APPENDH(a, v)) = HBOUND(a) + 1

        LENGTH(a) = HBOUND(a) $-$ LBOUND(a) + 1

        IS_IN(v, NEW(i)) = FALSE
        IS_IN(v, APPENDH(a, v1)) = (v = v1) | IS_IN(v, a)

        IS_DEFINED(a, j) = j $\geq$ LBOUND(a) & j $\leq$ HBOUND(a)

        FETCH(APPENDH(a, v), i) =
            if i=HBOUND(a)+1 **then** v **else** FETCH(a, i)

        SAME_BOUNDS(a1, a2) = (LBOUND(a1) = LBOUND(a2) &
            HBOUND(a1) = HBOUND(a2))

        REMOVEH(APPENDH(a, v)) = a

        ASSIGN(NEW(i), j, v) = NEW(i)
        ASSIGN(APPENDH(a, v), j, v1) = **if** j = HBOUND(a) + 1
            then APPENDH(a, v1)
            else APPENDH(ASSIGN(a, j, v1), v)

Figure 10.9 Specification of INT_ARRAY sort.

array that cannot be generated using only NEW and APPENDH. The axioms make it quite obvious that IS_DEFINED, LENGTH, and SAME_BOUNDS are extra observers. These functions were introduced simply to make it easier to read and write the interface specifications that appear later in this chapter and in chapter 11.

Note that this specification is not sufficiently complete. We have no way to reduce terms such as FETCH(NEW(1), 1) or REMOVEH(NEW(1)). As we shall see later, the incompleteness turns out to be harmless because the meanings of the interface specifications that use INT_ARRAY_SPEC are not affected by it. A related issue is the rather arbitrary choice of meaning for terms involving bounds errors, such as ASSIGN(NEW(i), j, v). Again, we shall see that this need not concern us because of the way we structure interface specifications that use INT_ARRAY_SPEC. Finally, note that the *partitioned by* contains only observers. This tells us that two INT_ARRAYs are equal if and only if they have the SAME_BOUNDS and FETCH returns the same value when applied to the same index.

10.2 Interface Specifications of Procedural Abstractions

We now turn our attention to interface specifications. We discuss how to specify both CLU procedures and CLU types. A discussion of how to reason with and about interface specifications is deferred to chapter 11.

A formal specification of a procedure has three parts:

1. A *header* giving its name, the types of its inputs, the types of its outputs, and any signals it is allowed to raise.

2. A *body* stating any requirements on its inputs and specifying the effects the procedure has when those requirements are met.

3. An *auxiliary specification*. In this chapter we shall assume that the auxiliary specification is the union of the traits associated with the types of the procedure's arguments and results.

The header provides the same information as would the header of an informal specification of the procedure. It specifies syntactic constraints that must be met by every implementation and use of the procedure.

The body provides the same kind of information as would the body of an informal specification, but there is greater rigor in the presentation. The body places constraints on the arguments with which the procedure should be called and defines the relevant aspects of the behavior of the procedure when it is called with appropriate arguments. The body is most conveniently thought of as two predicates, one giving constraints that must be met by calls of the procedure and one giving constraints that must be met by implementations. Their signatures are

Pre: Arguments → BOOL
Post: Arguments, Results → BOOL

A body of the form

requires Pre
effects Post

means that if the predicate Pre holds when a procedure is called, then the procedure terminates and the predicate Post must hold on termination. The specification is completely silent as to the behavior of the procedure when Pre does not hold; that is, the interpretation of the two predicates is exactly equivalent to the single predicate Pre ⇒ Post.

As discussed in chapter 5, in the interest of robustness we try to avoid abstractions that rely on conditions that must be established by callers. Consequently, the predicate in the requires clause is frequently TRUE; that is, nothing is required. In this case we shall omit the clause completely.

Three kinds of identifiers and function symbols may appear in requires and effects predicates:

1. locally bound logical variables,

2. formal arguments, and

3. identifiers and symbols defined in the auxiliary specification.

For example, in the specification

div = **proc** (i, j: int) **returns** (ans: int)
        requires j > 0
        effects i ≥ ans * j & ~∃ k: INT [k > ans & i ≥ j * k]

k is a logical variable, i, j, and *ans* are formals, and 0, >, ≥, *, and INT come from the trait associated with the type int (that is, the trait INT_SPEC). (Recall that by convention identifiers from auxiliary specifications are always uppercase.)

In the body of a specification, formal arguments may appear unqualified or qualified by *pre*, *post*, or *obj*. A formal qualified by *obj*, for example, x_{obj}, stands for the object bound to that formal. A formal qualified by *pre*, for example, x_{pre}, stands for the value of the object bound to that formal when the procedure is called. A formal qualified by *post*, for example, x_{post}, stands for the value of the object bound to that formal when the procedure returns. An unqualified argument formal is short for that formal qualified by *pre*. An unqualified result formal is short for that formal qualified by *post*.

The following specification defines a procedure that removes the elements from one array x, moves them to another array y, and sets the low bound of x to i:

> move = **proc** (x, y: int_Array, i: int)
>   **requires** $\sim(x_{obj} = y_{obj})$
>   **effects** $y_{post} = x$ & $x_{post} = \text{NEW}(i)$

This specification requires that the two formals x and y be bound to different objects; it says nothing about what *move* does when the same object is bound to both formals. Note that the equal sign is being used in two different ways in this specification. In the requires clause, the equal sign means object identity; that is,

$$x_{obj} = y_{obj}$$

if and only if x and y are bound to the same object. In the effects clause, we are comparing values of objects. This latter meaning of equality comes from the theory of the associated trait; for example,

$$y_{post} = x$$

means that the term that is the value of y when the procedure returns can be proved to be equal to the term that is the value of x when the procedure is called.

10.2.1 More Notation for Bodies of Procedure Specifications

To make the bodies of specifications easier to read and write, we introduce a stylized way to write predicates. We also introduce a set of five built-in predicates that will make it easier to describe notions that arise frequently in specifying CLU procedures.

The CLU-specific built-in predicates provide compact ways to state what old objects are not mutated by a procedure, what new objects are created, and when a procedure returns or signals. The predicate

> **modifies at most** $(o_1, \ldots, o_n)$

asserts that the procedure mutates no object except possibly some subset of $\{o_1, \ldots, o_n\}$. This predicate makes a statement about objects, so that we always use the *obj* qualifier, as in

> **modifies at most** (x_{obj})

Note that this predicate is really an assertion about all of those objects
that do not appear in the modifies list, rather than about those that do.
The predicate

modifies nothing

is equivalent to

modifies at most ( )

The predicate

new (o)

asserts that when the procedure returns, the object o has been added to
the universe of objects accessible in the environment of the caller. Like the
modifies predicate, the *new* predicate makes a statement about objects, so
that we use the *obj* qualifier. The predicate

returns

asserts that the procedure returns in the normal way; that is, it does not
abort, run forever, or raise a signal. The predicate

signals name

where *name* must be an exception listed in the header of the specification,
asserts that the procedure raises the exception *name*.

The following example uses the CLU-specific predicates to describe a
procedure that accepts as arguments two different *int_Arrays*. If they have
the same bounds, it stores their pairwise product in the second argument;
if they have different bounds, it raises an exception and does not modify
either argument:

array_mult = **proc**(x, y: int_Array) **signals** (bounds_mismatch)
    requires $\sim(x_{obj} = y_{obj})$
    effects
      ((SAME_BOUNDS(x, y) $\Rightarrow$
        (**modifies at most** (y_{obj}) & **returns** &
          $\forall$ i : INT [IS_DEFINED(x, i) $\Rightarrow$ FETCH(y_{post}, i) =
          FETCH(x, i) $*$ FETCH(y, i)]) &
      ($\sim$SAME_BOUNDS(x, y) $\Rightarrow$
        (**modifies nothing** & **signals** bounds_mismatch))

10.2.2 Syntax for Structuring Predicates

From now on we shall be using a somewhat richer syntactic structure for the bodies of specifications. We introduce this mechanism primarily to make it easier to read specifications that fall into the most common patterns. We begin by factoring a modifies predicate out of the effects clause. This again parallels the way we write informal specifications, so that a body of the form

> **requires** Pre
> **modifies** M
> **effects** Post

means

> Pre $\Rightarrow$ (**modifies** M & Post)

The modifies predicate specifies the set of objects that might be mutated by the procedure. It is either

> **modifies nothing**

or of the form

> **modifies at most** $(o_1, \ldots, o_n)$

The requires and effects predicates can have a significant amount of internal structure. They can be built up using

1. function symbols defined in the auxiliary part of the specification, such as IS_IN,

2. formal arguments and their qualifiers,

3. the normal connectives and quantifiers of first-order predicate calculus,

4. CLU-specific predicates, and

5. three special connectives: *if-then* and *if-then-else* (in both the requires and effects predicates) and *case* (in the effects predicate only).

The predicate "**if** P **then** Q **endif**" is equivalent to the predicate "P $\Rightarrow$ Q." The predicate "**if** P **then** Q **else** R" is equivalent to the predicate "(P $\Rightarrow$ Q) & ($\sim$P $\Rightarrow$ R)." These forms are particularly useful for dealing with nested conditions.

The general form of a *case* predicate is

normally PostPred
except signals $exception_1$ **when** $PrePred_1$ **ensuring** $PostPred_1$
    ...
    signals $exception_n$ **when** $PrePred_n$ **ensuring** $PostPred_n$

PostPred states a condition that must hold whenever the procedure returns normally, that is, when it does not raise an exception. The predicates $PrePred_i$ and $PostPred_i$ in each arm state conditions that must hold
whenever the $exception_i$ of that arm is raised. The $PrePred_i$ are like pre-
conditions that tell us when each exception may be raised. If none of the
$PrePred_i$ is true, the procedure must return normally. If exactly one of
them is true, the corresponding $exception_i$ must be raised. If more than
one of them is true, an exception associated with one of the true enabling
conditions must be raised, but the choice of which exception is nondeterministic. Each $PostPred_i$ is like a post-condition that tells us what must
be true whenever the associated $exception_i$ is raised.

To state this a bit more formally, a predicate of the form given above is
equivalent to the predicate

    (**returns** | **signals** $exception_1$ | ... | **signals** $exception_n$) &
    (**returns** $\Rightarrow$ PostPred & $\sim$($PrePred_1$ | ... | $PrePred_n$)) &
    (**signals** $exception_1$ $\Rightarrow$ ($PrePred_1$ & $PostPred_1$)) &
        ...
    (**signals** $exception_n$ $\Rightarrow$ ($PrePred_n$ & $PostPred_n$))

This semantics allows a nondeterministic choice of signals when this is
desired, as discussed. If instead we had the pre-conditions implying the
results, as in

    $PrePred_i$ $\Rightarrow$ **signals** $exception_i$ & $PostPred_i$

the nondeterminism would not be possible.

There is one more complication associated with the *case* predicate. For
each $exception_i$, we may leave out the *ensuring* clause. This is equivalent
to the appearance of "**ensuring** TRUE."

Figure 10.10 gives some sample specifications. The first shows how the
specification of *array_mult* can be rewritten using the new notation. The
next two examples illustrate the use of a qualifier to distinguish an object
from its value. (Recall that an unqualified argument identifier stands for
that identifier qualified by *pre*.) They also illustrate the use of the *new*
predicate. The *copy* procedure returns a new int_Array that is similar to

array_mult = **proc**(x, y: int_Array) **signals** (bounds_mismatch)
    requires $\sim(x_{obj} = y_{obj})$
    modifies at most (y_{obj})
    effects
        normally $\forall i$: INT [IS_DEFINED(x, i) $\Rightarrow$
            FETCH(y_{post}, i) = FETCH(x, i) $*$ FETCH(y, i)]
        except signals bounds_mismatch **when** $\sim$SAME_BOUNDS(x, y)
            ensuring modifies nothing

copy = **proc** (x: int_Array) **returns** (z: int_Array)
    modifies nothing
    effects new(z_{obj}) & (z = x)

share = **proc** (x: int_array) **returns** (z: int_array)
    modifies nothing
    effects $z_{obj} = x_{obj}$

share_and_copy = **proc** (x: int_array) **returns** (z: int_array)
    modifies nothing
    effects $z_{obj} = x_{obj}$ & **new**(z_{obj})

find_in_range = **proc** (a: int_array, x, y: int) **returns** (z: int)
        signals (els_too_small, els_too_big, wrong_xy)
    requires $\forall$ i, j: INT [(i $\leq$ j & IS_DEFINED(a, i) &
    IS_DEFINED(a, j)) $\Rightarrow$ FETCH(a, i) $\leq$ FETCH(a, j)]
    modifies nothing
    effects
        normally $\exists$ i: INT [FETCH(a, i) = z] & x < z & z < y
        except signals els_too_small
            when $\sim\exists$ i: INT [IS_DEFINED(a, i) & a[i] > x]
            signals els_too_big
            when $\sim\exists$ i: INT [IS_DEFINED(a, i) & a[i] < y]
            signals wrong_xy **when** $\sim$(x < y)

Figure 10.10 Some interface specifications for procedures.

the argument array; *share* is like *copy*, except that the returned array must be the input object.

The next example, *share_and_copy*, is a variation on the same theme. Here, however, we have written an inconsistent specification by asserting that the returned object is both new and the same as the input object. Inconsistent specifications cannot be implemented.

The final example illustrates the use of nondeterminism in a specification. Procedure *find_in_range* returns any element in the array *a* that is between the integers *x* and *y* and signals an appropriate exception if no such element exists. Note that under some conditions, for example, if *a* is empty, more than one exception may be appropriate. In such a case, any appropriate exception may be raised. If the array is not sorted, that is, if the requires clause is not satisfied, the procedure may do anything—loop, crash, modify *a*, return a value that is not in the proper range, return a value that is not even in the array, raise either signal, and so on.

10.3 Interface Specifications of Data Abstractions

Most of a data abstraction's interface specification is supplied by its auxiliary specification and the specifications of its operations. The specification has three parts:

1. A *header* specifying the name of the type, whether or not objects of the type are mutable, and the names of its operations.

2. A *link* associating types occurring in the specification with sorts occurring in the auxiliary specification.

3. A specification of each operation. These are interface specifications of procedures, as described in the previous section.

The information provided in the header is similar to that provided in the header of an informal specification. For example, the header for type *int_array* might be

> int_array **mutable type**
> **exports** [create, addh, remh, store, fetch, equal, similar]

The link takes the place of the overview section of our informal specifications. We include an auxiliary specification by giving a trait name. The link also associates the types appearing in the interface specification with sorts in the imported trait. It will often be the case that types will have names that are quite similar to the names of the sorts with which the type's values are associated. This should be regarded as a planned coincidence, with no formal meaning. Therefore we must state the association explicitly. For *int_array* we might have

> **based on sort** INT_ARRAY **from** INT_ARRAY_SPEC
> **with** [int **for** INT]

A specification of an *int_array* type is given in figure 10.11. This type will be used extensively in chapter 11. For the sake of brevity, we have not included all of CLU's built-in operations. Those that we have specified, however, have the same meaning as in CLU.

In reading the specification, take particular note of the way in which the incompletenesses of the trait are avoided. The specification of *fetch*, for example, guards against referring to terms in which the function FETCH

int_array **mutable type**
      exports [create, addh, remh, high, store, fetch, equal, similar]

      based on sort INT_ARRAY **from** INT_ARRAY_SPEC **with** [int **for** INT]

create = **proc** (i: int) **returns** (a: int_array)
      modifies nothing
      effects new(a_{obj}) & a = NEW(i)

addh = **proc** (a: int_array, v: int)
      modifies at most (a_{obj})
      effects a_{post} = APPENDH(a, v)

remh = **proc** (a: int_array) **returns** (v: int) **signals** (bounds)
      modifies at most (a_{obj})
      effects
            normally a_{post} = REMOVEH(a) & v = FETCH(a, HBOUND(a))
            except signals bounds **when** HBOUND(a) − LBOUND(a) < 0
            ensuring modifies nothing

high = **proc** (a: int_array) **returns** (i: int)
      modifies nothing
      effects i = HBOUND(a)

low = **proc** (a: int_array) **returns** (i: int)
      modifies nothing
      effects i = LBOUND(a)

store = **proc** (a: int_array, i: int, v: int) **signals** (bounds)
      modifies at most (a_{obj})
      effects
            normally a_{post}=ASSIGN(a, i, v)
            except signals bounds **when** ∼IS_DEFINED(a, i)
            ensuring modifies nothing

fetch = **proc** (a: int_array, i: int) **returns** (v: int) **signals** (bounds)
      modifies nothing
      effects
            normally v = FETCH(a, i)
            except signals bounds **when** ∼IS_DEFINED(a, i)

equal = **proc** (x: int_array, y: int_array) **returns** (b: bool)
      modifies nothing
      effects b = $(x_{obj} = y_{obj})$

similar = **proc** (x: int_array, y: int_array) **returns** (b: bool)
      modifies nothing
      effects b = (x = y)

Figure 10.11 Interface specification of an array of integers.

is applied to arguments for which it is undefined. This guard is embodied in the *except* clause calling for the exception *bounds*.

In this chapter, the trait INT_ARRAY_SPEC was presented before the specification of the type *int_array*. The order of presentation is arbitrary. In constructing this specification, we developed the interface and traits in tandem. We began with a preliminary trait. As we worked on the interface specification, we went back several times and modified the trait to make the interface specification easier to write.

10.4 Summary

In this chapter we introduced a formal language for writing specifications of CLU procedures and types. The basic structure of specifications written in this language follows closely that of the informal specifications used throughout this book. The most important point to carry away from this chapter is the way in which interface specifications are built on top of auxiliary specifications.

Auxiliary specifications are independent of the programming language. The unit of specification is the trait, which introduces functions and sorts and defines a theory. The theory contains a set of first-order formulas over sort-correct terms. These formulas follow from the axioms, *generated bys*, and *partitioned bys* of the trait.

It is important that an auxiliary specification be well-defined and sufficiently constraining. We defined two notions that can be used in evaluating traits. A trait is internally inconsistent if its theory contains TRUE = FALSE. Internal inconsistency should always be avoided because once TRUE = FALSE is in the theory, anything can be proved. A trait is sufficiently complete if the set of functions in the *generated by* is sufficient to generate all values of the sort. This notion can be used as an aid in constructing the trait because it gives us a set of candidates for the axioms. If an axiom for every candidate is included (and defined properly), the trait will be sufficiently complete. We may decide explicitly to omit one of the axioms, but starting with the candidates helps us avoid omitting an axiom by accident.

Interface specifications are dependent on the programming language. They define a set of program modules, that is, the set of all correct implementations of the specification. They deal explicitly with such issues as mutability and exceptions, which are not dealt with in auxiliary specifications. The connection between implementations and traits is made in an interface specification. The *based on* construct relates types to sorts.

The requires and effects clauses of procedure specifications constrain the behavior of procedures by relating the values of arguments and results to terms defined in traits.

We have now developed the machinery that allows us to specify procedures and types. In the next chapter we shall use specifications to reason about programs.

Further Reading

Bjørner, Dines, and Cliff B. Jones, 1982. *Formal Specification & Software Development*. Englewood Cliffs, N.J.: Prentice-Hall International.

Guttag, John V., James J. Horning, and Jeannette M. Wing, 1985. Larch in five easy pieces. Technical report 5, Digital Equipment Corporation Systems Research Center.

Exercises

10.1 Consider the specification of the bag abstraction in figure 10.6. Show that every variable-free term with outermost operator IS_IN is equal to a term in which no function with range INTBAG occurs.

10.2 Add an appropriate *partitioned by* to the specification in figure 10.6.

10.3 Add a function COUNT to the specification in figure 10.6. It should have the signature

COUNT: INTBAG, INT → INT

and return the number of occurrences of the integer in the bag.

10.4 Modify the specification in figure 10.6 so that DELETE removes all occurrences of the deleted element.

10.5 Write a complete specification of a *table* abstraction that maps strings to integers. It should have operations to create an empty table, add a string and its associated integer to the table, remove a string and associated integer from the table, look up the value of a string, and tell the number of mappings in the table. Tables should be mutable. (Hint: Use the STRING_TABLE sort developed in sections 10.1.1 and 10.1.2.)

10.6 Write a complete specification of a bounded stack type. It should have procedures with the following headers:

new = **proc** (i: int) **returns**(s: stack) **signals** (non_positive_ Size)
push = **proc** (s: stack, i: int) **signals** (overflow, duplicate)
pop = **proc** (s: stack) **returns** (i: int) **signals** (empty)

The integer passed to *new* is the maximum size of the stack. If this is not greater than zero, *new* signals *non_positive_ Size*. The procedure *push* signals *overflow* if pushing the integer passed to it would cause the stack to become too big. It signals *duplicate* if the integer passed to it is already in the stack. The procedure *pop* returns the last element inserted and removes it from the stack. (Hint: Use the INTSTACK sort developed in section 10.1.5, but modify it as needed.)

10.7   Write a natural-language specification that conveys all of the information contained in your answer to exercise 5 or 6. Do you believe that your natural-language specification would have been as complete if you had not written the formal one first?

10.8   Write a formal specification for poly (figure 4.3). As part of doing this specification, you will need to develop a trait for a related sort with appropriate functions.

10.9   Write a specification for a queue with first-in/first-out (FIFO) operations. As part of doing this specification, you will need to define a trait for a related sort with appropriate functions.

11

A Quick Look at Program Verification

This chapter is a short introduction to a complex topic. We hope to give the flavor of the process of program verification, without getting bogged down in the myriad details involved in actually constructing rigorous proofs of program properties. We believe that an understanding of what is involved in the formal verification of programs is an asset in informal reasoning about programs, and the most important thing to carry away from this chapter is thus simply a sense of the basic techniques involved in such formal reasoning. The structures of the proof rules and of the proofs are important; the details are not.

Program verification involves reasoning about program texts. This distinguishes verification from testing, which is always based upon observing computations. In verification we examine the program text and make inferences about the set of computations described by the program. We frequently refer to the set of computations to justify our method of reasoning, but we never construct this set or any part of it.

We start this chapter by introducing the basic program verification paradigm and reasoning about straight-line code. We then discuss programs containing **if** statements and **while** loops. Next we turn to the process of proving that procedures satisfy their specifications and of reasoning about programs that invoke procedures. Finally, we discuss reasoning about programs that contain clusters.

We shall generally avoid precise definitions in favor of illustrative examples. Moreover, we shall not treat all statement and expression forms of CLU; instead we shall rewrite code whenever necessary to make use of just the part of CLU that is covered. Other issues that will not be discussed include aliasing, recursive procedures, iterators, and exceptions.

11.1 Reasoning about Straight-Line Code

We shall use *total correctness formulas* to state what we wish to verify. Each formula includes two predicates, a *pre-condition* and a *post-condition*, and a segment of code. The domain of the predicates is the set of all states of computations. A total correctness formula has the form

$$P \ \{ \ S_1 \ S_2 \ldots S_n \ \} \ Q$$

Such a formula says that if the state of the computation before executing the statements S_1 S_2 ... S_n satisfies the pre-condition P, we want S_1 S_2 ... S_n to terminate and the state following execution to satisfy the post-condition Q. Of course, if the pre-condition is false in the initial state, then the whole formula is (vacuously) true. This is analogous to the rule of logic that

(FALSE $\Rightarrow$ P) = TRUE.

Here we follow the convention of using uppercase letters in formulas for identifiers, such as TRUE and FALSE, that do not appear in programs. Note that the pair of predicates in these formulas bears a striking resemblance to the body of a procedure specification. The pre-condition is like the requires clause, and the post-condition like the effects clause.

Consider an example:

TRUE
{ x := 0
   y := 1 }
y > x

How can we convince ourselves of the validity of this formula? That is, how do we know that this program fragment terminates, and that when it is finished, the value of y is greater than the value of x? We begin by arguing that exactly two statements get executed. After the second statement is executed, the value of y is 1 and the value of x is whatever it was after the first statement was executed, that is, 0. We complete the argument by observing that $1 > 0$. Note that in this argument we reasoned about

1. the flow of control through the program,

2. the meaning of the assignment statement, and

3. the meaning of the $>$ relation on integers.

The basic paradigm we shall use in all our reasoning about programs is as follows:

1. Locate all paths between the two predicates.

2. For each path, work backward from the final predicate to discover a derived predicate R that must be true prior to the first statement in the path if the final predicate is to hold. Do this one statement at a time.

3. Use our knowledge about the types of the values of the variables occurring in the predicates to convince ourselves that for each path the pre-condition implies the derived predicate R.

The simple proof just outlined can now be restated more carefully. Pushing $y > x$ back across the second assignment simplifies the program fragment to

TRUE $\{x := 0\}$ 1 $> x$

and pushing $1 > x$ across the remaining assignment gives

TRUE $\Rightarrow$ 1 > 0

which, by the rules of logic and what we know about integers, simplifies to TRUE. In constructing this proof, we used the following *rule of assignment* to push predicates backward across assignment statements:

> Let P be any predicate and e any side-effect-free expression: If P is to be true after the assignment $x := e$, P with e substituted for all free occurrences of x (that is, occurrences not bound to a quantifier) must be true before the assignment. That is to say, P with e substituted for all free occurrences of x is the *weakest pre-condition* that will ensure that P holds after the assignment $x := e$.

At first glance, it may seem a bit odd to be working backward through the program. The advantage of working this way is that our reasoning is goal directed: At each step, knowing what we wish to be true at the conclusion of some part of the program, we compute what needs to be true prior to that part. When we reach the start of the program, we have the weakest pre-condition sufficient to imply the desired post-condition ("weakest" in the sense that any other sufficient pre-condition implies it). We complete the proof by showing that the given pre-condition implies the weakest pre-condition.

If we were to work forward, we would begin with what was known to be true at the start of some part of the program and compute what must be true at the conclusion of that part. When we reached the end of the program, we would have the strongest post-condition implied by the known pre-condition ("strongest" in the sense that it implied all other valid post-conditions). We would complete the proof by showing that this strongest post-condition implied the given post-condition. The problem with this approach is that as we work forward through the program, we have no systematic way to prune irrelevant information. We might accumulate a large number of things that happen to be true but are totally unrelated to what we wish to prove.

We observed earlier that the two predicates in a total correctness formula are similar to the requires and effects clauses in procedure specifications.

There is, however, a crucial difference. In a procedure specification the effects predicate deals with two states, the one before and the one after the call. When we refer to the value of a formal, we qualify that formal (sometimes implicitly) with *pre* or *post*. In a total correctness formula each predicate deals with only one state. This presents a slight technical problem when we need to express a relationship between initial and final values.

As an example, consider a total correctness formula asserting that the statement $x := x+1$ increments x. The post-condition needs to refer to both the initial and final values of x. To do this, we use the trick of introducing a *fresh variable* x_0 in the pre-condition and asserting that it has the same value as x. It is "fresh" in the sense that it does not already occur anywhere in either the program text or the given pre- or post-condition. To assert that $x := x + 1$ increments x we write

$$x_0 = x \ \{ \ x := x + 1 \ \} \ x > x_0$$

To verify this formula, we push the predicate $x > x_0$ across the assignment statement to yield

$$(x_0 = x) \Rightarrow (x + 1 > x_0).$$

Substituting x_0 for x in the conclusion gives $x_0 + 1 > x_0$, which, assuming x is of type integer, simplifies to TRUE.

11.2 Reasoning about Programs with Multiple Paths

Thus far we have looked at programs that have only one path. Now consider a program with two paths through it, such as

if $x = 1$ **then** $y := 0$ **else** $y := x$ **end**

Here there is one path for the **then** clause and one path for the **else** clause. In reasoning about the **then** clause, we can assume that the predicate of the conditional is true; in reasoning about the **else** clause, that the predicate is false. In general, to reason about programs with **if** statements we use the following *case analysis rule:*

If the evaluation of b is side-effect-free, proving
  P {**if** b **then** s1 **else** s2} Q
is equivalent to proving both P & b {s1} Q and P & ~b {s2} Q.

A concise way of writing a *rule of inference* such as this is

side-effect-free(b), P & b {s1} Q, P & ~b {s2} Q
_____

P {if b then s1 else s2} Q

In this notation the clauses separated by commas above the line are called *hypotheses*. The single clause below the line is called the *conclusion*. Such a rule can be read as stating that demonstrating the truth of all the hypotheses is sufficient to prove the truth of the conclusion.

The case analysis rule is the first of a number of rules we shall use to reduce the verification of a complex total correctness formula to a set of simpler formulas or *verification conditions*. For example, to prove

$(x = 1) \mid (x = 0)$
   { if $x = 1$ then $y := 0$ else $x := y$ end }
$(y = 0)$

we need to prove three verification conditions:

1. side-effect-free $(x = 1)$
2. $((x = 1) \mid (x = 0))$ & $(x = 1)$ { $y := 0$ } $y = 0$
3. $((x = 1) \mid (x = 0))$ & $\sim(x = 1)$ { $y := x$ } $y = 0$

The validity of the second and third verification conditions can be shown using the rule of assignment. The validity of the first verification condition follows from the fact that the only procedure invoked by the expression is int$*equal*, which modifies nothing.

The kind of case analysis used above is applicable to any program segment with a relatively small number of paths through it. Unfortunately, as noted in chapter 9, most programs have a great many paths through them. Consider, for example,

$x \geq 0$ & $y = 0$
   { while $\sim(y = x)$ do $y := y + 1$ end }
$y = x$

There are an infinite number of paths through the program fragment in this formula, one for each possible value of x. If $x = 0$, the path contains no assignments; if $x = 1$, the path has one assignment; if $x = 2$, it has two assignments; and so forth. Clearly we cannot reason about this program by enumerating all paths through it. An ad hoc argument about the validity of this formula would probably go:

1. The pre-condition guarantees that $x \geq y$.

2. Every time through the loop the program increments y, thus bringing it closer to x.

3. Since there are only a finite number of integers between x and y, y will eventually equal x.

4. The program exits the loop and terminates as soon as $y = x$.

This is analogous to proving a theorem about all positive integers by ordinary induction. We first prove that the theorem holds for 1. We then show that if it holds for an arbitrary positive integer n, it must also hold for $n + 1$. We conclude that the theorem must hold for all positive integers. In reasoning about programs with loops, we resort to induction over the number of iterations through each loop.

A crucial step in constructing an inductive proof is choosing the induction hypothesis. In reasoning about loops, we call this hypothesis the *loop invariant*. Choosing a suitable loop invariant can be challenging. It must satisfy four conditions. Suppose we wish to prove that

     P   { **while** b **do** s **end** }   Q

We must be able to show that

1. The invariant is true at the start of the first iteration of the loop; that is, P implies the invariant.

2. If the invariant is true before evaluating the predicate controlling the loop, it is true after evaluating this predicate.

3. If the invariant and the predicate controlling the loop are true at the start of any iteration through the loop, the invariant is true at end of that iteration.

4. The conjunction of the loop invariant and the negation of the predicate controlling the loop implies Q.

The fourth condition implies that if the loop terminates in a state satisfying the invariant, Q will hold upon termination. The first three conditions guarantee that the invariant will indeed hold upon termination. Thus if we succeed in finding a loop invariant that satisfies these four conditions and in showing that the loop always terminates, we can conclude that the loop terminates in a state satisfying Q. This is expressed by the following *loop invariant rule*:

P ⇒ Inv,
Inv {b} Inv,
Inv & b {S} Inv,
(Inv & ~b) ⇒ Q,
while b **do** S **end** terminates

———————————————————————

P {**while** b **do** S **end**} Q

For our example, $x \geq y$ is a suitable loop invariant. It is good programming practice to include the loop invariant as a comment at the start of
loop. To prove that

x ≥ 0 & y = 0
    { % invariant x ≥ y
        while ~(y = x) **do** y := y + 1 **end** }
y = x

we show that the loop always terminates and demonstrate the validity of
each of the following:

1. The loop invariant holds on entry to the loop:

   (x ≥ 0 & y = 0) ⇒ x ≥ y

2. The loop invariant is maintained by evaluation of the test controlling
the loop:

   x ≥ y {~(y = x) } x ≥ y

3. The loop invariant is maintained by the body of the loop:

   x ≥ y & ~(y = x) { y := y + 1 } x ≥ y

4. The loop invariant and the negation of the test imply the post-condition:

   ((x ≥ y) & ~(~(y = x))) ⇒ (y = x)

The first and fourth conditions simplify immediately to TRUE. The second
follows directly from the fact that the evaluation of ~(x = y) is side-effect-
free. By pushing the final predicate across the assignment statement, we
can simplify the third verification condition to the predicate

   x ≥ y & ~(y = x) ⇒ (x ≥ y + 1)

Since $x \geq y$ & ~(y = x) implies $x > y$, this predicate also simplifies to
TRUE. Note, by the way, that this part of the proof is the inductive step.
We use the fact that the loop invariant holds on entry to the loop body to

show that it holds on completion of the loop body.

To complete our total correctness proof, we need to show that the loop always terminates. We do this by associating a *decrementing function* with the loop. This function maps a subset of the program variables to some well-ordered set. (A set is *totally ordered* if it is possible to compare every member of the set to every other member. A totally ordered set is *well-ordered* if every nonempty subset of it has a least element.) The nonnegative and positive integers are well-ordered. The set of all integers is not, since it has no least element. The set of nonnegative rationals is not, since there are nonempty subsets with no least element—for example, all positive rationals. In this chapter we shall use the nonnegative integers as the range of our decrementing functions.

The decrementing function must satisfy two conditions:

1. We must be able to show that every iteration of the loop reduces the value to which the decrementing function maps.

2. We must be able to show that the conjunction of the loop invariant and the fact that the decrementing function maps to the least element of the well-ordered set implies the negation of the predicate controlling the loop; that is, it implies that the loop will be exited.

If we can exhibit a decrementing function with these two properties, we can conclude that the loop always terminates. This is so because each time through the loop the decrementing function must map to a smaller value (condition 1); there are only a finite number of values between the initial value of the decrementing function and the least element of its domain (this is a property of well-ordered sets); and when the program reaches a state in which the decrementing functions maps to the least element, we exit the loop (condition 2). The result is analogous to the loop invariant rule: *Even though there may be infinitely many paths through a program, we can reason about the program's termination by examining only a finite number of them.*

For our example, $(x - y)$ is a suitable decrementing function. To prove that the loop in

```
x ≥ 0 & y = 0
    { % invariant x ≥ y
        % decrements (x − y), with range nonnegative integers
        while ∼(y = x) do y := y + 1 end }
y = x
```

always terminates we must show that

1. The decrementing function is decremented by the body:

$$(x - y) = d_0 \;\&\; \sim(y = x) \;\&\; x \geq y \;\{\; y := y + 1 \;\}\; (x - y) < d_0$$

2. The decrementing function is not incremented by the test:

$$(x - y) = d_0 \;\{\; \sim(x = y) \;\}\; (x - y) \leq d_0$$

3. When the decrementing function reaches 0, the loop is terminated:

$$(x - y) = 0 \;\&\; (x \geq y) \Rightarrow \sim(\sim(y = x))$$

All of these simplify easily to TRUE.

We started out by observing that the basic problem presented by a program with a loop is that it is not possible to enumerate all the paths through the program. This has two ramifications:

1. The proof of termination is no longer trivial.

2. It is not immediately obvious how to reduce arguments about the program's behavior to a finite number of steps.

We have dealt with these two issues separately. First, we argued that if the loop were to terminate, it would terminate in a state satisfying the post-condition. This is called a *partial correctness* argument, and the crucial step is inventing a loop invariant. We then argued that the loop would indeed terminate. The crucial step here was exhibiting a decrementing function. The combination of partial correctness and termination is called *total correctness*.

Students frequently have difficulty learning to write loop invariants. Fortunately, it is easy to tell whether or not one has an appropriate invariant. If the invariant is inappropriate, at least one step of the verification will fail. Most of the time we can quickly deduce the problem with the invariant by studying how the verification failed.

11.3 Reasoning about Procedures

The introduction of procedures complicates the proof system needed to reason about programs, but it also reduces the complexity of proofs constructed in that system. This is similar to the introduction of procedures, which complicates the definition of a programming language but simplifies the writing of programs in that language. Appropriate use of procedures makes it easier to understand a program and to reason formally about it. We reason separately about the correctness of a procedure's implementation and about parts of the program that call the procedure. To prove the correctness of a procedure definition, we show that the procedure's body satisfies its specification. When reasoning about invocations of a procedure, we use only the specification.

We shall split our discussion of procedures into two parts. First, we discuss procedures that return values but do not mutate their arguments. We then discuss procedures that mutate their arguments. The basic reasoning is the same for both, but there are differences in detail.

11.3.1 Reasoning about Procedures with No Side Effects

To prove that a procedure satisfies its specification, we first convert its body and its specification into a total correctness formula of the form we have been looking at,

P {body} Q

We then use the techniques discussed earlier to reason about this formula.

The pre-condition of the formula is based upon the requires clause of the specification. We construct it by first conjoining to the requires predicate a predicate $x = x_0$, where x_0 is a fresh variable, for each formal argument x. This "saves" the initial values of the arguments. We then conjoin the predicate $\sigma_0 = \sigma$. This is introduced to allow us to deal with the modifies clause of the specification. The reserved symbol σ stands for the current mapping from all objects to their values. The fresh variable σ_0 is introduced to save this mapping.

The post-condition is based upon the effects clause of the specification. We construct it by first substituting the saved initial value of each argument formal for every free occurrence of that formal in the effects predicate. (Recall that in the body of a specification the appearance of an unqualified argument formal stands for the initial value of that formal. Since in this section we are dealing only with procedures that do not modify their arguments, we defer our discussion of formals qualified by *post* or *obj* to section 11.3.2.) We then account for the *modifies nothing* by conjoining the predicate $\sigma_0 \subseteq \sigma$. By using a subset here, we permit the procedure to create new objects but guarantee that no existing objects are modified.

The statement part of the formula is constructed from the body of the implementation and the header of the specification. To transform the body we first delete the variable declarations. (Strictly speaking, we should not do this. In a completely formal treatment, all syntactic objects, including declarations, are dealt with in the proof rules. The rules associated with declarations are used to govern what axioms may be applied in reasoning

factorial = **proc** (x: int) **returns** (a: int)
    requires x ≥ 0
    modifies nothing
    effects a = x!

factorial = **proc** (x: int) **returns** (int)
  ans: int
  ans := 1
  while x > 0 **do**
    ans := ans * x
    x := x − 1
    end
  return (ans)
  end factorial

Figure 11.1 A specification and implementation of *factorial*.

about the values of variables. For example, the declaration

  x: int

indicates that we can use the theory associated with integers in reasoning about the value of x.) Next, we connect the value returned by the procedure with the formal, for example, *res*, used to represent the returned value in the specification; we do this by replacing any **return** statement of the form **return** (*ans*) by an assignment statement of the form *res* := *ans*. (We assume that there is only one **return** statement and that it is always the last statement in the procedure body.)

Figure 11.1 contains a specification and an implementation of a factorial function. When applied to this example, the transformations just described yield the verification condition

$$x \geq 0 \ \& \ x_0 = x \ \& \ \sigma_0 = \sigma$$
$$\{ \ ans := 1$$
$$\textbf{while } x > 0 \textbf{ do}$$
$$ans := ans * x$$
$$x := x - 1$$
$$\textbf{end}$$
$$a := ans \ \}$$
$$a = x_0! \ \& \ \sigma_0 \subseteq \sigma$$

From here we proceed as with any total correctness formula. We begin by annotating the loop with an invariant and a decrementing function. The invariant will include the predicate $\sigma_0 \subseteq \sigma$; we shall need this to show that the post-condition holds on exiting the loop. Pushing the final predicate across the assignment statement that replaced the **return** statement gives us

$x \geq 0$ & $x_0 = x$ & $\sigma_0 = \sigma$
    { ans := 1
      while $x > 0$ **do**
          % invariant ans $= (x_0!/x!)$ & $x \geq 0$ & $\sigma_0 \subseteq \sigma$
          % decrements x, range a nonnegative integer
          ans := ans $*$ x
          x := x $-$ 1
      end }
ans $= x_0!$ & $\sigma_0 \subseteq \sigma$

We can apply the loop invariant theorem to transform the above into four verification conditions:

1. The loop invariant holds on entry to the loop:

$x \geq 0$ & $x_0 = x$ & $\sigma_0 = \sigma$
    { ans := 1 }
ans $= x_0!/x!$ & $x \geq 0$ & $\sigma_0 \subseteq \sigma$

2. The loop invariant is maintained by the test:

ans $= (x_0!/x!)$ & $x \geq 0$ & $\sigma_0 \subseteq \sigma$
    { x > 0 }
ans $= (x_0!/x!)$ & $x \geq 0$ & $\sigma_0 \subseteq \sigma$

3. The loop invariant is maintained by the body of the loop when the test is true:

$x > 0$ & ans $= (x_0!/x!)$ & $x \geq 0$ & $\sigma_0 \subseteq \sigma$
    { ans := ans $*$ x
      x := x $-$ 1 }
ans $= (x_0!/x!)$ & $x \geq 0$ & $\sigma_0 \subseteq \sigma$

4. The loop invariant and the negation of the test imply the post-condition:

$((\text{ans} = (x_0!/x!)$ & $x \geq 0$ & $\sigma_0 \subseteq \sigma)$ & $\sim(x > 0)) \Rightarrow$
    (ans $= x_0!$ & $\sigma_0 \subseteq \sigma)$

Consider verification condition 1. Since the evaluation of the expression on the right-hand side of the assignment involves no side effects, we apply the assignment rule to yield

$$(x \geq 0 \ \& \ x_0 = x \ \& \ \sigma_0 = \sigma \ ) \Rightarrow$$
$$(1 = x_0!/x! \ \& \ x \geq 0 \ \& \ \sigma_0 \subseteq \sigma \ )$$

Since $x \geq 0$ occurs on both sides of the implication, and since $\sigma_0 = \sigma$ implies $\sigma_0 \subseteq \sigma$, we can reduce this to

$$(x \geq 0 \ \& \ x_0 = x \ ) \Rightarrow (1 = x_0!/x! \ )$$

Since $x_0 = x$ occurs on the left of the implication, we can reduce the right side to

$$x \geq 0 \Rightarrow 1 = (x!/x!)$$

that is, to TRUE. We discharge the second verification condition by observing that the test $x > 0$ is side-effect-free. Using the assignment rule twice, we simplify verification condition 3 to

$$(x > 0 \ \& \ ans = (x_0!/x!) \ \& \ x \geq 0 \ \& \ \sigma_0 \subseteq \sigma \ ) \Rightarrow$$
$$(ans * x = (x_0!/(x - 1)!) \ \& \ (x - 1) \geq 0 \ \& \ \sigma_0 \subseteq \sigma \ )$$

which, deleting $\sigma_0 = \sigma$ from both sides, reduces to

$$(x > 0 \ \& \ ans = (x_0!/x!) \ \& \ x \geq 0 \ ) \Rightarrow$$
$$(ans * x = (x_0!/(x - 1)!) \ \& \ (x - 1) \geq 0 \ )$$

Since $x > 0$ implies $(x - 1 \geq 0)$, we can reduce this to

$$(x > 0 \ \& \ ans = (x_0!/x!)) \Rightarrow (ans * x = (x_0!/(x - 1)!))$$

Since $ans = (x_0!/x!)$ occurs on the left of the implication, we can reduce this to

$$x > 0 \Rightarrow ((x_0!/x!) * x = x_0!/(x - 1)!)$$

which simplifies to

$$x > 0 \Rightarrow (((x_0! * x)/x!) = x_0!/(x - 1)!)$$

which, using the fact that for all v, $v > 0 \Rightarrow v! = v * (v - 1)!$, simplifies to TRUE. Finally, consider verification condition 4. Since $(x \geq 0 \ \& \ \sim(x > 0))$, we know that $x = 0$. The condition can therefore be reduced to

$$(ans = x_0!/0!) \Rightarrow ans = x_0!$$

Since $0! = 1$, this reduces to TRUE.

Suppose we had omitted the $x \geq 0$ part of the invariant. We would then have been able to simplify condition 4 to

$$(\text{ans} = x_0!/x! \;\&\; \sim (x > 0)) \Rightarrow \text{ans} = x_0!$$

At this point we would have concluded that this reduces to TRUE only when $x! = 1$, that is, when $x = 0$ or $x = 1$. We know that x is not equal to 1, since $\sim(x > 0)$. However, since we cannot deduce $x = 0$ from the hypothesis of the verification condition, we would be led to conclude that our loop invariant is too weak. Presumably this would lead us to strengthen it appropriately. Having changed the invariant, we would need to generate and prove the new verification conditions for the loop.

That completes our proof of partial correctness. To prove total correctness we use the decrementing function to show that the loop terminates, as follows:

1. The body reduces the decrementing function:

$$x = d_0 \;\&\; x > 0 \;\&\; \text{ans} = x_0!/x! \;\&\; x \geq 0 \;\&\; \sigma_0 \subseteq \sigma$$
   $\{\ \text{ans} := \text{ans} * x$
   $\quad x := x - 1\ \}$
$$x < d_0$$

2. The decrementing function is not incremented by the test:

$$x = d_0 \;\{\ x > 0\ \}\; x \leq d_0$$

3. The loop is exited when the decrementing function reaches 0:

$$(\text{ans} = x_0!/x! \;\&\; x \geq 0 \;\&\; \sigma_0 \subseteq \sigma \;\&\; x = 0\ ) \Rightarrow \sim(x > 0)$$

Using the assignment rule twice, we can simplify verification condition 1 to $(x = d_0 \Rightarrow (x - 1) < d_0)$, that is, to TRUE. Since $x > 0$ is side-effect-free, condition 2 simplifies to TRUE. Condition 3 also simplifies immediately to TRUE.

This completes our proof that the implementation of *factorial* satisfies its specification. The only thing that distinguishes this proof from those earlier in the chapter is the introduction of σ, which allowed us to reason about the absence of mutation in the same way we reasoned about other properties. In practice, it is often simpler to reason separately about mutability, since it is frequently possible to use some sort of meta-argument to show that the body of a procedure cannot possibly mutate anything. For *factorial*, we could have argued that since its only argument is an int, the body cannot access any mutable objects and therefore must modify nothing.

We use only the specification to reason about procedure invocations, and such reasoning is independent of the proof that a procedure satisfies its specification. For example, to reason about a program that invokes *factorial*, we assume *factorial*'s specification—that its requires implies its effects. It is convenient to reason about *factorial* using two different rules, each of which is valid because of *factorial*'s specification. First, since *factorial* is side-effect-free, we use it in formulas as an expression, as in

$$x = 3 \Rightarrow \text{factorial}(x) = 6$$

Such an expression stands for the value returned by the *factorial* procedure, just as the expression $a + b$ stands for the value returned by int$plus. We can reason about this value using the rule

1. $\forall$ w: INT [w $\geq$ 0 $\Rightarrow$ (factorial(w) = w!)]

where ! is a symbol defined in a trait. In addition, we sometimes need to reason about aspects of *factorial*'s behavior that are not captured by the returned value. To do this we use the rule

2. $x \geq 0$ & $\sigma_0 = \sigma$ { factorial(x) } $\sigma_0 \subseteq \sigma$

In this formula we treat *factorial*(x) as a command.

When we want to reason about a program fragment that invokes *factorial*, we can use these rules just as we use other components of our proof system, such as the assignment rule or the case analysis rule. (A slight technical problem arises if the first formula turns out to be unsatisfiable, that is, always false. However, in practice this is not a serious problem, and we shall ignore it.) Consider a proof of the formula

$x \geq 0$
    { **if** factorial(x) = 6 **then** c := **true else** c := **false end** }
$c = (x{=}3)$

Since rule 2 for *factorial* tells us that it has no side effects, we begin by applying the case analysis rule. This yields two verification conditions:

4. $x \geq 0$ & factorial(x) = 6 { c := **true** } c = (x{=}3)

5. $x \geq 0$ & $\sim$(factorial(x) = 6) { c := **false** } c = (x{=}3)

We use *factorial*(x) here to stand for the value returned by the *factorial* procedure when it is called with x as an argument. Pushing across the assignment statement of the first verification condition, we get

$$x \geq 0 \ \& \ \text{factorial}(x) = 6 \Rightarrow ((x = 3) = \text{TRUE})$$

or

$$x \geq 0 \ \& \ \text{factorial}(x) = 6 \Rightarrow (x = 3)$$

Rule 1 for *factorial* tells us that

$$\forall \ w: \text{INT} \ [\ w \geq 0 \Rightarrow (\ \text{factorial}(w) = w! \ ) \ ]$$

Since the hypothesis of the verification condition asserts that $x \geq 0$, we can use this rule to reduce the verification condition to

$$x! = 6 \Rightarrow x = 3$$

We then use the trait in which ! is specified to reduce this to TRUE. The second formula simplifies similarly.

As a second example, consider the *int_array$create* operation. Its specification (given in figure 10.11) is

create = **proc** (i: int) **returns** (a: int_array)
      modifies nothing
      effects a = NEW(i) & **new**(a_{obj})

Since *create* is side-effect-free, we have as our first rule

1. $\forall$ i: INT [array_int$create(i) = NEW(i)]

Here we use *array_int$create(i)* as an expression in a formula and explain how to simplify the formula; since *create* requires nothing, there is no left-hand side in this case. Note how we use the term NEW(i) from sort INT_ARRAY here. Rule 2 is

2. $\sigma_0 = \sigma\{ \ a := \text{array\_int\$create(i)} \ \}\sigma_0 \subseteq \sigma$

This rule allows σ to grow to include the new object but does not really define the effect of **new**. We would need to define this effect precisely only if we were doing proofs about aliasing. (Two variables that both refer to the same object are said to be *aliases*.) As mentioned earlier, we shall not treat aliasing in this chapter.

Now let us show that

$$x = 0 \ \& \ \sigma_0 \subseteq \sigma$$
$$\{ \ b := \text{array\_int\$create(x)} \ \}$$
$$b = \text{NEW}(0) \ \& \ \sigma_0 \subseteq \sigma$$

Rule 2 implies that the part of the post-condition dealing with σ holds. Since *create* is side-effect-free, the other part can be proved using the

sum_pointwise = **proc** (a1, a2: int_array)
    requires SAME_BOUNDS(a1, a2)
    modifies at most (a1$_{obj}$)
    effects SAME_BOUNDS(a1$_{post}$, a1) &
        $\forall$ i: INT[(IS_DEFINED(a1$_{post}$, i) $\Rightarrow$
            (FETCH(a1$_{post}$, i) = FETCH(a1, i) + FETCH(a2, i)))]

sum_pointwise = **proc** (a1, a2: int_array)
  i: int
  i := int_array\$low(a1)
  while i $<=$ int_array\$high(a1) **do**
    a1[i] := a1[i] + a2[i]
    i := i + 1
    end
  end sum_pointwise

Figure 11.2 Specification and implementation of *sum_pointwise*.

assignment rule to push the post-condition back across the statement. This yields

$$x = 0 \Rightarrow \text{array\_int\$create}(x) = \text{NEW}(0)$$

Then using rule 1, we get

$$x = 0 \Rightarrow \text{NEW}(x) = \text{NEW}(0)$$

which simplifies to TRUE.

11.3.2 Reasoning about Procedures That Mutate Arguments

Reasoning about procedures that mutate their arguments is similar to reasoning about those that do not. There are, however, important differences both in how we build verification formulas from the specifications and in how we incorporate the specification into our reasoning about invocations. As an example, consider a procedure *sum_pointwise* that computes the pointwise sum of two arrays. A specification and an implementation of *sum_pointwise* are given in figure 11.2. The specification uses the trait INT_ARRAY_SPEC that appeared in chapter 10.

As before, the first step in verifying that the implementation is consistent with the specification is building a verification condition:

SAME_BOUNDS(a1, a2) & $a1_0$ = a1 & $a2_0$ = a2 & σ_0 = σ
　{ i := int_array\$low(a1)
　　while i <= int_array\$high(a1) **do**
　　　a1[i] := a1[i] + a2[i]
　　　i := i + 1
　　end }
SAME_BOUNDS(a1, $a1_0$) &
　$\forall$ i: INT [IS_DEFINED(a1, i) $\Rightarrow$
　　(FETCH(a1, i) = FETCH($a1_0$, i) + FETCH($a2_0$, i))] &
　$(\sigma_0 - a1_{obj}) \subseteq \sigma$

This formula raises two issues that did not appear in the factorial example. Both are related to the translation of the predicates on two states that occur in specifications to the predicates on a single state that occur in total correctness formulas.

The first issue involves the qualifiers *pre* and *post* that appear (sometimes implicitly) in the specification. In the pre-condition of a total correctness formula a program identifier corresponds to the identifier subscripted by *pre* in a specification. In the post-condition a program identifier corresponds to the identifier subscripted by *post* in a specification. We use the trick of introducing fresh variables in the pre-condition of the formula to save the *pre* values of the formals so that they can be referred to in the post-condition.

The second issue is the handling of the *modifies at most* clause in the specification. It may appear that all we need do is to conjoin the predicate $a2 = a2_0$ to the final assertion. However, the *modifies at most* predicate asserts that all objects except *a1* have the same value on exiting the procedure as they had on entering. Not only is *a2* unchanged, but so is every other object except *a1*. Since this is an assertion about an unbounded number of objects, we cannot solve the problem merely by introducing some number of fresh variables. Instead we are forced to refer explicitly to σ. As in the factorial example, in the pre-condition we introduce a fresh variable σ_0 to save the initial mapping. In the post-condition, we introduce some new notation. Recall that σ is a mapping from objects to values. The notation $\sigma - x$ stands for a mapping whose domain does not include x but is otherwise identical to σ. The formula

$(\sigma_0 - a1_{obj}) \subseteq \sigma$

means that all objects except possibly the one bound to the identifier *a1* have the same value in σ_0 and in σ. (In addition, σ can contain new objects, as before.)

We now turn our attention to reasoning about calls of procedures that modify their arguments. Consider, for example, the verification condition

IS_DEFINED(a1, x) & $a1_0$ = a1 & σ_0 = σ
   { a1[x] := z }
FETCH(a1, x) = z &
$\forall$ j: INT [(IS_DEFINED(a1, j) & j $\neq$ x) $\Rightarrow$
   FETCH(a1, j) = FETCH($a1_0$, j)] & $(\sigma_0 - a1_{obj}) \subseteq \sigma$

Here we must reason about a call of *int_array$store*, which modifies its first argument. (Recall that $a1[x] := z$ stands for the procedure call *int_array$store(a1, x, z)*.)

In the *factorial* example, we began by transforming the specification into a formula that could be used as an assumption at any place in a proof. We cannot do the same thing here. There we treated *factorial(x)* as an expression, but since the procedure call of the array *store* operation will always occur as a command, that trick will not work. What we must do is introduce a rule for reasoning about procedure calls. The rule must govern

1. the transformation of the specification to a formula,

2. the replacement of the formal parameters that occur in the specification by the actuals that occur in the call, and

3. the elimination of the procedure invocation as we work our way backward through the program text generating verification conditions.

We begin by transforming the specification of *int_array$store* into the implication it denotes, yielding a formula of the form

Pre $\Rightarrow$ Post

The specification of the *store* operation (given in figure 10.11) is

store = **proc** (a: int_array, i: int, v: int)
    modifies at most (a_{obj})
    effects
        normally a_{post} = ASSIGN(a, i, v)
        except signals bounds **when** $\sim$IS_DEFINED(a, i) **ensuring**
           modifies nothing

Although we do not cover exceptions in this chapter, we shall cover calls of procedures such as *store* when the values of the arguments guarantee that exceptions cannot occur. The Pre of the formula for such a call is the conjunction of the requires predicate and the negation of the when clause for each signal. In this case the requires predicate is TRUE, so that we have

Pre = IS_DEFINED(a_{pre}, i_{pre})
Post = (**modifies at most** (a_{obj}) & a_{post} = ASSIGN(a_{pre}, i_{pre}, v_{pre}))

Note that those identifiers implicitly qualified by *pre* and *post* in the specification are explicitly qualified in these formulas.

Next, we transform Pre and Post by replacing references to the qualified formals by the actuals. We treat formals listed in the *modifies at most* clause differently from those that are not listed. For formals that can be modified, we use the associated actual for *post* and fresh variables for *pre*; for example, we replace references to the qualified formals a_{pre} and a_{post} by references to the variables $a1_1$ and $a1$. Note that we use $a1_1$ here rather than $a1_0$; whenever fresh variables are introduced, they must be distinct from both program variables and variables used in verification conditions. For formals that are not listed, we use the associated actual for both *pre* and *post*; for example, we replace references to i_{pre} and i_{post} by references to the actual *x*. We replace references to formals qualified by *obj* by the associated actual qualified by *obj*; for example, we replace references to a_{obj} by $a1_{obj}$. Finally, we add

$(\sigma_1 - a1_{obj}) \subseteq \sigma$

to Post to capture the effect of the *modifies at most* clause. Here again we use a fresh variable σ_1 that has not been used in the verification condition. This gives us the predicates Pre$'$ and Post$'$:

Pre$'$ = IS_DEFINED($a1_1$, x)
Post$'$ = (($\sigma_1 - a1_{obj}) \subseteq \sigma$) & (a1 = ASSIGN($a1_1$, x, v))

We now turn to the problem of eliminating the procedure invocation. We use a rule of inference of the form

hypothesis1, hypothesis2

---

pred1 { int_array\$store(a1, x, z) } pred2

The first hypothesis is of the form

pred1[$a1_1$ **for** a1, σ_1 **for** σ] $\Rightarrow$ Pre$'$

This states that what we know to be true prior to the procedure call is sufficient to imply the requirements for calling the procedure. Note that each identifier for which a fresh variable was substituted in Pre$'$ is replaced by the same fresh variable in pred1; for example, we rename occurrences of *a1* as $a1_1$ and σ as σ_1. This makes these identifiers coincide with the appropriate identifier in Pre$'$.

The second hypothesis is of the form

$$(\text{pred1}[a1_1 \text{ for } a1, \sigma_1 \text{ for } \sigma] \;\&\; \text{Post}') \Rightarrow \text{pred2}$$

This states that what we knew to be true before the call and what we know after the call are sufficient to prove the post-condition. No renaming of *a1* is done in pred2. When this predicate refers to *a1*, it is referring to the value of the actual on returning from the procedure call. This corresponds to the meaning of references to *a1* in Post$'$, where *a1* has been substituted for occurrences of $a1_{\text{post}}$ in Post.

In our example, the rule for *store* tells us that we first need to show that

$$(\text{IS\_DEFINED}(a1_1, x) \;\&\; a1_0 = a1_1 \;\&\; \sigma_0 = \sigma_1) \Rightarrow$$
$$\text{IS\_DEFINED}(a1_0, x)$$

This simplifies quickly to TRUE. The second hypothesis we need to discharge is

$$(\text{IS\_DEFINED}(a1_1, x) \;\&\; a1_0 = a1_1 \;\&\; \sigma_0 = \sigma_1 \;\&$$
$$(\sigma_1 - a1_{\text{obj}}) \subseteq \sigma \;\&\; a1 = \text{ASSIGN}(a1_1, x, z)) \Rightarrow (\text{FETCH}(a1, x) = z \;\&$$
$$\forall j: \text{INT} [(\text{IS\_DEFINED}(a1, j) \;\&\; j \neq x) \Rightarrow$$
$$\text{FETCH}(a1, j) = \text{FETCH}(a1_0, j)] \;\&$$
$$(\sigma_1 - a1_{\text{obj}}) \subseteq \sigma)$$

This is easily reduced to

$$(\text{IS\_DEFINED}(a1_1, x)) \Rightarrow$$
$$(\text{FETCH}(\text{ASSIGN}(a1_1, x, z), x) = z \;\&$$
$$\forall j: \text{INT} [(\text{IS\_DEFINED}(\text{ASSIGN}(a1_1, x, z), j) \;\&\; j \neq x) \Rightarrow$$
$$\text{FETCH}(\text{ASSIGN}(a1_1, x, z), j) = \text{FETCH}(a1_1, j)])$$

This can be reduced to TRUE using two lemmas that follow from trait INT_ARRAY_SPEC:

1. $\text{IS\_DEFINED}(a1_1, x) \Rightarrow \text{FETCH}(\text{ASSIGN}(a1_1, x, z), x) = z$

2. $\forall j: \text{INT} [(\text{IS\_DEFINED}(a1, j) \;\&\; j \neq x) \Rightarrow$
   $\text{FETCH}(\text{ASSIGN}(a1_1, x, z), j) = \text{FETCH}(a1_1, j)]$

Each of these lemmas can be proved by induction; the basis step proves the lemma for $a1_1 = \text{NEW}(k)$, and the induction step proves it for

$a1_1 = $ APPENDH(a, v). These proofs are identical in structure to the inductive proofs presented in chapter 10.

The proof rule for procedures that modify their arguments allows us to reason about calls of procedures such as *remh* that modify their arguments and return a result, provided the result is not an actual of the call. To simplify the presentation, we shall not present a rule that allows the result to be used. This is not a serious restriction as far as the built-in types of CLU are concerned because we can rewrite the code to get around it. For example,

  x := array_ int$remh(a)

can be rewritten as

  x := array_ int$top(a)
  array_ int$remh(a)

11.4 Reasoning about Data Abstractions

Like procedures, data abstractions force us to introduce a considerable amount of extra mechanism into our proof system. In return, they allow us to simplify proofs by factoring our reasoning. The proof that the implementation of a type satisfies its specification is completely separate from proofs about parts of the program that use the type. Among other things, this means that we can use the specification of the rep type when verifying the correctness of a cluster.

In this section we present the specification and implementation of a simple data abstraction, intset. We then sketch a verification of the correctness of the implementation. As always, the specification of a type is divided into a CLU-dependent interface specification and a CLU-independent auxiliary specification. The interface specification is shown in figure 11.3, and the auxiliary specification, INTSET_SPEC, is shown in figure 11.4.

The implementation of intset is shown in figure 11.5. This is similar to the implementation shown in chapter 4, but we have restricted ourselves to the subset of CLU for which we have presented proof rules, and we have used only **cvt** (and not **up** and **down**). Intsets are represented by the *int_array* abstraction defined in figure 10.11. Note that both the rep invariant NO_DUPS and the abstraction function SET_OF are defined using function symbols taken from the auxiliary part of the specification of the rep type. The abstraction function also uses function symbols taken from the auxiliary specification INTSET_SPEC. The implementation of

intset **mutable type exports** [create, insert, member, delete, choose, is_empty]

>   **based on sort** INTSET **from** INTSET_SPEC **with** [int **for** INT]

create = **proc** ( ) **returns** (s: intset)
>   **modifies nothing**
>   **effects new**(s_{obj}) & s = EMPTY

insert = **proc** (s: intset, i: int)
>   **modifies at most** (s_{obj})
>   **effects** s_{post} = INSERT(s, i)

member = **proc** (s: intset, i: int) **returns** (b: bool)
>   **modifies nothing**
>   **effects** b = MEMBER(i, s)

choose = **proc** (s: intset) **returns**(i: int)
>   **requires** SIZE(s) > 0
>   **modifies nothing**
>   **effects** MEMBER(i, s)

delete = **proc** (s: intset, x: int)
>   **modifies at most** (s_{obj})
>   **effects** s_{post} = DELETE(s, i)

is_empty = **proc** (s: intset) **returns** (b: bool)
>   **modifies nothing**
>   **effects** b = (SIZE(s) = 0)

end intset

Figure 11.3 Interface specification of the data abstraction intset.

intset uses an internal procedure *getind*; the specification of this procedure is shown in figure 11.6.

Verifying that an implementation of a type is consistent with its specification is similar to verifying that each procedure in a cluster is consistent with its specification. The new wrinkles involve the rep invariant, the abstraction function, and **cvt**.

Specifications for a type's operations are expressed in terms of abstract values. For example, the value of the result formal *s* of intset$*create* is constrained by

s = EMPTY

Here we are using value EMPTY from sort INTSET. In verifying the implementation of *create*, however, we must have a way to treat *s* as an *int_array*; that is, we want its value to come from sort INT_ARRAY. We do this very simply: All formals of the abstract type that are **cvt**'d on call or return

trait INTSET_SPEC

   imports INT

   introduces
      EMPTY: → INTSET
      INSERT: INTSET, INT → INTSET
      DELETE: INTSET, INT → INTSET
      MEMBER: INT, INTSET → BOOL
      SIZE: INTSET → INT

   constrains INTSET **so that**
      INTSET **generated by** [EMPTY, INSERT]
      INTSET **partitioned by** [MEMBER]
      for all [s: INTSET, i, j: INT]

        MEMBER(i, EMPTY) = FALSE
        MEMBER(j, INSERT(s, i)) = **if** i = j **then** TRUE
          else MEMBER(j, s)

        SIZE(EMPTY) = 0
        SIZE(INSERT(s, i)) = **if** MEMBER(i, s) **then** SIZE(s)
          else 1 + SIZE(s)

        DELETE(EMPTY, i) = EMPTY
        DELETE(INSERT(s, i), j) = **if** i = j **then** DELETE(s, j)
          else INSERT(DELETE(s, j), i)

Figure 11.4 The INTSET_SPEC trait.

intset = **cluster is** create, insert, member, delete, choose, is_empty

   rep = int_array

   % rep invariant NO_DUPS(r)
   %   ∀x, y: INT [IS_DEFINED(r, x) & IS_DEFINED(r, y) ⇒
   %   ((FETCH(r, x) = FETCH(r, y)) ⇒ x = y)]

   % abstraction function SET_OF(r)
   %   **if** LENGTH(r) = 0 **then** EMPTY
   %   **else** INSERT(SET_OF(REMOVEH(r)), FETCH(r, HBOUND(r)))

   create = **proc** ( ) **returns** (**cvt**)
      return (rep$create(1))
      end create

   insert = **proc** (s: **cvt**, i: int)
      if getind(s, i) > rep$high(s) **then** rep$addh(s, i) **end**
      end insert

Figure 11.5 Implementation of *intset* (continues on next page).

```
getind = proc (s: rep, i: int) returns (int)
    j: int
    found: bool
    j := rep$low(s)
    found := false
    while j <= rep$high(s) & ~found do
        % invariant found = ∃ k: INT [ IS_DEFINED(s, k) & k < j &
        % FETCH(s, k) = i ]  &  s = s₀  &  i = i₀
        % decrements HIGH(s) − j + 1, range is the nonnegative integers
        if s[j] = i
            then found := true
            else j := j + 1
            end
        end
    return (j)
    end getind

member = proc (s: cvt, i: int) returns (bool)
    return (getind(s, i) <= rep$high(s))
    end member

delete = proc (s: cvt, i: int)
    j: int
    j := getind(s, i)
    if j <= rep$high(s)
        then s[j] := s[rep$high(s)]
            rep$remh(s)
        end
    end delete

choose = proc (s: cvt) returns (int)
    return (s[rep$low(s)])
    end choose

is_empty = proc (s: cvt) returns (bool)
    return (rep$size(s) = 0)
    end is_empty

end intset
```

Figure 11.5 (continued)

```
getind = proc (s: int_array, i: int) returns (j: int)
    modifies nothing
    effects if ∃ k: INT [IS_DEFINED(s, k) & FETCH(s, k) = i]
        then IS_DEFINED(s, j) & i = FETCH(s, j)
        else j = HBOUND(s) + 1
```

Figure 11.6 Specification of *getind*.

will be considered to be of the rep type in the verification conditions for the operation implementations. (Formals that are not **cvt**'d will continue to be of the abstract type.) This treatment matches the CLU semantics well: For operations that use **cvt**, every object of abstract type outside the cluster is of rep type inside.

This simple treatment does not provide a way to reason about operations that use **up** and **down** explicitly. To simplify the proof rules, we shall not treat explicit use of **up** and **down**. (To verify an operation that uses **up** and **down**, we suggest rewriting it in a form that uses only **cvt**.)

Since formals are considered to be of the rep type in the verification conditions, we need a way to talk about values of the abstract type. We do this by using the abstraction function. For example, in the post-condition of the verification condition for the implementation of *create*, we assert that

SET_OF(s) = EMPTY

Here s is a concrete value, and we produce the associated abstract value by using the abstraction function SET_OF.

As mentioned in chapter 4, all operations are required to preserve the rep invariant. For each procedure that accepts an argument of the cluster's type, we may assume that the rep invariant holds for that argument on entry to the procedure. To show that the rep invariant is preserved we must prove:

1. The invariant holds for any value of the rep type that is **cvt**'d on return.

2. On exit from the procedure, the invariant holds for each argument of the abstract type that was **cvt**'d on entry.

This proof is trivial for procedures that do not modify the rep of an abstract object or return objects of the abstract type.

The verification condition for the implementation of the procedure intset$*create* is

1. $\sigma_0 = \sigma$
         { s := **rep**$create(1) }
   SET_OF(s) = EMPTY & NO_DUPS(s) & $\sigma_0 \subseteq \sigma$

Here we assume in the post-condition that s is of type *int_array*. To guarantee that *create* establishes the rep invariant, we require that the rep invariant hold for s in the post-condition. We use the abstraction function to relate s to the desired result.

Since *int_array$create* is side-effect-free, we know that $\sigma_0 \subseteq \sigma$ is true in the post-condition, and we can use the assignment rule to reduce the verification condition to the predicate:

2. SET_OF(**rep**$create(1)) = EMPTY & NO_DUPS(**rep**$create(1))

Next we use the specification of *int_array*$ *create* to replace **rep**$*create*(1) by its value NEW(1), obtaining

3. SET_OF(NEW(1)) = EMPTY & NO_DUPS(NEW(1))

To prove NO_DUPS(NEW(1)), we replace NO_DUPS by its definition to get

4. ∀ x, y: INT
   [IS_DEFINED(NEW(1), x) & IS_DEFINED(NEW(1), y) ⇒
      ((FETCH(NEW(1), x) = FETCH(NEW(1), y)) ⇒ x = y)]

The auxiliary specification INT_ARRAY_SPEC implies that

5. ∀ x: INT [IS_DEFINED(NEW(1), x) = FALSE]

so that 4 simplifies to TRUE, and *s* therefore satisfies the rep invariant. We now use the definition of the abstraction function to transform the first conjunct of 3 to

6. [**if** LENGTH(NEW(1)) = 0
     then EMPTY
     else INSERT(SET_OF(REMOVEH(NEW(1))),
                FETCH(NEW(1), HBOUND(NEW(1))))]
   = EMPTY

Because the auxiliary specification INT_ARRAY_SPEC implies that

7. LENGTH(NEW(1)) = 0

step 6 simplifies to TRUE. This completes our verification that the implementation of intset$ *create* is consistent with its specification. The verification conditions for the remaining operations of intset are given in figure 11.7. The proofs of these verification conditions introduce nothing new and are not presented.

In these formulas, the translations of the modifies clauses of the specifications take into account the difference in viewpoint between the specifications and implementations of the procedures in the cluster. In the specification of a procedure, the clause *modifies nothing*, for example, states that as far as users of the procedure can observe, the procedure does not mutate any object. It does not state that the procedure cannot modify the representation of an abstract object. For example, the *modifies nothing* clause of *member*'s specification has been transformed into

$$(\sigma_0 - s_{obj}) \subseteq \sigma \ \& \ \mathrm{SET\_OF}(s) = \mathrm{SET\_OF}(s_0)$$

in the post-condition of *member*'s verification condition. This post-condition does not preclude modifications of the rep as long as it continues to represent the same intset. This permits implementations that involve benevolent side effects.

Note that in building the verification conditions for the operations, we assume the invariant in the pre-condition and must show that it is preserved across the body of the procedure. This is true even for operations such as *member* and *is_empty* that modify nothing. Since even these operations are allowed to modify the rep of an abstract object as long as the change is not visible outside the cluster, we must show that the rep invariant holds on exit. This proof will be trivial if the operation, in fact, does no modification.

The most important thing to observe about these verification conditions is how use of the rep invariant and abstraction function captures the distinction between a type and a mere collection of procedures. The abstraction function bridges the gap between the objects manipulated by the implementation and the specification that must be satisfied. The invariant captures the way the implementations of the different procedures cooperate to attain the desired effects. For example, NO_DUPS captures the interdependence of the implementations of *insert* and *delete*. It is the introduction of the rep invariant into the pre-condition of *delete*'s verification condition that allows us to prove the correctness of its implementation. Otherwise we could not show that removing one copy of the element was enough to delete it from the set.

11.5 A Few Remarks about Formal Reasoning

From the foregoing it should be obvious that formal reasoning about programs involves a large amount of tedious symbol manipulation. This is not a process that people usually enjoy or do very well. Clearly this is a perfect job for a machine. People do need to provide

1. a specification of the program to be verified,

2. a putative implementation of that program,

3. specifications of the procedures and types used in the implementation,

4. a loop invariant and decrementing function for each loop, and

Verification condition for insert:

$s_0 = s$ & NO_DUPS(s_0) & $\sigma_0 = \sigma$
    { **if** getind(s, i) > **rep**\$high(s) **then** **rep**\$addh(s, i) **end** }
NO_DUPS(s) & SET_OF(s) = INSERT(SET_OF (s_0), i) &
$(\sigma_0 - s_{obj}) \subseteq \sigma$

Verification condition for member:

$s = s_0$ & NO_DUPS(s_0) & $\sigma_0 = \sigma$
    { b := (getind(s, i) <= **rep**\$high(s)) }
NO_DUPS(s) & b = MEMBER(i, SET_OF(s)) & $(\sigma_0 - s_{obj}) \subseteq \sigma$ &
SET_OF(s) = SET_OF(s_0)

Verification condition for delete:

$s = s_0$ & NO_DUPS(s_0) & $\sigma_0 = \sigma$
    { j := getind(s, i)
    if j <= **rep**\$high(s)
      then s[j] := s[**rep**\$high(s)]
        rep\$remh(s)
    end }
NO_DUPS(s) & SET_OF(s) = DELETE(SET_OF(s_0), i) &
$(\sigma_0 - s_{obj}) \subseteq \sigma$

Verification condition for choose:

$s = s_0$ & NO_DUPS(s_0) & $\sigma_0 = \sigma$ & SIZE$(s_0) > 0$
    { i := s[**rep**\$low(s)] }
NO_DUPS(s) & MEMBER(i, SET_OF(s)) & $(\sigma_0 - s_{obj}) = \sigma$ &
SET_OF(s) = SET_OF(s_0)

Verification condition for is_empty:

$s = s_0$ & NO_DUPS(s) & $\sigma_0 = \sigma$
    { b := (**rep**\$size(s) = 0) }
NO_DUPS(s) & b = (SIZE(SET_OF(s)) = 0) & $(\sigma_0 - s_{obj}) \subseteq \sigma$ &
SET_OF(s) = SET_OF(s_0)

Figure 11.7 Verification conditions for remaining intset operations.

5. a rep invariant and abstraction function for each cluster.

It is important not to think of this information as being generated during an ex post facto program verification phase. Our whole approach to programming is based on developing specifications during program design. Invariants, decrementing functions, and abstraction functions should all be included as part of a program's documentation.

    Given the above information, the second stage of program verification—the construction of verification conditions—is algorithmic and therefore best left entirely to a program. Much of the symbol manipulation done so laboriously in this chapter involved the generation of verification conditions.

The final stage of the verification—the simplification of verification conditions—can be partially automated. Almost all of the simplification done in this chapter can be done by theorem-proving programs. In general, however, we shall not always be so fortunate. A program verifier can fail to prove that a program satisfies its specification under a number of conditions:

1. The deductive power of the simplifier is insufficient to allow it to derive a valid predicate as a theorem. There are a number of interesting domains for which it is impossible to write theorem provers that can answer all relevant questions.

2. The program being verified satisfies its specification, but the specifications of internal procedures are too weak to allow this to be shown.

3. The program being verified satisfies its specification, but the loop invariants, decrementing functions, rep invariants, or abstraction functions are inappropriate.

4. The correctness of the program depends upon some fact about the application area. Ideally, one should incorporate such information into the specification or other documentation.

5. The program being specified does not satisfy its specification. The program, the specification, or both may be "wrong."

The first of these causes of failure is a serious problem in program verification. It can generally be overcome by guiding the verifier, giving it appropriate lemmas to prove. Unfortunately, deducing the lemmas that need to be supplied is often difficult if one is unfamiliar with the theorem-proving technology used by the verifier. The second, third, and fourth causes often represent failures in the design or documentation of the program. Discovering such failures is an important benefit of program verification. The fifth cause represents a failure in either the specification or the implementation of the program. There is no a priori way to know whether it is the program or the specification that is "wrong." Uncovering such problems is the primary purpose of program verification.

11.6 Summary

Throughout this chapter we treated programs as text and reasoned about them as such. We associate with a programming language a set of proof rules that allow us to reduce a total correctness formula of the form

Pre-condition {program text} Post-condition

to a predicate whose meaning is independent of the programming language. In order to make this possible, we embed various kinds of auxiliary information in the program text.

To deal with loops, we introduce loop invariants and decrementing functions. Decrementing functions are used to prove termination. Loop invariants reduce reasoning about the arbitrarily large number of paths through a program containing a loop to reasoning about a relatively small number of paths.

In dealing with procedures, we rely heavily upon specifications. Although procedures have complicated proof rules (simplified versions of these rules were presented here), they play a vital role in modularizing reasoning about programs. The key point is that reasoning about the implementation of a procedure is independent of reasoning about its invocations.

Specifications also play a prominent role in reasoning about types. Like procedures, types help to modularize our reasoning. The proof that the implementation of a type satisfies its specification is completely separate from proofs about parts of the program that use the type. The abstraction function and rep invariant play essential roles in reasoning about the correctness of an implementation of a type. The abstraction function bridges the gap between the abstract objects referred to in the specification of a type and the concrete objects used in the implementation. The rep invariant captures a relationship between the implementations of the operations of the type. It allows us to reason separately about the implementation of each operation.

The usefulness of loop invariants, decrementing functions, rep invariants, and abstraction functions is not limited to formal verification. They are useful to the implementer since they are statements of assumptions on which the implementation depends. They are useful during program testing as a source of test cases, and during debugging as evidence of errors. They also provide useful program comments for those who must try to understand the program later.

Understanding a program means reasoning informally about its meaning. Informal reasoning is like formal reasoning; it follows the same steps and uses the same information. Thus it is no surprise that concepts developed for formal reasoning are useful for understanding programs. Perhaps the most important outcome of work on formal verification has been the identification and definition of these concepts.

Further Reading

Gries, David, 1981. *The Science of Programming.* New York: Springer-Verlag.

Hoare, C. A. R., 1969. An axiomatic basis for computer programming. *Communications of the ACM* 12(10): 576–583. Reprinted in *Programming Methodology, A Collection of Articles by Members of IFIP WG2.3*, edited by David Gries (New York: Springer-Verlag, 1978).

Jones, Cliff B., 1980. *Software Development, A Rigorous Approach.* Englewood Cliffs, N.J.: Prentice-Hall International.

London, Ralph L., Mary Shaw, and William A. Wulf, 1981. Abstraction and verification in Alphard: A symbol table example. In *Alphard: Form and Content*, edited by Mary Shaw (New York: Springer-Verlag), pp. 161–190.

Exercises

11.1 Give a loop invariant for the *sum_pointwise* procedure implemented in figure 11.2. Prove the correctness of this implementation.

11.2 Complete the verification of the implementation of intset by proving the validity of each of the formulas in figure 11.7.

11.3 Specify a procedure, *max_elem*, that accepts a nonempty array of integers as an argument and returns the largest element in the array. Specify a procedure, *max_array*, that accepts two nonempty arrays of integers as arguments and returns the array that contains the largest element. Implement *max_array* using *max_elem* and prove that your implementation satisfies the specification. (Note: You need not implement *max_elem*.)

11.4 Specify a procedure that reverses the order of the elements of an array of integers. Implement a procedure that satisfies that specification, and then prove the total correctness of your procedure.

11.5 Write a specification of an unbounded stack type. It should have operations with the following headers:

```
new = proc ( ) returns (s: stack)
push = proc (s: stack, i: int)
pop = proc (s: stack) returns (i: int)
```

Give an implementation of the type you specified, including the rep invariant and abstraction function. Prove that your implementation satisfies the specification.

11.6 Discuss the relationship between program verification and the material on test data selection presented in section 9.1.

12

A Preamble to Program Design

12.1 The Software Life Cycle

So far we have concentrated on the specification, implementation, and validation of program modules. The next three chapters deal with issues related to programs as a whole and with the process of program development.

Program development is usually broken up into a number of phases: requirements analysis; design; implementation and testing; acceptance testing; production; and modification and maintenance. Typically, the process begins with someone we shall call the *customer* who wants a program to provide a particular service. Sometimes the service is well understood and described a priori in a complete and precise manner, but this is quite rare. More often customers do not fully understand what they want the program to do. Even if the desired service is well understood, it is probably not described precisely enough to serve as a basis for constructing a program. The purpose of the requirements analysis phase is to analyze the needs of the customer and produce a *requirements specification* of a program that will meet those needs.

The specification that results from requirements analysis is the input to the design phase. In this phase a modular decomposition of a program satisfying the specification is developed. In the next phase the individual modules are implemented and then tested to ensure that they perform as intended. As discussed in chapter 9, we use two kinds of tests: unit tests, in which individual modules are tested in isolation, and integration tests, in which modules are tested in combination.

At best, integration testing shows that the modules together satisfy the implementer's interpretation of the specification. The implementer may have misinterpreted the specification or neglected to test some portion of its behavior, though, and the customer therefore needs some other basis for deciding whether or not the program does what it is supposed to do. This typically takes the form of *acceptance tests*. Acceptance tests provide an evaluation of the program behavior that is independent of the design, and they are generally performed by an organization different from the one that worked on the design and implementation. They should include both trial runs under conditions approximating those the customer will actually encounter and tests derived directly from the requirements specification.

When the program has passed the acceptance tests, it enters the production phase and becomes a product that the customer can use. The useful life of the program occurs during this phase, but the program is unlikely to remain unchanged even here. First, it almost certainly harbors undetected errors that must be corrected during production. Correcting such errors is called *program maintenance*. Second, the customer's requirements are likely to change. Responding to such changes requires *program modification*.

Errors are a problem throughout the entire development process. Those made in the earlier stages, however, are likely to be harder to fix and to result in more wasted effort. An error made in the requirements analysis can lead to a totally useless program, since it is not the program the customer wants. If this is not discovered until the acceptance tests, an enormous amount of design and implementation may have to be redone. An error made during design can result in the implementation of unusable modules and failure to create needed modules. By contrast, an error made in implementing a single module affects only that module and can be corrected simply by reimplementing it.

The earlier an error is detected, the less serious its consequences. If an error in the requirements analysis is caught during that phase, it may be necessary to rethink parts of the requirements specification, but not to discard design work. Similarly, if an error in design is caught before implementation begins, it may be necessary to rethink a number of design decisions, but not to discard work done in implementing and testing unneeded modules. It is clearly important to make use of error-detection methods and techniques during requirements analysis and design, and not just during implementation.

It is worth noting once again that there are no guarantees of correctness. Program verification can ensure that a program meets its specification, but it is not generally practical today, and even if it were, it would still find errors much too late in the program development process. As was just noted, the errors made in the early phases matter most, and it is important to find these errors quickly to minimize wasted effort. One can never be absolutely certain that a requirements specification describes the exact product the customer wants. Nor are there existing techniques that can tell us whether the modular decomposition produced by the design phase will satisfy the specification when the modules are implemented. There are methods that can be used to catch errors in these early phases; they depend on explicit documentation of decisions and careful review of these decisions. While better than nothing, they are far from adequate. Identification and invention of better methods is an important area for research in programming methodology.

Although we have emphasized the importance of doing the first three phases in the proper order, it is important to recognize that inevitably a certain amount of iteration through these phases will be necessary. During design we often uncover problems with the requirements specification. When this happens, it is necessary to redo part of the requirements analysis. Similarly, if a problem with the design is found during implementation, the relevant parts of the system must be redesigned.

This chapter focuses on the requirements analysis. We shall give an overview of the issues that must be addressed during this phase and provide a short example. The topics covered are complicated ones. Our discussion of them is abbreviated, oversimplified, and intended to serve only as an introduction.

12.2 Requirements Analysis Overview

The original product description produced by a customer is unlikely to be either complete or precise. The purpose of requirements analysis is to analyze the customer's needs so that we can identify and carefully describe the customer's requirements. This analysis must involve consultation, since the customer is the ultimate judge of what is wanted.

Often, a good way to get started on requirements analysis is to examine how the customer does things at present. If a system has been in place for a while, it will have developed methods for normal processing and for coping with errors and a variety of contingencies. The present system can be a source of ideas not only about how to do things, but about what needs to be done. Studying the customer's organization can also help ensure that the program developed will fit smoothly into that organization. It must be compatible with other systems already in use and also with the current practices of the organization.

The requirements analysis must consider both normal situations and errors. Studying normal case behavior means defining the effect of all nonerroneous user interactions with a program under the assumption that the program is in a normal state. Normal case behavior is often the easiest part of the problem, so that focusing on it is a good way to get started. A good approach is to work out scenarios of typical interactions with the system. We shall give an example of this in the next section.

Cases with no errors represent only a small part of program behavior, and it is essential to consider and describe how a program behaves in the presence of errors. This part of the analysis should never be neglected or underemphasized. The analyst must try to uncover all possible errors that might occur and develop the appropriate responses for each case. Errors come from two sources: users interacting with the program, and hardware and software malfunctions. Scenarios are a good way to study user errors, since they allow us to pinpoint the ways in which these errors might be made. Sometimes proper interface design can prevent user errors—for example, by presenting an interactive user with a menu containing only legal commands. Another possibility is for the system to recognize erroneous input and reject it. Of course, it is not possible to avoid all errors. For example, if a bank clerk makes an error entering an account holder's deposit, it might not be noticed until the account holder examines a monthly statement. The analyst must identify all situations like this and decide what should be done to handle them. Often these decisions require consultation with the customer, since business policy is involved (for example, deciding how to compensate an account holder whose check has bounced because of a bank error).

As far as software errors are concerned, the analyst must make a decision about the amount of effort to be expended during program design and implementation to avoid, remove, or limit the scope of such errors. For example, if it is important to limit the scope of software errors, the output of critical modules can be checked for reasonableness, and the system shut down if the checks fail. The analyst must also decide what to do about hardware failures. It may be important for a system to be *highly available*; that is, it should be highly likely that the system is up and running all the time (or at certain times). Satisfying such a requirement may involve the use of redundant hardware and software. A related requirement is that the system be *highly reliable*. Here we are concerned with avoiding loss of information because of failures. Determining how ambitious the system must be in trying to recover from hardware and software malfunctions is an important aspect of requirements analysis.

In addition to functional requirements, a program must satisfy efficiency requirements. Time and space efficiency should be considered together since it is often necessary to trade one off against the other. The first thing to do is to find out whether or not there is a hard limit in either dimension. For example, the program might have to run on a microcomputer with 128K of memory and one floppy disk. Alternatively, it might have to satisfy some real-time constraint (for example, computing the current altitude of a plane every tenth of a second). In considering time efficiency, it is important to distinguish between *throughput* and *response time*. Throughput refers to the amount of data processed by a system over an interval of time. Response time refers to the amount of time between interactions with the system. Often, optimizing one of these characteristics has a deleterious effect on the other. A communications network, for example, might maximize throughput at the cost of response time by batching messages.

The customer's space and time requirements must be checked for compatibility with the functional requirements, the hardware the customer intends to use, and the price the customer is willing to pay for the system. For example, the customer may want some activity to satisfy efficiency requirements that are either not possible given the hardware or can be satisfied only with very sophisticated software. If there is an incompatibility, negotiations may be necessary to produce new requirements.

These issues affect the content of the requirements specification directly. There are a number of other issues that should be considered during requirements analysis, not because they affect the specification, but because they provide useful input for the designers. Two such issues are *modifiability* and *reusability*. Usually there will be areas with fixed requirements and others in which changes are likely. Information about likely changes is useful because a design can be shaped in such a way as to make certain changes easy. This information is often best obtained by careful study of earlier systems designed to do a similar job. Changes are likely in places where those systems differ from the current one. The specified system might also be intended to be the first of many systems that will be similar but will differ in details. The job of building the additional systems can be simplified if all or part of the software produced for the first system is designed to be reusable. If the designers know what the similarities are, they can shape the design to accommodate them. An outstanding example of this is the isolation of target-machine dependencies in a compiler.

Pinning down constraints on the delivery schedule is another important part of the requirements analysis. Knowing that the customer is in a hurry, for example, may encourage the designer of the software to trade extendability for simplicity. Knowing that a program providing a proper subset of the functionality of the entire system should be available early would have an impact on the implementation schedule as well as on the design.

Ideally, the requirements specification should be precise and complete. A precise specification is one that defines behavior unambiguously, while a complete specification is one that describes all relevant aspects of that behavior. Since requirements specifications tend to be written in an informal language, ambiguities and incompleteness are unavoidable and hard to notice. The analyst should look hard for ambiguities and incompleteness in all parts of the specification.

The requirements specification can be the input to two activities in addition to the design. It can be used to produce acceptance tests and as a basis for a user's manual. The user's manual is something that must be produced anyway, but its production can provide an independent check on the suitability of the specification. If the system is hard to use, this may be evident when the manual is written. Also, by reading the manual, the customer may notice deficiencies in the specification that were overlooked earlier.

In addition to the specification, the analyst should also produce a document explaining the decisions made during analysis and, if reasonable, the alternatives that were rejected (and why they were rejected). This information is useful when requirements must be rethought because of errors or changing customer needs.

12.3 A Sample Problem

To illustrate this discussion, we now investigate an example, a program for generating KWIC indexes for sets of titles, with the goal of developing a partial requirements specification for this program. KWIC indexes are used to scan a database of titles to find titles of interest. In general, the database is restricted to titles in a particular domain, such as computer science research. The purpose of the index is to allow someone scanning the titles to identify those of interest quickly. Some words in a title are more useful than others for this purpose. Words like "the" and "and" are not useful at all, whereas words like "compiler" and "CLU" are clearly useful. The words that are useful are called *keywords*. In the output the titles are presented to the user sorted by keywords. The user can then scan the sorted list, looking for keywords related to the topic of interest. Once a set of titles containing a relevant keyword has been found, the user can choose specific titles to consider in more detail.

There are two ways to present the index: Keywords can be presented "in context" or "out of context." In a KWIC (Key Word In Context) index the keywords stay in their position in the title, but are highlighted. For example, a KWIC index for the titles

A Simple Database Language for Personal Computers

A Practical Tool Kit for Making Portable Compilers

Triply Redundant Databases

Compiling Made Easy

would begin with

A Practical Tool Kit for Making Portable <u>Compilers</u>

<u>Compiling</u> Made Easy

A Simple Database Language for Personal <u>Computers</u>

A Simple <u>Database</u> Language for Personal Computers

Triply Redundant <u>Databases</u>

We want a program to construct a KWIC index for a database of titles supplied by the user. It should go through the database to find all keywords, using information supplied by the user, and should print the index. This description is typical of the kind of problem description that exists prior to the requirements analysis phase. Enough information has been given to provide a general idea of what is wanted. However, as soon as we try to define what is wanted in any detail, we can see that the problem description is vague and imprecise.

We begin by considering the normal case behavior. A good way to start is to enumerate all of the information that must be communicated between the user and the program. The program must find out from the user:

1. where to find the database of titles,

2. which words are keywords, and

3. where to put the KWIC index it has produced.

The user must find out from the program

1. what information the program wants him to provide and

2. whether or not it has succeeded in building a KWIC index.

As mentioned, a good way to determine how this information should be communicated is to sketch a sample session. Before doing this, however, we should decide what style of interaction to use. In general, there are two methods of interacting with a program: The user either waits for the program to demand information or offers it spontaneously—for example, as arguments when the program is called. Let us make the provisional decision that our KWIC program will request responses from the user.

We now start to sketch a sample session. It seems reasonable to assume that the user must start by invoking the program. How this is done almost certainly depends upon the computing environment in which KWIC is to run. Let us assume that KWIC is invoked by typing

```
>run kwic
```

(Here > is the prompt from the operating system. We shall follow the convention of printing input from the user in roman and messages sent by the program in italic.)

Note that we have already made a decision to invoke KWIC without passing it any arguments. There are two reasons behind this choice. First, it preserves the narrowest possible interface between the operating system and KWIC; for example, we need not concern ourselves with the syntax used for passing arguments to programs or even with whether or not it is possible to pass arguments to programs. This will make it easier to move our program from one system to another and will make it more robust to changes in the environment. The second reason is that we have decided to let user interaction be demand driven.

Next the program should ask for some of the information it needs. It could ask about how to find the database of titles or keywords. Since there is no obvious a priori reason to start with either one, we should ask whether getting one kind of information first might make it easier to get the other. Knowing something about keywords is unlikely to help us find out where the database is, but having access to the database may help with keywords. Let us assume, then, that the program asks the user where the database can be found, and the user supplies this information. We shall also assume that the database is in a file that can be accessed by its name:

>run kwic
Enter name of database to be indexed: cs.titles

Here the user enters the file name followed by a carriage return to indicate that the name is complete.

The program now needs information that will allow it to decide which words are keywords. One possibility is to ask the user to list the keywords and assume that all words not listed are not keywords. Alternatively, the program could ask the user to list the nonkeywords. Of these alternatives, it would probably be better to supply a list of nonkeywords. In either case, errors of omission seem inevitable. When errors cannot be avoided, we should choose the alternative in which the consequences of an error are less severe. Forgetting to list one nonkeyword only results in some titles being listed in a uninteresting way. On the other hand, if one interesting keyword is forgotten, then the user of the listing may miss a title of interest. Assuming that KWIC asks for nonkeywords, our sample session might continue:

Please enter the list of nonkeywords.
The words should be separated by a comma or carriage return, and the
    list terminated by esc.
Nonkeywords: a, as, at, ...

As we construct this sample session, it becomes clear that compiling this list is going to be a long and not terribly easy process, and errors of omission are going to be quite a problem. We should clearly consider other possibilities. For example, if the user were to supply both a list of keywords and a list of nonkeywords, the program could look up every word occurring in a title in these lists. If it found a word that was not on either list, it could then ask the user which list to put it on. In this way we could avoid errors of omission.

The question now arises as to why the user should enter lists at all. Perhaps it would be better for the program to go through the titles and every time it encounters a new word to ask the user which list to put it on. This approach has two important assets: It eliminates a class of user errors and it reduces the amount of typing a user must do. If we adopt this approach, our sample session might continue

For the following sequence of words, type k if the word is a keyword and
    n if it is not.
A: n
Simple: n
Database: k
Language: k
For: n
Personal: k
Computers: k
Practical: k
    .
    .
    .
    .

In looking at this sample session we discover that if our database has a lot of different words, the enumeration will be tedious. It would be particularly annoying if the user were producing more than one index and had to repeat the process for each set of titles. A little thought, however, should convince us that this last problem is easily surmountable. KWIC could store the lists of keywords and nonkeywords in such a way that they can be reused. It could ask at the start of a run whether it should use

a predefined dictionary, and if so in what file it can be found. When it
finds a word that is not in the dictionary, it could ask whether or not it
is a keyword and then augment the dictionary with that information. At
the end of a run, KWIC could ask the user whether or not to save the
dictionary, and if so in what file. A sample session might then look like

```
>run kwic
```
Enter name of database to be indexed: `cs.titles`
Enter name of initial dictionary: `cs.dict`
*For the following sequence of words, enter k if the word is a keyword
    and n if it is not.*

.

.

.

If you want to save the amended dictionary, enter y else enter n: `y`
Enter name of file in which to store new dictionary: `mycs.dict`

We now have several ways to obtain information about keywords, so this
is probably a good point at which to consult with the customer. We can
prepare a summary of the advantages and disadvantages of each approach
and present that and the sample sessions to the customer. Let us assume
that the customer prefers the last alternative discussed.

Another point to discuss is how sophisticated the program should be
at recognizing variants of words. If the user has already indicated that
"assembler" is a keyword, should the program ask about "assemblers," or
should it assume that "assemblers" is also a keyword? If so, how about
"assembled," "assembly," "assembly-line," and so forth? Similarly, if the
user tells the program that "is" is not a keyword, should the program
assume the same thing about "are," "was," "were," and so forth? This is a
difficult and open-ended problem. There is a clear trade-off between ease of
use and the difficulty and expense of building the program. The customer
should be apprised of this trade-off and asked to make a decision. We shall
assume that the unsophisticated approach to recognizing words is chosen.

This gets us to the final piece of information that must be supplied by
the user: where to put the output of KWIC. Let us assume that the output
is to be written to a file and that the user will be asked for the name of
that file:

Enter name of output file: `cs.titles.out`

The user can print this file or read it online using whatever utilities are
provided by the operating system.

We have now completed a pass outlining how information is to be communicated from the user to KWIC. To complete our normal case requirements analysis, we still need to clarify the contents of the database of titles (assuming that it has no errors in it) and the form of the output (assuming that no errors occurred during processing). We shall not define these formats here. The key question to ask about the database is how much control we have over its contents. Is it an existing database with contents and organization beyond our control or is it to be built up in a way we may specify? As far as the output is concerned, its format is an important issue, but one that is difficult to deal with before the event. It is hard to anticipate what formats users will like or dislike. Often the best approach is to do something quite simple initially, but to design the program in such a way that it is easy to respond to suggestions for improvements.

We now consider errors. One possible error is for the user to supply an invalid file name, perhaps one that KWIC is unable to open in the appropriate mode. In response we can simply ask the user to supply another name.

Another error might be imperfect data in the database, such as titles with misspelled words. Suppose KWIC is not allowed to alter this database, so that it is not possible for the user to correct the error immediately. What should be done about further processing of the input and about production of the output in this case? The user should have the option of continuing to process the input, so that other errors can be found. To decide about the output, we need to know how much computation would be lost if we discarded it, and also how useful it would be given the errors. KWIC indexes are usually run over very large collections of titles, and the reading in of those titles is a substantial part of what the program does. Furthermore, an error in one word is unlikely to affect the processing of surrounding words or titles. The information in the index about the title containing the error may be misleading, but information about other titles should be correct. Therefore let us make the following decision. If the user decides to continue, we shall skip the word containing the error insofar as producing keyword listings in the index is concerned, but continue normal processing of subsequent words. When input is finished, we shall give the user the option of canceling the output through a further interaction:

If you want to produce the index, enter y else enter n:

Finally, we arrange for titles containing skipped words to appear at the end of the output (if any). The advantage of appending these titles to the regular output is that someone using the index can look at them too.

How does the user indicate that processing should stop? Probably a special user response should be permitted at every interaction with KWIC. Termination should happen gracefully; for example, the user should be given the option of producing the new dictionary and the output.

What about error checking of the dictionary of keywords and nonkeywords? Since the dictionary is produced by KWIC, in theory there should be no errors in it, but, of course, this may not be the case. The user might supply a file that was not produced by KWIC, a file that was originally produced by KWIC but has since been altered accidentally, or a file produced by a version of KWIC that had a bug in it. How easily such problems can be detected will depend upon the format of the dictionary. This format should not be part of the requirements specification, but should be left to the designers. However, we shall require that the dictionary be an ASCII file so that users can inspect it easily. We should probably also require that an auxiliary program be available that allows users to inspect a dictionary and repair errors. (An existing text editor may be sufficient for this purpose.) The effects of an error in the dictionary are likely to be serious. They will lead to either spurious or missing entries in the index. Once an error in the dictionary has been discovered, it should be reported to the user, who should be given the option of providing another dictionary.

Finally, suppose KWIC gets into an unanticipated state because of a hardware or software malfunction. In this case, it should try to write out the amended dictionary to a temporary file (this will not always be possible) and then stop processing. If we have reason to believe that hardware or software malfunctions are going to be moderately frequent, KWIC should probably be required to save the dictionary periodically in a temporary file.

To complete the requirements specification for KWIC, we must define the details that have been omitted and specify efficiency constraints. Potential modifications should include easy reformatting of the output.

12.4 Summary

We began this chapter with a description of the software life cycle. While conceding that it is indeed a cycle, we emphasized the importance of doing a careful requirements analysis before starting design and a careful design before starting implementation. The bulk of the chapter was devoted to a discussion of the requirements analysis. The next chapter takes up design.

We usually start the requirements analysis with an incomplete understanding of what the customer really needs. The goal of analysis is to deepen our understanding and describe the needed system in a moderately complete and precise requirements specification. Customers should be consulted during analysis because they are the ultimate judges of what is needed.

A number of issues must be considered during requirements analysis:

1. The program's behavior must be defined for both correct and incorrect inputs.

2. Issues related to hardware and software errors, such as availability and reliability constraints, must be explored.

3. Constraints on time and space efficiency must be pinned down. Efficiency is not an add-on feature. It must be designed in from the start.

4. Scheduling constraints must be addressed. The customer's desired delivery schedule for the software or part of the software may well have an impact on the design and implementation.

5. It is useful to try to pin down those parts of the requirements that are most likely to change.

6. It is useful to know whether the system being specified is the first of a number of similar systems, so that it can be designed in a way that will allow components to be reused.

The requirements analysis process is a difficult one. We suggested an approach based on sample sessions or scenarios and illustrated that approach with an example. Our discussion was far from thorough. Involving customers in a productive way is perhaps the most difficult and critical aspect of requirements analysis. It involves many issues that are beyond the scope of this book.

Further Reading

Boehm, Barry W., 1981. *Software Engineering Economics.* Englewood Cliffs, N.J.: Prentice-Hall.

De Marco, T., 1979. *Structured Analysis and System Specification.* Englewood Cliffs, N.J.: Prentice-Hall.

Heninger, Kathryn L., 1980. Specifying software requirements for complex systems: New techniques and their application. *IEEE Transactions on Software Engineering* SE-6(1): 2–13.

Exercises

12.1   Prepare a complete and precise requirements specification for KWIC. Be sure to address issues related to the format of its input and output and the handling of errors. State reasonable efficiency requirements. You may assume that there is enough space to fit all of KWIC's data in main memory, assuming that space is not wasted.

12.2   Program *xref* produces an index for a document: For each word containing more than one letter, it lists the word followed by the lines in which it appeared, for example,

compiler   3, 17, 25, . . .

Prepare a complete and precise requirements specification for *xref*.

13

Design

In preceding chapters we have discussed the specification and implementation of individual abstractions. We have emphasized abstractions because they are the building blocks out of which programs are constructed. We now discuss how to invent abstractions and how to put them together to build good programs.

13.1 An Overview of the Design Process

The purpose of design is to define a program structure consisting of a number of modules that can be implemented independently and that, when implemented, will together satisfy the requirements specification. This structure must provide the required behavior and also meet the performance constraints, including space and time efficiency, availability, and reliability. Furthermore, the structure should be a good one: it should be reasonably simple and avoid duplication of effort (for example, it should not contain two modules where one suffices). Finally, the structure should support both initial program development and maintenance and modification. The ease of modification will depend on the sort of change desired. A particular decomposition cannot accommodate all changes with equal ease. This is why likely modifications should be identified prior to design—so that a decomposition can be developed that facilitates them.

We start with the requirements specification. Sometimes it describes a single abstraction, sometimes several that together make up the system. We pick one of these abstractions to start work on. This becomes the initial *target abstraction*. In designing a program, we carry out three steps for each target:

1. Identify *helping abstractions*, or *helpers* for short, that will be useful in implementing the target and that facilitate decomposition of the problem.

2. Specify the behavior of each helper.

3. Outline a program to implement the target in terms of the helpers.

The first step consists of inventing a number of helping abstractions that are useful in the problem domain of the target. The helpers can be thought of as constituting an abstract machine that provides objects and operations tailored to implementing the target. The idea is that if the machine were available, implementing the target to run on it would be straightforward. Next, we define the helpers precisely by providing a specification for each one. Usually when an abstraction is first identified, its meaning is a bit hazy. The second step involves pinning down the details and then documenting the decisions in a specification that is as complete and unambiguous as is feasible. Once the behavior of each helper is defined precisely, we can use them to write programs. In principle, the target can now be implemented, but we generally do not do this at design time. Instead we merely outline enough of an implementation to convince ourselves that an acceptably efficient and modular implementation can be constructed. If the abstract machine existed, we would now be finished, but in reality the helpers need to be considered in more detail. The next step is to select one helper and design its implementation. This process continues until all the helpers have been studied.

For large programs, we usually do not know in advance what the structure of the program ought to be. Instead, the discovery of this structure is a major goal of the design. As the design progresses, sometimes a choice must be made between a number of structures, none of which is well understood. Later, the decision may be found to be wrong. This is especially likely for choices made early in design, when the structure of the program is least understood and least constrained by other decisions, and when the effect of an error on the global structure is largest.

When errors are discovered, we must correct them by changing the design. Usually we must discard all later work that depends on the error. That is why we delay the start of implementation until after we have a complete design. Of course, no matter how careful we are about our design, during implementation we are likely to uncover problems with it. When this happens, we must rethink the part of the design related to the problem.

Our discussion so far has outlined how design occurs, but has neglected a number of questions, such as:

1. How is decomposition accomplished? That is, how do we identify subsidiary abstractions that will help to decompose the problem?

2. How do we select the next target?

3. How do we know whether we are making progress? For example, are the helpers easier to implement than the target that caused them to be introduced?

4. How are performance and modification requirements factored into a design?

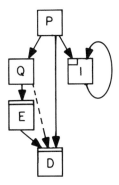

Figure 13.1 A module dependency diagram.

5. How much decomposition should be done?

These and other, similar questions will be addressed as we carry out an example. First, however, we discuss how to document a design.

13.2 The Design Notebook

The decisions made during design must be recorded. This documentation should be done in a systematic manner and kept in a *design notebook*. The notebook contains an introductory section describing the overall design of the program and a section for each abstraction.

13.2.1 The Introductory Section

The introductory section lists all the abstractions identified so far and indicates where information about them can be found in the notebook. It also indicates which helpers are to be used in the implementations of which targets. This documentation takes the form of a *module dependency diagram*, which identifies all the abstractions encountered during design (those present in the requirements specification or introduced as helpers) and shows their relationships.

A module dependency diagram consists of nodes and arcs. The nodes represent abstractions, and each contains the name of its abstraction. There are three kinds of nodes, one for each of the three kinds of abstractions we use. Nodes like the ones labeled P and Q in figure 13.1 represent procedures, ones like E and D represent data types, and ones like I represent iterators.

The arcs show which abstractions are to be used in implementing which other abstractions. There is an arc from a target to each helper. We say that the target *uses* or *depends on* the helpers. A target uses a procedure or iterator by calling it. A data type is used if one or more of its operations is called. The diagram does not show the specific operations used; the use of any operation is sufficient to set up the dependency. To record more detailed information would result in an overly cluttered diagram that would be hard to use. If this extra information is needed, it is recorded in notes accompanying the diagram.

Sometimes, an abstraction uses an object of a type without calling any of the type's operations. For example, it might receive an object as an argument and simply pass the object on to another abstraction. In such a case we say that the target *weakly uses* the type. A special dashed arc is used to indicate this kind of dependency.

The dependencies determine the level of the nodes. An abstraction A is at a higher level than B if A depends on B and B does not depend on A (that is, there is no direct or indirect recursion). If A and B depend on each other, they must appear at the same level.

In figure 13.1 procedure P is to be implemented by the three helpers Q, D, and I. Q is another procedure; its implementation uses E and weakly uses D. E and D are data types; the implementation of E uses D, but D uses no helpers. I is an iterator; it is implemented in terms of itself; that is, it is recursive. Note that the entire structure is a directed graph. It is not a tree, since one abstraction can be used in implementing several others. Note also that the graph may have cycles in it. Cycles occur when there is recursion. If an abstraction is to be used directly in implementing itself, there is an arc pointing from that node to itself (as is the case with I in figure 13.1). Cycles involving more than one arc represent mutual recursion. In general, recursion is reasonable for a program if the problem being solved has a naturally recursive structure. We saw examples of natural recursion in the implementations of list and sorted_list in chapter 4. However, if the cycles get very long and involve many arcs, it is wise to be suspicious of the program structure.

The module dependency diagram is useful when errors are detected. A design error shows up as a flaw in an abstraction; for example, an efficient implementation becomes impossible or needed arguments are missing. The potential impact of the error can be determined from the diagram. Implementations of all abstractions that use the erroneous one must be reconsidered. (That reconsideration may find some of those abstractions erroneous, and so on.) An abstraction that weakly uses an erroneous type is affected if the type disappears entirely, but not if the type changes or is replaced by another type. For example, in figure 13.1, if we found a problem with D, we would have to rethink the implementations of E and P, but Q and I would not be affected. Of course, if rethinking E prompted us to change its specification, we would be forced to reexamine the implementation of Q.

The introductory section of the notebook should also contain information about the overall process of implementing the design. It should suggest how many people should work on the implementation, the order in which various abstractions should be implemented, and how long various phases of the implementation should take. It should also propose a set of tests to be used on the completed implementation. Chapter 14 discusses strategies to be used in the implementation process.

13.2.2 The Abstraction Sections

We partition the notebook entry for an individual abstraction into five pieces:

1. a specification of its functional behavior;

2. a description of efficiency constraints it must observe;

3. information about how it is to be implemented;

4. information about how it is to be tested; and

5. miscellaneous information that does not fit into any of the previous four categories.

The specification is, of course, the most important item in this list. However, we have already said a great deal about such specifications and will say no more about them here.

Efficiency constraints are generally propagated top-down through a program. The requirements specification may constrain the time the program may take to perform certain tasks and the amount of primary and secondary memory it may use. To convince ourselves that the implementation of a target will meet its efficiency constraints, we have to make assumptions about the efficiency of its helpers. These assumptions show up as efficiency constraints in the parts of the design notebook detailing each helper.

Efficiency constraints can be expressed in a variety of ways. Frequently we express them as functions of the size of the input. We might, for example, write as part of the description of a *sort* procedure with formal argument x of type list[int]

efficiency
>   worst case time $=$ Order(length(x) $*$ log(length(x)))
>   main memory added $=$ Order(length(x))
>   max temporary main memory $=$ Order(length(x))

This says

1. The time required to sort list x should never exceed some constant multiplied by the product of the length of x and the log (base two) of the length of x.

2. When *sort* returns, the total amount of main memory currently allocated for the program in which *sort* is embedded should not have been increased by more than a constant multiplied by the length of x.

3. During the execution of *sort*, the amount of main memory allocated for *sort* should not exceed some constant multiplied by the length of x.

For many applications, it suffices to supply relative efficiency constraints such as those just given. Sometimes, however, it is useful to bound the multiplicative constants or even to impose an absolute bound. An absolute upper bound on time is needed in real-time applications. Analogously, we put absolute bounds on the space used in programs that are to be run on small machines or machines that do not support virtual memory. If *sort* were to be embedded in a real-time system that was intended to run on a machine with no secondary memory, we might write

efficiency
>   worst case time $=$ 1 second
>   main memory added $=$ Order(length(x))
>   max temporary main memory $=$ Order(length(x))
>   secondary store added $=$ 0
>   max temporary secondary store $=$ 0

(For *sort*, an absolute constraint on time would be satisfiable only if the requires clause of its specification put some upper bound on the length of the input list x.)

The section of the notebook entry containing information about how the abstraction is to implemented should include a list of its helpers. For data types, the entry should describe the representation, representation invariant, and abstraction function associated with the intended implementation. This section might also include a description of how the implementation works. This description can be omitted if the implementation is straightforward, but it is necessary if the implementation is clever. Frequently it is helpful to give a sketch of the intended algorithm, such as

Implement sort(x) by a recursive algorithm of the form:
> **if** length(x) = 1
>> **then** x
>> **else** merge(merge_ sort(first_ half(x)), merge_ sort(second_ half(x)))

Alternatively we might give an appropriate reference, such as

> Implement sort using one of the merge sorts in section 5.2.4 of *The Art of Computer Programming, Volume 3*, by Donald Knuth, Addison-Wesley, 1973

An important criterion for evaluating a design is how easy or hard it will be to test its implementation. The notebook entry for each abstraction should contain information on how the designer believes its implementations should be tested. This information might include suggested test data, a description of a driver that could be used to supply the test data and check the results, and suggestions about stubs that could be used to replace the helpers.

As its name implies, the section of the abstraction entry labeled "miscellaneous" might contain almost anything. Typical items are

1. a justification for the decisions documented elsewhere in the entry,

2. a discussion of alternatives that were considered and rejected,

3. potential extensions or other modifications of the abstraction, and

4. information about the context in which the designers expect the abstraction to be used.

In closing this section we should note that if the design is very large, it may be useful to structure the notebook by introducing subsidiary notebooks. In a module dependency diagram, any subgraph can be viewed as an independent subsystem. However, the most convenient choice is a subgraph in which only one node is used from outside.

13.3 Problem Description

In the remainder of this chapter, we carry out a complete design of a simplified text formatter. This program is small and is intended primarily to serve as an illustration of the design process. This section gives a specification of the formatter; subsequent sections describe its design. An implementation is given in appendix B. The input to the text formatter consists of a sequence of unformatted text lines mixed with command lines.

Each line (except perhaps the last) is terminated by a newline character, and command lines begin with a period to distinguish them from text lines. An example is

```
Justification only occurs in "fill" mode.
In "nofill" mode, each input text line is output without
modification.
The .br command causes a line-break.
.br
Just like this.
```

The program produces justified, indented, and paginated text:

```
Justification  only occurs in "fill" mode. In "nofill" mode,
each input text line is output without modification. The .br
command causes a line-break.
Just like this.
```

The output text is indented 10 spaces from the left margin and is divided into pages of 50 text lines each. A header of 5 lines is output at the beginning of each page. The header consists of 2 blank lines, a line of the form

```
Page n
```

where n is the page number, and 2 more blank lines.

An input text line consists of a sequence of words and word-break characters. The word-break characters are space, tab, and newline; all other characters are constituents of words. Tab stops are considered to be every eight spaces. The tabs and spaces are accumulated in the current output line along with the words. For example, if two spaces occur in the input between two words and those words appear on the same output line, they should be separated by at least two spaces. If two words appear on different output lines, the spaces between them need not be preserved.

The formatter has two basic modes of operation. In "nofill" mode, each input text line is output without modification. The formatter is initially in "fill" mode, in which it produces output lines of 60 characters and accepts input until no more words can fit on the current output line (newline characters are treated as spaces). The line is then justified. Justification is performed by adding spaces between words until the last word has its last character in the rightmost position of the line. Spaces should be added as evenly as possible, in a way that also avoids "rivers" of white space in the text. (This could occur, for example, if all extra spaces were added to the

format = **proc** (ins, outs, errs: stream) **signals** (badarg(string))
    modifies *ins, outs, errs*
    effects If *ins* is not open for reading, or *outs* or *errs* is not open for
        writing, *badarg(s)* is signaled, where *s* identifies an argument that
        was opened improperly (e.g., *badarg* ("input stream")). Otherwise
        format proceeds as described in the text, taking input from *ins* and
        producing output on *outs* and error messages on *errs*. *Ins, outs*, and
        errs are closed before returning.

Figure 13.2 Specification of the formatter.

same side of each line.) Spaces can be added only between words to the right of all tabs. If there is no place to add spaces, then no justification is performed and no error message is produced. Thus if a word is encountered that is longer than the line length, it is output as given on a line by itself with no error message.

Any input line that starts with a word-break character causes a *line break*: If the current output line is not empty, it is neither filled nor adjusted, but output as is. If the current output line is empty, a line break has no effect. An empty input text line (one containing no words, tabs, or spaces) causes a line break and then causes a blank line to be output.

The formatter accepts three commands:

.br   causes a line break
.nf   causes a line break and changes the mode to nofill
.fi   causes a line break and changes the mode to fill

An unrecognized command name causes an error message and is otherwise ignored. The message describes the error and gives the line number on which it occurred.

The specification of the formatter is given in figure 13.2. Since the formatter is intended for general use, we have not made any assumptions about its inputs. (That is, it has no requires clause.) The specification describes the effects by referring to the text of this section; a specification that appears in the design notebook would contain this text. The efficiency constraints are as follows: The time used should be Order(n), where n is the number of characters in *ins*. The space added should be Order(n) (this is the storage for *outs*), but the temporary space used should be small, much less than Order(n). Finally, the design must support such potential modifications as additional commands (for example, for figures and different fonts) and different kinds of output devices. And it must be reasonably simple and modular.

13.4 Getting Started on a Design

In the next several sections, we illustrate the design process by developing a CLU program to implement the formatter. Our presentation idealizes the design process; for example, no design errors are made. We discuss this point further in section 13.9.

When design begins, we have a module dependency diagram containing the abstractions identified during requirements analysis. The graph may contain arcs, since some of the identified abstractions may depend on some of the others. (For example, the requirements specification for an accounting system might have identified a command processor—a procedure—and an account database—a data type; the command processor depends on the database since it makes calls on the database operations.) We begin our design by entering this diagram and the specifications of the abstractions in the design notebook. Then we choose our first target. Since in the case of the formatter there is just one abstraction, *format*, this automatically becomes our first target.

The first step is to invent helpers. There is one main heuristic to guide us in this process: Let the problem structure determine the program structure. This means that we should concentrate on understanding the problem being solved (namely, how to implement the target) and use the insights we gain into the problem to help us develop the structure of the program.

A good way to study the problem structure is to make a list of the tasks that must be accomplished. Here is a list for *format*:

1. Read input.

2. Interpret input.

3. Produce output.

We do not assume that the final program will have subparts that correspond to the listed tasks. Listing the tasks is just a first step toward a design. The next step is to use the list as a guide in inventing the abstractions that will determine the program structure.

Although we have listed the tasks in approximately the order in which they might be carried out, we do not assume that this order will exist in the final program. In fact, for the tasks just given we have a major choice to make. We could complete each task before starting the next, but this problem does not require that all input be interpreted before output can be produced. We can thus do the job incrementally, interpreting the input and producing the output as we go. Since an incremental solution is needed to satisfy our space constraints, we shall make this choice.

In looking for abstractions, we seek to hide details of processing that are not of interest at the current level of the design. Although we can use procedures and iterators to hide details, data types are most useful for this. We shall look for types in likely places: input, output, internal and external databases, and individual data items. For example, we can use a type to hide details of output processing (task 3). First we have to decide what level of encapsulation to provide. We could have a line abstraction that hides the formatting of a single line or a page abstraction that hides the formatting of a single page. Neither of these completely encapsulates the output, though, and complete encapsulation is a good idea, because as production of output becomes fancier (a likely modification), it would be useful if we only had to change one abstraction (and not *format*). Our solution is to introduce a data type, *doc*, that is informed about details of the input and produces the output accordingly. Doc is a kind of abstract output device, with high-level operations well matched to formatting.

Just as doc encapsulates details of output processing, we would like to encapsulate the details of processing input. Here we could invent a token abstraction to represent the meaningful items in the input text, such as words and tabs. However, to produce the proper kind of token, we must know whether we are at the start of a line, since a period is interpreted differently in this position. In fact, the problem as stated is line oriented. If we adopt this problem structure in our program, we can capture knowledge about current position in a line in a natural way. We therefore define a procedure, *do_line*, that processes a single line of input. *Format* will be responsible for processing all the lines of the input in order; *do_line* will be responsible for processing individual lines.

Finally, we must consider task 2, interpreting input, and the relationship between *do_line* and doc. A simple solution is to let *do_line* interpret the input and call appropriate doc operations. This solution requires that we pass a doc object as an argument to *do_line*. *Format* is responsible for creating and finishing the output; doc provides a *create* operation to start the output and a *terminate* operation to finish it.

Now we are ready for the second step of the design, namely, firming up and documenting these abstractions. In the process of producing the documentation, we often uncover loose ends. Some are simply details that need to be worked out, but some might be design errors that require modifications. That is to say, a lot of design work is done during documentation. For example, in specifying *do_line*, we must decide exactly where the input stream is positioned when *do_line* is called and when it returns. We must also decide which module is responsible for checking that the various argument streams are in the proper mode. Furthermore, we must decide

do_ line = **proc** (ins: stream, d: doc, errs: stream) **signals** (all_done)
    requires *can_read(ins)* & *can_write(errs)* &
       d has not been terminated.
    modifies *ins, errs, d*
    effects If *ins* is empty, signals *all_done*. Otherwise processes one input
       line from *ins* to *d* as defined in the specification of *format*, writing
       error messages on *errs*. The entire line is processed including the
       end-of-line character.

    efficiency The time and space taken to scan and parse a line should be
       proportional to the number of characters in the line.

Figure 13.3 Specification of *do_line*.

whether *do_line* writes error messages and how *format* finds out that all
input has been processed.

The specification of *do_line* is shown in figure 13.3, which also shows
the efficiency constraints. This specification contains the answers to our
questions. *Do_line* is responsible for error messages, but not for checking
the mode of *ins* and *errs* or for closing them; *ins* is positioned at the start of
a line both when *do_line* is called and when it returns; and *do_line* notifies
format of the end of input by raising an exception.

The specification of doc is given in figure 13.4. Specifications are given
only for *create* and *terminate*; more operations will be specified later. Note
that doc relies on *format* to check that *outs* is writable. Also, doc requires
that other modules in the formatter not modify *outs*; this requirement is
needed so that doc has control over what is written to the stream.

Our picture of doc is rather incomplete at this point. When inventing
data abstractions, we define only the operations used by the implementation we are studying at the time. Additional operations will be defined
when we study other implementations that use the type. Note that the
specifications of doc and *do_line* both refer to that of *format*. It is typical to
have such references in the specifications of helpers. References are better
than copying text, since they lead to shorter specifications that highlight
how an abstraction fits into the overall structure.

The extended module dependency diagram is shown in figure 13.5. The
diagram contains nodes for doc and *do_line* and arcs that indicate use. At
this stage of the design, we know that *format* uses doc and *do_line* and also
that a doc object is passed to *do_line*. If an abstraction takes an object of
some type as an argument or result, then it uses that type. Here *do_line*
uses doc, and there is thus an arc from *do_line* to doc and *do_line* appears
at a higher level in the diagram than doc.

A sketch of the implementation of *format* is shown in figure 13.6.

doc = **data type is** create, terminate

Requires

> The output stream bound to the argument *outs* of operation *create* must not be modified outside of doc between the time the doc object is created and the time it is terminated.

Overview

> Doc is an abstract output device that produces formatted output (as defined in the specification of *format*) corresponding to the information passed to it via its operations. A doc object is connected to the output stream passed to the *create* operation. All output will be written to this stream by the time the *terminate* operation returns.

Operations

> create = **proc** (outs: stream) **returns** (doc)
> > **requires** *can_write(outs)*.
> > **modifies** *outs*
> > **effects** Prepares to produce the document on *outs*, starting on page 1, in "fill" mode.
>
> terminate = **proc** (d: doc)
> > **requires** *terminate(d)* has not been called previously.
> > **modifies** *d*
> > **effects** Finishes writing information in *d* to the connected output stream and then closes that stream.

end doc

efficiency time = Order(n), where n is the number of characters written to the connected output stream. Space added = Order(n). Temporary space used much less than Order(n).

Figure 13.4 Partial specification of doc.

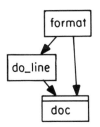

Figure 13.5 Extended module dependency diagram.

check modes of arguments
d: doc := doc$create(outs)
call do_line(ins, d, errs) repeatedly until it signals all_done
terminate d and close ins and errs

Figure 13.6 Sketch of the *format* implementation.

13.5 Discussion of the Method

In constructing our design, we have used a single method for focusing our attention on what needed to be done—we broke the work into subtasks and then investigated how we might accomplish the subtasks. We did not simply introduce a procedure for each subtask, however. Instead we looked for abstractions, especially data types, to take care of the details of the tasks. We believe that this results in a better design, by making it easier to hide details until later stages of the design and to see and exploit connections among different tasks.

We introduced abstractions to hide details that we deemed inappropriate at the current level. This raises the question of how to decide what is appropriate at a given level. Such a decision is largely a matter of judgment, but there are some guidelines. An implementation should accomplish something, but it should not do too much. Our goal is to end up with small modules; implementations of procedures and iterators, both within and outside of data types, should be no more than a page or two in length, and usually shorter. If in *format* we had decided to take care of the details of input and output, we would have ended up with too large a module. In addition, the different parts of the implementation of a module should be at roughly the same level of detail. Even though it deals with input line by line, while output is one big abstraction, *format* is reasonably well balanced in this regard because these decisions match the structure of the problem.

The design of *format* is typical of the design of a top-level abstraction in a system. The implementations of such abstractions are concerned primarily with organizing the computation, while the details of carrying out the steps are handled by helpers. Introducing partially specified data types like doc is also typical. In the early stages of design we frequently know that two modules must communicate with each other, but we do not know exactly how this communication is to take place. In particular, we know that the modules are to communicate through objects of some type, but we do not know what operations on those objects will be useful. Therefore the specification of the shared type is necessarily incomplete. It will be completed as we continue with the design.

do_text_line = **proc** (ins: stream, d: doc)
> **requires** *can_read*(*ins*) & ~*empty*(*ins*) & *d* has not been terminated.
> **modifies** *ins*, *d*
> **effects** Reads one line from *ins*, including the end-of-line character, and processes it as text, as defined in the specification of *do_line*.

do_command = **proc** (ins: stream, d: doc) **signals** (error(string))
> **requires** *can_read*(*ins*) & ~*empty*(*ins*) & *d* has not been terminated.
> **effects** Reads one line from *ins*, including the end-of-line character. Discards the initial character and processes the rest of the line as a command line as defined in the specification of *do_line*. If the command is erroneous, signals *error*(*s*), where *s* describes the error; however, the entire line is processed even if the command is erroneous.

Figure 13.7 Specifications of *do_text_line* and *do_command*.

13.6 Continuing the Design

We now have a module dependency diagram containing several abstractions. How do we select the next target? The first thing to notice is that not all abstractions are suitable as the next target; those that are suitable we shall call the *candidates*. Clearly, *format* itself is not a candidate since we have already designed its implementation. In addition, doc is not a candidate, since it is incomplete. (Actually, we shall occasionally select an incomplete abstraction as a candidate—see section 13.9.) Therefore we have a single candidate, *do_line*.

Do_line must process either a text line or a command line. The processing is quite different in the two cases, and we shall provide two procedures, *do_text_line* and *do_command*, to handle them. Although the work of processing command and text lines could be done directly in *do_line*, providing these two procedures may simplify future modifications by localizing what would have to be changed if, for example, a new command were added.

Specifications for the two procedures are given in figure 13.7. Note that both *do_text_line* and *do_command* assume that *ins* is positioned at the first character of the line when they are called. This means that *do_line* must only peek at and not read the first character of the line; that is, it should use stream$*peekc* rather than stream$*getc*.

The new module dependency diagram is given in figure 13.8. Now there is a dashed arc from *do_line* to doc because *do_line* does not call any doc operations. Before studying *do_line*, we did not know whether its use of doc would be weak or strong, so we assumed it would be strong. We are now making the same assumptions about *do_text_line* and *do_command*.

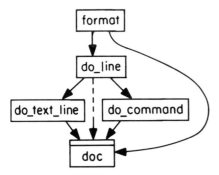

Figure 13.8 Module dependency diagram after design of *do_line*.

Now we have two candidates, *do_text_line* and *do_command*. How do we choose between them? There are no hard and fast rules here; either candidate could be studied. However, there are several reasons why we might prefer one of several candidates:

1. We are uncertain about how to implement it. For example, we might need to implement an abstraction very efficiently, and we are not sure how to achieve this efficiency or even whether it can be achieved.

2. We are uncertain about its appropriateness.

3. It may be more central to the design than another, so that studying its implementation will provide more insight into the design.

4. We may have been working on some area of the design and would like to finish that area.

The first two rules concern the investigation of areas considered to be questionable; choosing an abstraction for one of these two reasons is likely to expose design errors quickly. If such an uncertainty exists, it is almost always best to investigate it immediately. The other two rules are really opposites: Finishing up an area is unlikely to lead to further insights, but it may nevertheless be useful (not to mention psychologically comforting) to get part of the design fully out of the way.

In our example, there is little difference between the two candidates, so we shall arbitrarily choose *do_text_line*. The main question here is how to convey information about the input to doc. One possibility is to pass a token that encodes the information. However, such an approach involves unnecessary work: *Do_text_line* must encode information in a token that must then be decoded in doc. A better approach is to provide a doc operation for each case. Such a solution will be more efficient in both space and time.

The following kinds of information must be passed to doc: words, spaces, tabs, end of line, line break (if the first character in the line is a space or tab), and empty line (if the entire line is empty). Each of these will be handled by a separate operation: *add_word*, *add_space*, *add_tab*, *add_newline*, *break_line*, and *skip_line*. This scheme has the advantage of being easily extendable.

Another question concerns the representation of words: Should words be strings, or do we need a data abstraction? We should use a data abstraction if it will ease potential modifications, but that appears not to be the case here. For example, a word abstraction would not help in adding different fonts, since a font change typically affects more than one word at a time. Therefore we shall just use strings for words. The rest of *do_text_line* involves scanning the input, but this is relatively straightforward and no helpers are needed.

Do_command must scan the command from the stream and discard the rest of the line. The main decision here is whether it should recognize the command or just pass the string to a doc operation. Either decision is satisfactory for now. However, as more commands are added, it is likely that fancy parsing will be required (since commands may have arguments). It is appropriate for *do_command* to do this parsing, since we should keep doc free of such considerations. Therefore *do_command* will have to recognize the commands. It can then call specific doc operations. The doc operations needed are *break_line*, *set_fill*, and *set_nofill*.

The specification of doc is given in figure 13.9. The module dependency diagram is still that shown in figure 13.8.

13.7 The Doc Abstraction

Now doc is the only candidate, and we begin by considering its rep. To decide on the rep, we must make a fundamental decision: Do we produce output incrementally or wait until the end? Since it is not necessary to wait, we shall choose the incremental approach. Again, this choice is necessary to support the space constraints of the formatter.

The rep must contain at least the following: the connected output stream, the current page number, the current line number, the current mode (fill or nofill), and a buffer. The buffer is needed because output cannot be completely incremental; when we are in fill mode, we must collect a line's worth of characters before doing the output.

Doc must perform two different kinds of tasks: processing lines and arranging lines in pages. Most of the complications are involved in line

doc = **data type is** create, add_word, add_ space, add_tab, add_ newline,
     break_ line, skip_ line, set_fill, set_ nofill, terminate

Requires

The actual output stream bound to the argument *outs* of operation *create* must not be modified outside of doc between the time the doc object is created and the time it is terminated.

Overview

Doc is an abstract output device that produces formatted output (as defined in the specification of *format*) corresponding to information passed to it via its operations. A doc object is connected to the output stream passed to the *create* operation. All output will be written to this stream by the time the *terminate* operation returns.

Operations

create = **proc** (outs: stream) **returns** (doc)
     requires *can_write(outs)*.
     modifies *outs*
     effects Prepares to produce the document on *outs*, starting on page 1, in "fill" mode.

add_word = **proc** (d: doc, w: string)
     requires *d* has not been terminated.
     modifies *d*
     effects Adds *w* to *d*.

add_ space = **proc** (d: doc)
     requires *d* has not been terminated.
     modifies *d*
     effects Adds a space to *d*.

add_tab = **proc** (d: doc)
     requires *d* has not been terminated.
     modifies *d*
     effects Adds a tab to *d*.

add_ newline = **proc** (d: doc)
     requires *d* has not been terminated.
     modifies *d*
     effects Adds an end-of-line to *d*.

skip_ line = **proc** (d: doc)
     requires *d* has not been terminated.
     modifies *d*
     effects Adds an empty line to *d*.

Figure 13.9 Complete specification of doc (continues on next page).

break_line = **proc** (d: doc)
>**requires** *d* has not been terminated.
>**modifies** *d*
>**effects** Adds a line break to *d*.

set_fill = **proc** (d: doc)
>**requires** *d* has not been terminated.
>**modifies** *d*
>**effects** Adds a line break to *d* and causes "fill" mode to be entered.

set_nofill = **proc** (d: doc)
>**requires** *d* has not been terminated.
>**modifies** *d*
>**effects** Adds a line break to *d* and causes "nofill" mode to be entered.

terminate = **proc** (d: doc)
>**requires** *d* has not been terminated.
>**modifies** *d*
>**effects** Finishes writing information in *d* to the connected output stream and then closes that stream.

end doc

efficiency time = Order(n), where n is the number of characters written to the connected output stream. Space added = Order(n). Temporary space used much less than Order(n).

Figure 13.9 (continued)

processing, namely, filling and adjusting a line when in fill mode. If we introduce a *line* abstraction, we may be able to deal with the details of line processing separately. By encapsulating these details, we can avoid considering them in the implementation of doc.

We could encapsulate the details of line processing at several different levels. One possibility is a high-level abstraction that justifies and outputs the characters of the line automatically when the line becomes full. Such an abstraction would have to know the current mode of operation (fill or nofill) and also the maximum line length. Note that outputting a line cannot be completely invisible; the doc abstraction must be informed when line output happens so that it can take care of page formatting. An alternative approach is to have a low-level abstraction that keeps track of words and the spacing between them, but does not do justification. However, such an abstraction provides little benefit over simply keeping the information in an array.

We shall use a third approach, a line abstraction at a level between the two just discussed. This type will provide justification, but only on command. Information about the current processing mode and line length will be kept in doc, not in line. This decision seems appropriate because such information concerns the document as a whole; that is, it is information at the document level. Note that we have achieved a good separation of concerns: Line takes care of all details of managing lines, while doc is concerned with all other details of formatting the document.

The intermediate-level abstraction has potential for future flexibility. For example, we might want to vary the size of the left margin, or to prepare several lines in advance of output. These possibilities can be accommodated provided we decouple the line abstraction from physical lines; in particular, outputting a line should not happen automatically, and should not include producing the left margin.

To determine the line operations, we must look at the implementations of doc operations. To take care of spaces and tabs, line$*add_space* and line$*add_tab* will be useful. However, most of the interaction between line and doc takes place in doc$*add_word*. A rough sketch of what this operation does is

if fill mode and no room for word on line
    then justify and output line **end**
add word to line

Given this sketch, we can define the line operations we need. We need operations for adding a word to the line and for finding out whether a word can fit in a line. We must also be able to justify and output the line. Justification and output should be separate operations, so that later we could, for example, arrange to justify several lines before outputting any of them. The output operation should write the line directly onto the stream rather than return a string, since this may be more efficient for some line representations (for example, if the rep is not a string).

Justification must be done in such a way as to avoid rivers of white space in the text; since rivers concern adjacent lines, this is an issue we must consider in doc. It is not always possible to add spaces evenly, so some pairs of words will have more spaces added between them than will others. We can avoid rivers by adding these extra spaces alternately from the right and left in subsequent lines. We must remember in the rep of doc where the extra spaces should go for the current line. We can begin at the right at the start of each page.

One last question concerns the creation of line objects. Obviously we need a *create* operation to get started. After we have output a line, we could create another object for the next line, but such an approach is wasteful because it creates garbage (old lines) that may need to be collected. It is better to reinitialize the old object and reuse it. Operation *clear* will allow us to do this. A specification of line is given in figure 13.10.

Doc$*add_word* is not the only doc operation that causes lines to be output; operations *add_newline*, *break_line*, *skip_line*, and *terminate* also do this. Furthermore, outputting lines involves page formatting. A procedure *output_line* is useful for handling these details. Since *output_line* is unlikely to be useful outside the implementation of doc, it can be internal to doc. This means that we do not add it to the module dependency diagram, but instead design its implementation as part of designing the implementation of doc.

Output_line is in charge of all page formatting, so we must make decisions about how to keep track of page numbers and how to recognize the need for a new page. A reasonable choice is to store the number of the current page and also the number of lines on the current page in the rep. When we have produced all lines on this page, we can prepare for the next. However, we should not produce the page break and page header for the next page unless there is something on it.

The implementation of doc is described in figure 13.11. Note that we have included not only the rep, rep invariant, and abstraction function, but also the specification of the internal routine *output_line*. The extended module dependency diagram is shown in figure 13.12.

13.3 The Line Abstraction

As usual when designing the implementation of a data type, we begin our study of line by considering its rep. The major issue here is to store the words, spaces, and tabs in a way that makes justification relatively straightforward and efficient; in particular, we should avoid moving characters around any more than necessary. Since *justify* is the operation most affected by our choice, we should study its implementation first.

Justification consists of the following tasks:

1. Find first space subject to justification.

2. Enlarge spaces.

A plausible approach is to keep words, tabs, and spaces separate in the rep, so that *justify* can locate easily the first space subject to justification and all the spaces it must enlarge. For example, we could use an

line = **data type is** create, add_word, add_ space, add_tab, length, justify,
            output, clear

Overview

A line is a mutable line of text consisting of the characters corresponding
to words, tabs, and spaces. Operation *justify* justifies the line to a given
length, and *output* writes the line to a given output stream.

Operations

create = **proc** ( ) **returns** (line)
        effects Returns a new, empty line.

add_word = **proc** (l: line, w: string)
        modifies *l*
        effects Adds the characters of *w* to the end of *l*.

add_ space = **proc** (l: line)
        modifies *l*
        effects Adds a space character to the end of *l*.

add_tab = **proc** (l: line)
        modifies *l*
        effects Adds a tab to the end of *l*. This causes from 1 to 8 spaces to
            be added to the end of *l*; afterward $length(l_{post})$ mod $8 = 1$.

length = **proc** (l: line)
        effects Returns the number of characters in *l*.

justify = **proc** (l: line, len: int, from_right: bool)
        modifies *l*
        effects Justifies *l* as described in the specification of *format* to con-
            tain *len* characters if possible. Justification occurs by enlarg-
            ing spaces to the right of the rightmost tab. Spaces are added
            as evenly as possible, with extra spaces added from the right
            if *from_right* is true and otherwise from the left. If justifica-
            tion is impossible (e.g., because *l* has just one word), *l* is left
            unchanged.

output = **proc** (l: line, outs: stream)
        requires *can_write(outs)*.
        modifies *outs*
        effects Writes the characters in *l* to *outs*.

clear = **proc** (l: line)
        modifies *l*
        effects Reinitializes *l* to be empty.

end line

efficiency The space added should be Order(n), where n is the number of
    characters in the line. The time used to justify and output should be
    Order(n). The time used for other operations should be constant.

Figure 13.10 Specification of line.

% Doc produces output incrementally, making use of a buffer consisting of a
% single line object.

   rep = record[line: line,   % The current line
               fill: bool,   % True if in fill mode
               from_right: bool, % True if extra spaces should be added
                               %   from the right
               lineno: int,   % Number of lines on current page
               pageno: int,   % Number of the current page
               outs: stream]   % The output stream

   lines_per_page = 50   % kept as a constant so it can be changed easily

% A typical doc object is a sequence of characters of all the lines completed
%   so far, followed by the characters in the current line.
% The abstraction function is:
%   Characters in d.outs || Characters in d.line

% The rep invariant is
%   can_write(r.outs)
%   & 1 < r.lineno <= lines_per_page
%   & 0 <= r.pageno
%   & r.page_no > 0 =>
%   The current page has the proper header & r.line_no lines, not
%   counting the header
%   & ∀ previous pages p (i.e., pages with numbers 1, ..., r.page_no − 1)
%   p has the proper header
%   & p has lines_per_page lines, not counting the header
%   & p is followed by a page break

% Internal routine *output_line* takes care of the page formatting needed
%   when lines are produced.

output_line = **proc** (d: **rep**)
      modifies *d*
      effects Writes *d.line* and an end-of-line character on *d.outs*, increments
         d.lineno and *d.pageno* appropriately, and takes care of page headers
         and page breaks.

Figure 13.11 Implementation documentation for doc.

array[item], where

   item = variant [word: string, tab, space: int]

Here we are keeping information about adjacent spaces in the same item
in the array; the int for a tab or space keeps track of how many blank
characters should be output. For efficiency, the current line length should
be stored redundantly in the rep. Also, justification will be faster if we keep
track of the rightmost tab as tabs are added to the line. This optimization

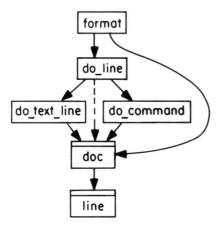

Figure 13.12 Module dependency diagram including line.

is worthwhile if tabs are rare, which is likely. The implementation of line is described in figure 13.13.

13.9 Review and Discussion

The design of *format* was accomplished in stages. At each stage we selected a new target and investigated its implementation. There were two main activities in this investigation. First, we invented some helping abstractions that were useful in the problem domain of the target. Then we defined the properties of the helpers precisely and described those properties in specifications.

The hardest part of program design, and the place where the greatest creativity is required, is in the invention of the helpers. The primary method used here is to study the problem structure and derive from it a structure for the program. Our study of the problem structure takes the form of a list of tasks to be accomplished. However, we do not simply invent a program structure that carries out those tasks one after another. Instead we use the list as a way of focusing attention on what needs to be done, and then invent abstractions to accomplish the needed work.

Abstractions are introduced in each case to hide detail. For example, doc hides the details of the way in which the output is formatted, while *do_line* hides the processing and interpretation of input. It is desirable to hide detail for two reasons: control of complexity and modifiability. We control complexity in order to delay making decisions and thus limit the number of concerns that must be dealt with at any particular level of the design.

% The rep of a line keeps spaces, tabs, and words separately, although adjoining
%   spaces are merged.
% Tabs and spaces are kept as a count of the number of spaces to output.
% The line length and the position of the rightmost tab are also kept.

item = variant[word: string, tab, space: int]
    rep = record[length: int,   % the current line length
                right_tab: int,   % index of rightmost tab or 0 if none
                stuff: array[item]]   % the words, tabs, and spaces

    max_tab_width = 8   % kept as a constant to ease program modification

% A typical line is a sequence of characters.
% The abstraction function is:
%   if l.length = 0 then " "
%   else chars(l.stuff[1]) $\|$ ... $\|$ chars(top(l.stuff))
% where chars(item) is
%   if the item is a word, then the characters in the word
%   otherwise n spaces, where n is the integer in the tab or space item

% The rep invariant is
%   r.length = the sum of the ints in the tab and space items + the
%   sizes of the words
%   & r.right_tab >= 0
%   & low(r.stuff) = 1
%   & r.right_ stuff > 0 => r.stuff[r.right_tab] is
%   a tab & there are no tabs higher in r.stuff
%   & there are no adjacent spaces in r.stuff
%   & the integer in each space and tab token is > 0
%   & the integer in each tab token is <= max_tab_width
%   & for each tab item in r.stuff, count mod max_tab_width = 1,
%   where count is the number of characters from the beginning
%   of r.stuff up to and including this tab

Figure 13.13 Implementation documentation for line.

We support modifiability because the hidden details can be changed later
without affecting other abstractions.

In developing the design we are guided by several factors:

1. Knowledge of what abstractions are already available. This includes
abstractions present in the programming language and in any available
library of programs as well as those already identified during this design.

2. Knowledge of preexisting algorithms and data structures. It is necessary
to know about the methods that have already been developed. For example, the designer should be familiar with existing sorting and searching
techniques and not have to reinvent them.

3. Knowledge about related programs and their structure. As we get deeper into the design, this knowledge is likely to become more and more useful, since even radically different applications are likely to have some similar subtasks.

For example, our decision to process input and output incrementally was based on our knowledge that one-pass algorithms are usually more space efficient than two-pass algorithms. Also, we relied on the existence of CLU types and made explicit choices between them and data abstractions—for example, strings versus a word abstraction and streams versus an output abstraction. We did not use knowledge of data structures in the formatter, but we would do so in any program that needed to store large amounts of data. For example, if we had many formatting commands, we might store information about them in a command table that provides fast lookup.

Types usually have the most impact on the shape of the design. Of course, procedures and iterators are also important, but they tend to appear as operations of types rather than independently. (The formatter was unusual in not having any iterators in its design.) We look for types in four areas: input, output, internal data structures, and individual data items. The design of *format* contains examples of two of these four uses. Doc is an abstraction of output, while line is an abstract data structure. The word abstraction would have been an abstract item, but we decided it was not needed. However, a command abstraction might be useful in the future, especially if users are permitted to define their own commands.

Encapsulating input and output permits us to consume input or produce output in terms of abstract quantities. Input is often handled with procedures or iterators, since typically we just read the next item. The buffering provided by streams is then adequate. A data abstraction allows fancier buffering and also permits us to produce or consume the next item whenever it is convenient. We made use of both these features in doc.

As the design of the implementation becomes better understood, we can use sketches of the program as a guide to our study of abstractions. These sketches are written in English and contain descriptions of subtasks to be accomplished. We can then note for each subtask what procedure or iterator, or operation of a type is to be used and what its arguments are. Unlike the list of tasks used to start the design, the order in which subtasks are carried out in the sketch is significant. We made such a sketch in studying the doc *add_word* operation.

In inventing helpers we are concerned only with concepts; the details of those concepts are usually left ill-defined. In the second step of design, we firm up those details and document them by means of specifications.

During this activity we commonly uncover cases that have been overlooked, especially error cases and end cases (for example, the empty input). We may discover missing arguments or results, or find that arguments or results are of the wrong type. In the case of data types, we may discover missing operations. Moreover, if an abstraction has become overly complex, this may be evident when we try to write the specification. A complicated specification leads naturally to the question of how things can be simplified. A better, simpler abstraction is often the result.

After we have specified all the helpers precisely, we document our design decisions in the notebook by extending the module dependency diagram, adding or extending definitions of the helpers, and sometimes recording a description of the implementation of the target. Such information can be omitted for procedures and iterators unless the implementation is clever; for data types a sketch of the rep, rep invariant, and abstraction function should always be given.

One issue we have not yet discussed is how thorough the design must be. At each stage, the investigation must be thorough enough that all helpers are identified and specified precisely. It is not necessary to go further than this. For example, in *do_text_line* we did not discuss scanning, since it was clear it could be done easily using stream operations. By contrast, the design of doc was carried out in more detail. This extra detail was not really necessary; for example, we need not have considered the details of producing pages. If we had stopped the design sooner, the rep, rep invariant, and abstraction function could only have been sketched. In such a case, more design (but of a localized nature) would be needed during implementation—for example, to define the exact meanings of *lineno* and *pageno* in the rep of doc, and possibly even to identify and specify the internal routine *output_line*. (In fact, the implementations of doc and line shown in appendix B both use additional internal routines.)

As soon as one target has been studied, the next step is to select a new target. First, the candidates are identified; these are abstractions whose implementations have not yet been studied but whose using abstractions have all been analyzed. One candidate is then chosen as the next target. While there are guidelines for making this choice, there are no rules. For example, there is no requirement that all abstractions at one level in the design must be studied before those at lower levels. Instead, common sense should be used, with the goal of finishing the entire design as quickly as possible. This is why we look at questionable abstractions as soon as possible and even study them before they are complete.

If, while studying the implementation of some target, we discover an error in the abstraction itself, we must correct the error before proceeding further with the current target. We use the arcs in the module dependency diagram to discover all implementations affected by the error. Then we correct the design for those implementations. In the process, we may discover more errors that must also be corrected.

For example, suppose we had forgotten to pass the argument *from_right* to line$*justify*. This error might have been discovered while we were figuring out what to do with the extra spaces. By consulting the module dependency diagram, we would discover that the implementation of doc must be reconsidered. So we would back up, correct the error (perhaps we had forgotten to include the information in the rep of doc), correct the specification of *justify*, and then continue with the implementation of line.

When there are no more candidates, then ordinarily the design is finished. There is one special case, however. When two or more abstractions are mutually recursive, then none can be a candidate, since each is used by another abstraction that has not yet been studied. When dealing with mutually recursive abstractions, we must proceed with caution. One must be selected as a candidate and studied first. Since this candidate is used by another abstraction that has not been studied, we may discover later that its behavior is not what is needed. This problem is another indication of why mutual recursion must be viewed with suspicion.

The result of a finished design is a design notebook containing the completed module dependency diagram and sections for each abstraction. It is important that this notebook contain not only the decisions that were made but also the reasons for them. Ideally, this part of the documentation will explain both what problems are being solved by a particular structure and what problems are being avoided that would have been introduced by other structures studied during the design.

Because such documentation is difficult and time consuming to produce, it is often neglected. It is important in the later stages of program development, however, whenever a situation arises in which a design decision must be reconsidered or changed. The new decision can best be made by someone who fully understands the design. The documentation makes the needed information available, both to people other than the original designers and even to the original designers, who will forget it as time goes by.

Note that design is a top-down process; that is, we always reason from what is wanted to how to achieve it. In this way we keep our goal (the required program) firmly in view and are free to use our intuition about programs to guide us to a solution. Through experience in writing programs, programmers develop intuition about what is implementable with what efficiency. They also come to know what program structures are appropriate to various problems. This knowledge can be used to good effect in top-down design.

An alternative to top-down design is bottom-up design, which starts with what is known to be implementable and somehow proceeds from there to achieve what is wanted. Bottom-up design is not really a tenable process for any but the smallest programs. For large programs the gap between what is available and what is wanted is large and must be bridged by the introduction of many abstractions. This gap is easier to bridge top-down because we can concentrate on the thing that is least understood (the program that is wanted) and use what we know to help us. With top-down design, our intuition about what is implementable tells us whether we are making progress: The helpers should be easier to implement than the target. With bottom-up design, it is much harder to measure progress, since it is hard to evaluate how close we are to what is wanted.

In reality, we tend to go back and forth between top-down and bottom-up design. For example, we may investigate the implementation of key abstractions or questionable abstractions even when all uses are not yet understood. However, it is important that the design be driven from the top and that we avoid prematurely implementing any abstractions; doing implementation too early is truly a waste of effort, since the chances of all the details being right are small.

13.10 Summary

This chapter has discussed program design. Design progresses by modular decomposition based on the recognition of useful abstractions. We discussed how this decomposition happens and illustrated the design process with an example. We also discussed a method of documenting a design in a notebook.

The example used was a simple text formatter, and the resulting program was small, containing only six modules. In addition, the presentation of the design process was unrealistic—we did not make errors as we went along. In the real world, any design, even of a simple program like the formatter, requires a great deal of iteration, and many errors will be introduced and corrected as it progresses. Nevertheless, the basic methods we presented still apply, even to much larger programs. We have used it in such programs ourselves.

We do not claim that the formatter design is the best possible. In fact, the goal of the design process is never a "best" design. Instead it is an "adequate" design, one that satisfies the requirements and design goals and has a reasonably good structure. We discuss this issue in the next chapter.

Further Reading

Bentley, Jon L., 1982. *Writing Efficient Programs*. Englewood Cliffs, N.J.: Prentice-Hall.

Hester, S. D., David L. Parnas, and D. F. Utter, 1981. Using documentation as a software design medium. *Bell System Technical Journal* 60(8): 1941–1977.

Kernighan, Brian W., and P. J. Plauger, 1981. *Software Tools in Pascal*. Reading, Mass.: Addison-Wesley.

Parnas, David L., 1972. On the criteria to be used in decomposing systems into modules. *Communications of the ACM* 15(12): 1053–1058.

Exercises

13.1   Design and implement the KWIC program specified in chapter 12 and exercise 1 of chapter 12.

13.2   Design and implement the *xref* program specified in exercise 2 of chapter 12.

13.3   Form a team of three and design and implement a moderately large program. One possibility is the Trivicalc program discussed in appendix C.

14

Between Design and Implementation

In this chapter we discuss briefly the two considerations that arise between the completion of a design and the start of implementation, namely evaluation of the design and choice of a program development strategy. We also discuss the relationship of abstraction to efficiency.

14.1 Evaluating a Design

Periodically during the design of a large program it is worthwhile to step back and attempt a comprehensive evaluation of the design so far. This process is called *design review*. Design reviews should always be conducted by a team composed of some people involved in the design and others who are not. Those not involved in the design should be familiar both with the program requirements and with the technology that will be used in implementing the design. They should also be familiar with the design itself.

It is important that both the designers and the outside reviewers understand that the point of a design review is to find problems with the design. While the designers will inevitably find themselves attempting to justify their design decisions to the outside reviewers, they should not view this as their primary goal. In addition, both the reviewers and the designers must keep in mind that the purpose of the review is only to find errors, not to correct them. Errors should be recorded in an error log, and then the review should continue (unless so many errors have been found that continuing is no longer productive).

It is useful for the designers to present not only the design but also the alternatives that were considered and rejected. This will give the outside reviewers a context for evaluating the chosen design. It may also help the reviewers to find flaws in the chosen design. A common problem is failure to apply design criteria uniformly. Explaining that an alternative was rejected because it failed to meet some criterion may well prompt the reviewers to notice that some other part of the design fails to meet that same criterion.

There are three critical issues to address in evaluating a design:

1. Will all implementations of the design exhibit the desired functionality? That is, will the program be "correct"?

2. Are there implementations of the design that will be acceptably efficient?

3. Does the design describe a program structure that will make implementations reasonably easy to build, test, and maintain? Also, how difficult will it be to enhance the design to accommodate future modifications, especially those identified during the requirements phase?

14.1.1 Correctness and Efficiency

Earlier in this book we discussed two approaches, testing and verification, to increasing our confidence that a program will behave as desired. Unfortunately, neither of these approaches can be applied to designs. Since designs cannot be run, testing is out of the question. If designs were presented completely in a formal language, some verification might be possible. However, formal specification of designs of large programs is beyond the current state of the art.

While there are no completely rigorous techniques for reviewing a design, it is important that design reviews be systematic. They should examine both local and global properties of the design. Local properties can be examined by studying the specifications of individual modules; global properties, by studying how the modules fit together.

Two important local properties are restrictiveness and generality. These were discussed in chapter 8. Another important local property is efficiency. The first step in estimating the overall efficiency of a system is to construct for each abstraction an expression relating its running time and storage consumption to its arguments. How accurately this can be done depends upon the completeness of the design. Consider the *sort* procedure:

> sort = **proc** (a: int_array)
>> **modifies** a
>> **effects** Permutes the elements of a so that a is sorted in increasing
>>> order.

If the design does not specify any efficiency constraints for *sort*, relatively little can be said about the efficiency of its implementations or about the efficiency of abstractions to be implemented using *sort*. The problem is that implementations of *sort* span a wide range with respect to efficiency. Considerably more can be said about the efficiency of implementations if the design includes the following criterion:

efficiency worst case time = n * log(n) comparisons, where n is the
   size of a.

and more yet if the design states:

efficiency worst case time = n * log(n) comparisons, where n is the
   size of a. Maximum temporary main memory allocated is a small
   constant. No main memory added.

An important function of a design review is to discover places where the
design needs to be more specific about what is required of implementations.

After each module has been evaluated in isolation, we examine the design as a whole. A good way to begin is by tracing paths through the
design that correspond to various uses of the program. We select some
test data and then trace how both control and data would flow through
an implementation based on the design. This tracing process is sometimes
called a *walk-through*. The test data are chosen in much the same way as
described in chapter 9. However, since "testing" a design is much more
labor intensive than testing an implementation, we must be more selective
in choosing our test data.

Since the point of tracing the design is to convince ourselves that all implementations of the design will have the desired functionality, the success
of this method is related to the completeness of the test cases. We are
using the test cases to carry out an informal verification process. During
the design review it is also important to discuss the completeness of the
process, that is, to argue that all cases have been considered. Normal and
exceptional cases should be considered separately.

Picking test cases for a design review is somewhat simplified by the fact
that the data can be symbolic. We need only identify properties that the
test data should have; we do not need to invent data with those properties.
As an example, consider the *format* program designed in chapter 13. We
start with normal input; that is, we assume that all of *format*'s arguments
are in the correct mode. Since the initial contents of the output and error
files are irrelevant, we need only consider what the contents of the input
file, *ins*, should look like. The test should include cases involving various
combinations of

1. short and long words, including words too long to fit on a single output
line,

2. single and multiple spaces between words,

3. tabs and spaces at different positions on output lines—for example, the
front, the middle, and the end,

4. short and long input lines, including input lines that are longer than output lines,

5. output lines where no justification is necessary,

6. output lines where more spaces have to be inserted than there are word breaks in the line, and

7. output lines containing exactly one word.

The following symbolic data cover some of these cases:

wd1 space wd2 space wd3 space wd4

where the sum of the lengths of wd1 and wd2 is less than the length of an output line, but the sum of the lengths of wd1, wd2, and wd3 is greater than the length of an output line.

We use these data to walk through the design of *format*. The walk-through is, in effect, a hand simulation of the design. The main thing we want to examine is the flow of information through the program. Here is how we might start a walk-through based upon the previous data:

1. After checking the modes of its arguments, *format* creates a new doc, *d*, by calling doc$*create*(*outs*). The specification of doc$*create* tells us that this doc is in "fill" mode. Nothing is written to *outs* by doc$*create*.

2. *Format* next calls *do_line*(*ins, d, errs*). At this point we note that *do_line* does not have direct access to *outs* and ask ourselves whether this will present a problem. It does not seem to.

3. *Do_line* peeks at the first character of the first line of *ins*, discovers that it is not a period, and then calls *do_text_line*(*ins, d*). We note that *do_text_line* is not being passed *errs*. We note also that it has no way to pass information about any errors back to *do_line*. This presents no problem for *format* as currently specified, but is likely to present a problem if the specification is modified to allow something other than words, for example, information about changes in type face, within text lines. We should note this in our design documentation.

The walk-through continues in this fashion until the behavior of the entire program has been explored. In the process we estimate the efficiency of each module, so that we can construct estimates of worst-case and average efficiencies for the whole program.

Walk-throughs are a laborious and imprecise process. Experience indicates that designers are seldom able to examine their own designs adequately. The process works best when it is performed by a team of people including, but not dominated by, the designers.

Tracing through the entire design with a few sets of inputs helps us to uncover gross errors in the way the abstractions comprising the design are fit together. A good next step is to work bottom-up through the module dependency diagram, isolating subsystems that can be meaningfully evaluated independently of the context in which they will be used. Since these are likely to be considerably smaller than the system as a whole, we can trace more sets of test data through them.

The modifiability of the design should be addressed explicitly during the review. There should be a discussion of how the design must be changed to accommodate each expected modification. A plausible measure of how well the design accommodates modifications is how many abstractions must be reimplemented or respecified in each case. The best situation is one in which only a single abstraction needs to be reimplemented.

A walk-through forces us to look at the design from a different perspective than the one that characterized the design process. During design, we focused on identifying abstractions and specifying their interfaces. These abstractions arose from considering what steps were to be carried out, but our attention was focused on parts of the program separately. Now we go back over the steps carried out by the whole program as it uses the abstractions, and this exercise forces us to address the question of whether the abstractions can be composed to solve the original problem.

14.1.2 Structure

The most important structural issue to address in evaluating a design is the appropriateness of the module boundaries. There are two key questions to ask:

1. Have we failed to identify an abstraction that would lead to a better modularization?

2. Have we grouped together things that do not belong in the same module?

We can provide no formula for answering these questions. What we can provide is a list of symptoms that occur when a program has been badly modularized. We shall concentrate on local symptoms, that is, problems that can be detected by looking at a single module or at the interface between two modules.

Coherence of Procedures

Each module in a design should represent a single coherent abstraction. The coherence of an abstraction can be examined by looking at its specification. A procedure should perform a single abstract operation on its arguments. (Our discussion applies to iterators too. An iterator maps its inputs to a sequence of items; this mapping should be a single abstract operation.)

Some procedures have no apparent coherence. They are held together by nothing more than some arbitrarily placed bracketing mechanism. In the early days of "structured programming" many people failed to understand that good program structure is basically a semantic notion. They looked for simple syntactic definitions of "well-structured." Many of these simplistic definitions included an arbitrary upper bound, such as one page, on the size of procedures. Another example of an arbitrary size restriction occurs in programs that must manage their own memory and are divided into modules to facilitate overlays. Such arbitrary restrictions frequently led to procedures totally lacking in coherence.

A second cause of lack of coherence is hand optimization of programs. An eagle-eyed programmer may notice that some arbitrary group of statements appears several times. In an attempt to save space, the programmer may bundle these statements into a procedure. In the long run, however, such optimizations are generally counterproductive, since they make the program harder to modify.

There are two reliable indicators of a total lack of coherence. If it seems that the best way to specify a procedure involves describing its internal structure (that is, how it works), the procedure is probably incoherent. The second good tip-off is difficulty in finding a suitable name for a procedure. If the best name we can come up with is "procedure1," there is probably something wrong with our design. If there is no apparent coherence to a procedural abstraction, we should rethink the design with the goal of eliminating that procedure.

Conjunctive coherence is a step up from no coherence at all. It is indicated by a specification of the form

A & B & C & ...

Conjunctive coherence usually occurs when a sequence of temporally contiguous actions is combined into a single procedure. A typical example is an abstraction whose job is to initialize all data structures. The specification of such an abstraction is likely to be a conjunction:

initialize A & initialize B & ...

Note that such a structure can make it more difficult to identify data abstractions, since part of the job of each type is taken over by the procedure.

In an environment in which procedure calls are unduly expensive, con-

get_end = **proc** (a: int_list, j: int) **returns** (i: int) **signals** (empty)
    requires $0 < j < 3$
    effects if $size(a) < 1$ then signals *empty*
        else if $j = 1$ then $i = first(a)$
        else if $j = 2$ then $i = last(a)$.

Figure 14.1 An example of disjunctive coherence.

junctive coherence may be useful, since it can eliminate some calls. However, unless there is a strong logical connection between the actions, it is generally better not to combine procedures. The more we put into a procedure, the harder it will be to debug and maintain it. Furthermore, as we maintain a program, we are likely to discover occasions when it would be useful to perform some subset of the conjuncts. If this happens, the appropriate thing to do is probably to break up the original procedure. What people often do instead, however, is to add another procedural abstraction. This leads to more code to debug and maintain, and to a program that occupies more space at runtime.

Disjunctive coherence is indicated by a specification with an effects clause of the form

A | B | ...

often in the guise of an if-then-else or a conjunction of implications. A robust procedure is likely to involve some disjunctive coherence, to separate the normal return from exceptions. However, if the specification of what happens when the procedure returns normally involves disjunction, one should be concerned. Consider the specification in figure 14.1. The abstraction *get_end* can raise the signal *empty* or it can return normally and do one of two different things. Each of the these two would have been a perfectly reasonable abstraction in its own right; that is to say, we could have had the two abstractions of figure 14.2. Combining these two procedures into one has no advantage and several disadvantages. First, a call of the form *get_end*(a, 1) is likely to be harder for a reader to understand than a call of the form *get_first*(a). Second, a new class of errors is possible—for example, a call of the form *get_end*(a, 3). Third, a program using *get_end* is less efficient than one using the two abstractions of figure 14.2. Whenever a call to *get_end* is made, the caller knows which of the procedures is wanted. However, this information must be encoded into the second argument of the call, and *get_end* must test this argument to figure out what to do. This extra work requires both time and space. Also, we may implement *get_end* using subsidiary abstractions such as *get_first* and *get_last*, thus increasing the number of procedure calls that get executed.

get_first = **proc** (a: int_list) **returns** (i: int) **signals** (empty)
    effects if $size(a) < 1$, signals *empty*; otherwise $i = first(a)$.

get_last = **proc** (a: int_list) **returns** (i: int) **signals** (empty)
    effects if $size(a) < 1$, signals *empty*; otherwise $i = last(a)$.

Figure 14.2 Two coherent procedures.

Disjunctive coherence often arises from a misguided attempt to generalize abstractions. When a program design contains two or more similar
abstractions, it is always worthwhile to consider whether a single more
general abstraction might replace all or some of the similar ones. If successful, generalization saves space and programmer effort with little cost
in execution speed or complexity in the implementation of the generalized abstraction. However, if the result is an abstraction with disjunctive
coherence, then it is usually better not to do the replacement.

The procedure *justify* in type line in the design of *format* is an example of successful generalization. We could have had two procedures,
justify_from_left and *justify_from_right*, but this would have bought us little, except a larger program and more code to maintain. While there is
a disjunction in *justify*'s specification, it is buried rather deeply. Most of
the specification, and most of the code, is independent of the value of the
argument *from_right*.

A generalization we did not make in line was combining *add_word*,
add_space, and *add_tab* into a single procedure, such as

add_token = **proc** (l: line, t: token)

where *token* is an appropriate one_of type. This would not have saved any
code, and it would have been slower than having the three operations.

Occasionally the appearance of excessive disjunctive coherence indicates
failure to introduce appropriate data abstractions into the design. In such
cases combining several distinct functions into one procedure may be an
attempt to encapsulate representation information that should have been
encapsulated in a missing type. In effect, the type is implemented by a
single procedure, and extra arguments are used to identify the operation
being called.

Coherence of Types

Each operation of a type should be a coherent procedure or iterator. In addition, a type should provide an abstraction that its users can conveniently
think of as a set of values and a set of operations intimately associated with
those values. One way of judging the coherence of a type is to examine each

operation to see whether it really belongs in the type. As discussed in chapter 4, a type should be *adequate*, that is, should provide enough operations so that common uses are efficient. In badly designed types one frequently finds operations that do not seem particularly relevant to the abstraction and whose implementation can take little or no advantage of direct access to the representation. It is generally better to move such operations out of the type. If fewer operations have access to the representation, it is easier to modify the representation if it becomes desirable to do so.

Consider, for example, a stack type containing the operation *sqrt_top*:

sqrt_top = **proc** (s: stack) **returns** (i: real) **signals** (empty)
    effects if $size(s) = 0$, signals *empty*; otherwise $i = sqrt(top(s))$.

Sqrt_top has little to do with stacks, and most of its implementation will not depend upon the representation of stacks. Therefore this operation should be moved out of the stack abstraction.

Communication between Modules

A careful examination of how much and what kind of information is exchanged between modules can uncover important flaws in a design. Throughout this book we have stressed the importance of narrow interfaces: A module should have access to only as much information as it needs to do its job. CLU is designed to encourage narrow interfaces—for example, by not allowing procedures to refer to global variables—but it is still possible to pass too much information to a module.

A module may be passed too much information because a type has not been identified. In the absence of a type, all modules that would have have communicated in terms of the abstract objects instead communicate in terms of the representation. The result is modules that have much wider interfaces than necessary; instead of being related only through the type, they share knowledge of how that type is implemented. Note that this includes the abstraction function and representation invariant in addition to the representation itself. Note also that all using modules must be considered in reasoning that the implementation of the (missing) type is correct. Furthermore, if the implementation of the type changes, every using module must also change.

Even if all needed types have been identified and are implemented by their own clusters, some interfaces may still be wider than necessary. Well-designed programs frequently have types that include a great deal of information. Some modules may not need to access all of this information. Yet many designs call for passing the entire abstract object when a small piece of it would be sufficient. For example, we might have a type *student_record* that includes, among other things, a student's name, social security number, residence, and transcript. A procedure, *print_address*, that prints an address label might need only a student's name and residence. Such a procedure should not be passed the whole student record; instead the information it needs should be extracted from the student record by its caller and passed to it explicitly. There are several good reasons for this:

1. If *print_address* is passed an object of type *student_record*, its implementor will have to know how to extract the needed information—that is, what operations to call. If the specification of *student_record* is changed, the implementation of *print_address* may have to be changed. None of this would be necessary if the needed information were passed explicitly.

2. The implementation of *print_address* may have a bug that causes it to mutate the *student_record*. Such bugs can be very hard to find.

3. If *student_record* is a CLU record (or struct), rather than a recordlike abstraction, then whenever extra fields are added to the record, *print_address* will have to be recompiled.

14.2 Ordering the Program Development Process

Throughout this book we have stressed a program development strategy that is basically top-down. While we recognize that program development is an iterative process, we have argued that in each iteration, specification and design should precede implementation. We have not yet discussed how to order the process of going from a design to an implementation. The basic choice is between a top-down strategy and a bottom-up strategy.

The traditional mode of development is bottom-up. In bottom-up development we implement and test all modules used by a module M before we implement and test M. Consider, for example, implementing a design with the module dependency diagram shown in figure 14.3. We might begin by implementing and unit testing D and E. We might then implement and test B and C. In testing B and C we might use D and E in order to avoid writing stubs. Doing this implies that we are no longer performing unit testing, since we are not testing a single module. Finally, we would implement and test A. Note, by the way, that this is not the only possible bottom-up order. We might equally well have used the order D, B, E, C, A.

In top-down development, we implement and test all modules that use a module M before implementing and testing M. Possible top-down orders for the above example include A, B, C, D, E and A, C, E, B, D. Just as bottom-up development reduces our dependence on stubs, top-down development reduces our dependence on drivers. It is important that top-

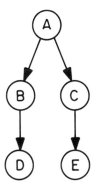

Figure 14.3 A simple module dependency diagram.

down development be accompanied by careful unit testing of all modules. If we tested B only as it is used by A, we might see only part of B's specified behavior. Were we to change A later, new bugs might be revealed in B. Therefore if we choose to use A as a driver for B, we must make sure that A tests B thoroughly.

Neither development strategy strictly dominates the other. Most of the time, it seems better to work top-down, but there can be compelling reasons to pursue a bottom-up approach. We advocate a mixed strategy in which one works top-down on some parts of the system and bottom-up on others.

Top-down development has the advantage of helping us to catch serious design errors early on. When we test a module, we are testing not only the implementation of that module but also the specifications of the modules that it uses. If we follow a bottom-up strategy, we might easily spend a great deal of effort implementing and testing modules that are not useful because there is a problem with the design of one of their ancestors in the module dependency diagram. A similar problem can occur in top-down development if we discover that some crucial descendent module is unimplementable or cannot be implemented with acceptable efficiency. Experience indicates, however, that this problem occurs less often. This may be because lower-level abstractions are often similar to things we have built before, whereas higher-level abstractions tend to be more idiosyncratic.

In top-down development it is always possible to integrate one module at a time. We merely replace a stub by the module it is intended to simulate. In bottom-up development, on the other hand, we tend to integrate several modules at once; in most cases a single higher-level module corresponds not to one but to several drivers. For example, when A is integrated into the program of figure 14.3, it will be replacing drivers for both B and C.

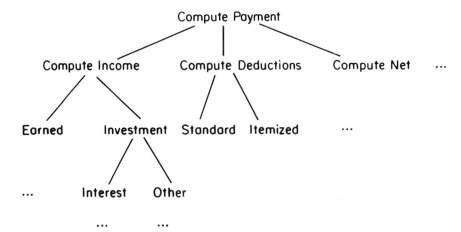

Figure 14.4 Module dependency diagram for an income tax program.

Since system integration tends to proceed more smoothly if we can add one interface at a time, top-down development has a significant advantage in this regard.

Top-down development also increases the likelihood of bringing up useful partial versions of the program being developed. Suppose a program to compute income tax payments has the (partial) module dependency diagram of figure 14.4. By working top-down, it is possible to bring up many useful partial implementations of this design. A system without *investment* and *itemized* procedures, for example, might be quite useful to some people.

Even if partial versions of the system cannot be used productively by the customer, bringing partial versions up early in the development process may have some important benefits. If implementers are able to demonstrate something early, they can get feedback that may reveal problems with the requirements for the program. Furthermore, on lengthy projects the ability to produce a series of working partial systems seems to help the morale of both the customer and the developers.

While bottom-up development delays the point at which one has a working partial system, it does lead to the earlier completion of useful subsystems. These subsystems generally have wider applicability than do the partial systems we get when working top-down. This is particularly true of low-level subsystems (for example, input/output subsystems). In an environment in which several related systems are being developed simultaneously, it can be helpful to bring up shared subsystems by working bottom-up. Moreover, it is sometimes easier to build a low-level subsystem than to build a stub to simulate it.

Another potential advantage of bottom-up development is that it may place less of a load on machine resources than does top-down development. This is particularly true early in the development process. Top-down development tends to use more machine time because we are running a larger part of the system during many phases of development. Typically, full tests of the system are developed and then run each time a new module is added. Such tests may exercise many other parts of the system besides the module being added, and thus consume much time and space that really is not needed for this step of the integration. On the other hand, such a systematic approach to testing is bound to uncover more errors and be more thorough than a more ad hoc approach.

A related advantage of bottom-up development is that it may allow us to proceed in the absence of some computational resource. Consider building a system that is to run on a machine with 512K of memory. If that machine has not yet arrived, but there is a similar machine with 128K available, we might be able to do quite a bit of bottom-up development before memory becomes a problem. We would probably run out of memory much sooner working top-down.

In deciding when to work top-down and when to work bottom-up, we must consider nontechnical as well as technical issues. Sometimes large implementation efforts must begin before an adequate team of programmers has been assembled. In such cases top-down development seems to work best. Since module dependency diagrams are always narrower at the top than at the bottom, the need for programmers tends to grow as we get deeper into the implementation. If, on the other hand, the programming team is already assembled and ready to go to work, bottom-up development may be better.

Generally, we recommend a mixed strategy in which top-down development is favored but not followed entirely. In working out this strategy, both technical and nontechnical factors should be taken into account. The most important point is that a development strategy should be defined explicitly before implementation begins. Some insight into implementation and testing order can be gained by considering *format*:

1. *Format* and *do_line* should probably be implemented first. They are simple to implement and together can be used as drivers to test lower-level abstractions. They can be tested easily since stubs for *do_text_line*, *do_command_line*, and the two doc operations *create* and *terminate* need only print the arguments with which they are called.

2. *Do_command* and *do_text_line* should probably be implemented next.

Again, they can be easily tested, since stubs for the procedures they call need only print their arguments.

3. At this point we have only two abstractions left, doc and line. A strictly top-down development would lead us to implement doc next. There are two things arguing against this. First, many of the procedures of doc (for example, *add_word*) do nothing more than call corresponding procedures in line. Second, it will be difficult to test doc without line$*clear* and line$*output*, and writing stubs to replace these is probably no easier than actually implementing them. For these reasons it probably makes sense to consider implementing and testing parts of each of doc and line at the same time. The *create*, *add_word*, *add_space*, and *add_tab* operations of each plus line$*clear* and line$*word* would be a reasonable choice. It is possible to implement these types in stages because their reps and rep invariants have already been chosen.

4. We conclude the implementation by finishing up doc and then line. The implementation of line$*justify* should probably be the last thing done. It will be the most complicated procedure to implement and require the most testing. Testing it will be much easier once the rest of the program is in place.

14.3 The Relationship of Abstraction to Efficiency

Throughout this book we have argued that well-chosen abstractions simplify programming, testing, debugging, and maintenance. We have put less emphasis on their impact on program efficiency.

Most of the time, well-chosen abstractions have a salutary effect on efficiency. They can be a big help in designing efficient algorithms. Moreover, by increasing the modularity of programs, they make it far easier to introduce modifications. This, in turn, helps us to eliminate bottlenecks and respond to changes in efficiency requirements. Occasionally, however, the introduction of abstractions can have an adverse impact on efficiency. An example of a serious problem is a procedure that could take advantage of accessing the rep of some type. When such a problem arises, there is nothing to do but consider reworking the design.

Less serious problems arise when the programming language being used imposes an unacceptably high cost on the use of abstractions. The style of programming we advocate leads to programs that make a large number of procedure calls. Calling a procedure always involves some runtime overhead; how much depends upon the programming language and the compiler. For some programming languages (not CLU), calling a proce-

Q = **proc** ...
   ...
   if ~stack$empty(x)
      then P(x)
      end
   ...
   end Q

P = **proc** (s: stack)
   if ~stack$empty(s)
      then x: int := stack$pop(s)
         % use x and return
      end
   end P

Figure 14.5 A procedure call of P by Q.

Q = **proc** ...
   ...
   if ~stack$empty(x)
      then % start of the body of P
         if ~stack$empty(x)
            then P_x: int := stack$pop(x)
               % use P_x
            end % of the body of P
      end
   ...
   end Q

Figure 14.6 Replacing P's call by its body.

dure is an inherently time-consuming operation. Even if the procedure call is not expensive, there may be overhead stemming from the introduction of redundant computation. The problem is that for any particular use of a procedure there may be information available that is not available for all uses. In figure 14.5, for example, the check ~stack$*empty(s)* in the body of P is redundant for Q's call of P.

The solution to this problem is an optimization called *inline substitution.* Doing an inline substitution means replacing a procedure call with the body of the procedure being called. This involves replacing occurrences of arguments in the body by the corresponding actuals in the call. If the body has local identifiers that conflict with identifiers in the calling environment, these must be renamed. Figure 14.6 shows the result of applying inline substitution to the code in figure 14.5. Given the code in figure 14.6, a reasonably good optimizing compiler will remove the redundant code.

Some compilers perform inline optimization automatically, but most do not. Since inline substitution is a source-level optimization, it can be done by programmers. Hand optimizations of this sort can be useful when applied on a small scale to critical sections of code. Attempting to do it on a large scale, however, is not a good idea. If this is desired, the best approach is to write a program to do it automatically. Unfortunately such a program is moderately difficult to write.

14.4 Summary

In this chapter we discussed some things that should be done between the completion of a design and the start of implementation. The key points to take away are the importance of conducting a systematic evaluation of the design and developing a precise plan specifying the order of implementation and testing of the modules comprising the design.

In a design review one considers whether or not implementations of the design will exhibit the desired behavior and efficiency, and whether or not the program structure described by the design will be reasonably easy to build, test, and maintain. We suggested conducting design reviews by tracing the path of symbolic test data through the design. We also suggested several criteria that could be used in evaluating structural issues. All of the criteria were related to the appropriateness of module boundaries.

We did not give fixed rules for picking an order of implementation and testing. We discussed the relative merits of top-down and bottom-up development and testing. Our conclusion was that most of the time it is best to follow a mixed strategy, but with an emphasis on proceeding top-down.

In our discussion of design reviews we presented an extremely abbreviated review of the text formatter designed in chapter 13. The reader should not infer from this that design reviews should be conducted only after a design is complete. For larger programs it is imperative that we conduct careful reviews during the design phase. It is also important to start considering how the implementation effort is to be organized. A need for early completion of subsystems, for example, can have a significant impact on design decisions.

Further Reading

Lampson, Butler W., 1984. Hints for computer system design. *IEEE Software* 1(1): 11–28.

Myers, G. J., 1979. *The Art of Software Testing.* New York: John Wiley & Sons.

Exercises

14.1 Perform a design review for some program that you have designed. Be sure to include a discussion of the structure and modifiability of the program as well as a discussion of its correctness.

14.2 Define an implementation strategy for a program that you have designed.

14.3 A student proposes that in the design of *format* (chapter 13) the line operations *justify*, *clear*, and *output* should be replaced by a single operation that does all three things. A second student objects that this would lead to one of the bad forms of coherence. What kind of coherence would result? What are the advantages and disadvantages of merging the three operations, or of merging just *clear* and *output*?

14.4 Suppose that in the design of *format* (chapter 13) we had invented a *command* data type that encapsulated all information about commands, including their format and their meaning. We could then change the doc abstraction so that it would have a single operation in place of *set_fill* and *set_nofill*. Evaluate this alternative design with respect to both the current specification of the formatter and future modifications involving more commands.

14.5 Suppose that in the design of *format* (chapter 13) we had not introduced the line abstraction. Evaluate this alternative design with respect to both the current specification of the formatter and future modifications.

14.6 Suppose that in the design of *format* (chapter 13) we had introduced neither line nor doc. Evaluate this alternative. Discuss the coherence of the design and also the issues of module communication discussed in section 14.1.2.

14.7 The map abstraction was discussed in exercise 5 of chapter 5. It provided an operation *insert* to add a string with its associated element to the map, and another operation *change* to change the element associated with the string. Suppose these two operations were replaced by a single operation that added the association if it did not already exist, and changed it if it did. Discuss the coherence of this modified abstraction. How does the modified abstraction compare with the original?

15

Using Other Languages

This book has presented a methodology for program construction. Our approach uses abstractions, particularly data types, in program design and implementation. Not all programming languages support all the kinds of abstractions we use, nor do they support abstractions of our exact form. When faced with such a situation, we do not abandon the methodology. Instead we bridge the gap between the methodology and the language in two ways:

1. Adapt the forms of abstractions to better fit the language.

2. Invent conventions for implementing abstractions in the language.

Both methods were illustrated in chapter 7, which explained how to use the methodology with Pascal. The following techniques were used:

1. Procedural abstractions were subdivided into procedures, which return no results, and functions, which return a single result. Furthermore, function and procedure specifications defined the parameter-passing method (call by value or call by reference) for each argument.

2. Since there are no exceptions in Pascal, we used instead functions that return an enumerated type whose value indicates the kind of termination. In addition, we implemented a *failure* procedure that could be called in situations where the *failure* exception would be raised in CLU.

3. Iterators were replaced by special types called generators, with operations to start the iteration, to determine whether iteration is complete, and to return the next item.

4. Parameterization was done with text editing. A parameterized module was written in a form that could not be compiled. Each instantiation was carried out by text editing.

5. Data abstractions were implemented by following a set of conventions. These conventions governed the way in which the type and its operations were named and the way in which objects were represented. They also prohibited access to the rep from outside the code that implemented the type.

6. Finally, entire programs were assembled by following conventions on the placement of modules.

The first three of these techniques tailor the methodology to Pascal, while the second three are conventions for the use of Pascal.

A similar approach can be used for any language, although the details change. Before discussing other languages, however, we show a different way to implement data types in Pascal. The approach shown in chapter 7 was very close to CLU: All objects belonging to abstract types reside in the heap. By contrast, storage for objects of most built-in types in Pascal is allocated on the stack. An approach that allocates storage for objects of abstract types on the stack will be described. This approach is only suitable for objects of fixed size; objects that grow dynamically must use the heap approach.

15.1 The Stack Approach

Suppose we wanted to define integer sets that contained a maximum of 100 elements. Let us name such a type *intset_100*. Since the maximum size of an intset_100 object is known in advance, it is possible to allocate space for its representation on the stack. A possible representation is

```
type intset_100 = record
    size: 0..100;
    els: array [1..100] of integer
    end;
```

Here, as in chapter 7, we use the convention of defining the rep by a type definition:

```
type typename = {definition of rep type goes here};
```

In this case the real storage representation is defined on the right-hand side of the equal sign. By contrast, in section 7.2 the right-hand side was a pointer to a type named *typename_rep* whose definition specified the real storage representation.

Types for which object storage is allocated on the stack differ in several ways from those that use the heap. They have neither create nor destroy operations, but instead have an initialization operation. Also, all operations of a stack-based type take objects of the type by reference.

Stack-based types have no create operations because storage for their objects is allocated whenever a declaration is evaluated. For example, the declaration

```
var t: intset_100;
```

causes room for a new variable *t* to be allocated on the stack. The amount
of space allocated depends on the type of the new variable. In this case,
the type of *t* indicates that space is needed for a record with an integer
component and an array component. Of course, the storage thus allocated
cannot be a meaningful representation because it has no data stored in it.
We solve this problem by providing one or more initialization operations
whose purpose is to initialize the rep and thus ensure that it satisfies the
rep invariant. For example, *intset_100_init* does initialization of intset_100
objects. Unfortunately, there is no guarantee that an initialization operation will ever be called, and therefore no guarantee that the rep invariant
holds when the other operations are called. This problem is one of the
weaknesses of the stack approach in Pascal.

The intset_100 argument of *intset_100_init* is passed by reference, since
intset_100_init must modify *s*. The other intset_100 operations also take
their intset_100 arguments by reference, both to permit modifications
where needed and because call by reference is more efficient than call by
value for large objects. Finally, intset_100, like other stack-based types,
has no destroy operation because storage for an intset_100 variable is deallocated when the procedure that declared that variable returns.

The specification of intset_100 is given in figure 15.1. Note that although all operations take their intset_100 argument by reference, only
some operations are allowed to modify that argument. An implementation
of intset_100 is given in figure 15.2.

It is worth noting that the above approach can be used in Pascal only if
the sizes of objects are known at compile time. If object sizes are defined
at creation time, the objects must be allocated on the heap in Pascal even
if their sizes do not vary dynamically.

Iteration over a stack-based collection such as intset_100 can be provided
by a generator type. This type is similar to those discussed in section 7.4,
but some of the details differ. For example, figure 15.3 gives a specification
of a generator *s100_ elems* for intset_100 objects. Note that, like intset_100
itself, s100_elems initializes rather than creates its objects, and there is no
destroy operation. Creation is again done by declaration; for example,

var g: s100_ elems;

creates an uninitialized generator object, which must then be initialized by
calling *s100_elems_init*. This procedure and the functions *s100_elems_next*
and *s100_elems_done* all take the generator object by reference.

Also note that *s100_elems_next* and *s100elems_done* take the set being
iterated over as an extra argument. This is done for efficiency; it gives

intset_100 = **data type is** intset_100_init, intset_100_insert, intset_100_delete,
        intset_100_size, intset_100_member

Overview

Intset_100 objects are mutable sets of integers that can contain a maximum
of 100 elements. They are allocated on the stack.

Operations

procedure intset_100_init (**var** s: intset_100)
    modifies s
    effects Initializes s to contain no elements.

function intset_100_insert (**var** s: intset_100, x: integer):
            intset_100_insert_exceptions
    modifies s
    effects The type intset_100_insert_exceptions = (normal, noroom).
    If $size(s_{\mathrm{pre}} \cup \{x\}) > 100$, returns *noroom* and does not modify
    s. Otherwise modifies s to include x as an element and returns
    normal.

procedure intset_100_delete (**var** s: intset_100, x: integer)
    modifies s
    effects Modifies s so that x is not an element.

function intset_100_size (**var** s: intset_100): integer
    effects Returns a count of the number of elements in s.

function intset_100_member (**var** s: intset_100, x: integer): boolean
    effects Returns true if x is an element of s; otherwise returns false.

end intset_100

Figure 15.1 Specification of intset_100.

these operations access to the intset_100 object. The generator could store
a copy of this object in its rep (*s100_elems_init* would make the copy), but
making and storing the copy would be too costly. An implementation of
the s100_elems generator is shown in figure 15.4.

15.2 Choosing an Approach

In adapting a programming language to the methodology (and vice versa),
the approaches described for Pascal in the last section and in chapter 7
can be used as a guide. Obviously, the approach selected will be heavily
influenced by the language; wherever possible, the features of the language
should be used to best advantage.

316 Using Other Languages

{ Rep definition for intset_100. }

type
   intset_100 = **record**
     size: 0..100;
     els: **array** [1..100] **of integer**
     end;

   s100_insert_exceptions = (normal, noroom);

(∗ The rep invariant is
   r.els(1), ..., r.els(r.size) contain no duplicates
The abstraction function is
   A(r) = {r.els[i] | 1 <= i <= r.size} ∗)

{ Operations of intset_100. }

procedure intset_100_init (**var** s: intset_100);
   begin
     s.size := 0
   end { intset_100_init };

function intset_100_insert (**var** s: intset_100; x: **integer**):
       intset_100_insert_exceptions;
   var i: **integer**;
   not_found: **boolean**;
   begin
     i := 1;
     not_found := **true**;
     while (i <= s.size) **and** not_found **do**
       if s.els[i] = x **then** not_found := **false else** i := i + 1;
     if not_found
       then if s.size < 100
         then begin s.size := s.size + 1;
           s.els[s.size] := x;
           intset_100_insert := normal
         end
         else intset_100_insert := noroom
       else intset_100_insert := normal
   end { intset_100_insert };

Figure 15.2 Implementation of intset_100 (continues on next page).

For example, Ada provides good support for data abstractions, exceptions, and polymorphic abstractions. For Ada we would proceed as follows:

1. Use both functions and procedures and define their interfaces in terms of the calling mechanisms of Ada (**in**, **out**, and **inout**).

2. Use Ada exceptions, but require (by convention) that callers catch all exceptions listed in the interface plus *failure*, and signal *failure* explicitly

```
procedure intset_100_delete (var s: intset_100; x: integer);
   var i: integer;
   not_found: boolean;
   begin
      i := 1;
      not_found := true;
      while (i <= s.size) and not_found do
         if not(s.els[i] = x) then i := i + 1
            else begin
               s.els[i] := s.els[s.size];
               s.size := s.size − 1;
               not_found := false
            end
   end { intset_100_delete };

function intset_100_size (var s: intset_100): integer;
   begin
      intset_100_size := s.size
   end { intset_100_size };

function intset_100_member (var s: intset_100; x: integer): boolean;
   var i: integer;
   found: boolean;
   begin
      i := 1;
      found := false;
      while (i <= s.size) and not(found) do
         if s.els[i] = x then found := true else i := i + 1;
      intset_100_member := found
   end { intset_100_member };

{ End of the intset_100 implementation. }
```

Figure 15.2 (continued)

in all places where it would be signaled implicitly (or explicitly) in CLU.
This convention is necessary because Ada propagates unhandled exceptions automatically; in the absence of such a convention, there is no way
to guarantee that a procedure signals only those exceptions listed in its
interface.

3. Implement data abstractions with packages and private or limited private types.

4. Replace iterators with generators. A generator for a type would be
implemented in the same package that implements the type.

5. Implement parameters with generics.

6. Assemble programs using the Ada programming environment.

s100_elems = **generator type is** s100_elems_init, s100_elems_next,
                    s100_elems_done

Overview

   s100_elems is a generator for intset_100 objects.

Operations

   procedure s100_elems_init (**var** s: intset_100, **var** g: s100_elems)
         modifies g
         effects Initializes g to be a generator that can be used to iterate
            over the elements of s.

   function s100_elems_done (**var** s: intset_100, **var** g: s100_elems): boolean

            requires g is a generator over s, and s has not been modified since g
               was created.
            effects Returns true if all elements of s have been produced by
               previous calls of *s100_elems_next*; otherwise returns false.

   function s100_elems_next (**var** s: intset_100, **var** g: s100_elems): integer
            requires g is a generator over s, and s has not been modified since g
               was created, and not all elements of s have been "yielded."
            modifies g
            effects Returns an arbitrary element of s that has not been
               produced previously, and modifies g to record the production of
               this element.

end s100_elems

Figure 15.3 Specification of s100_elems generator.

   In Ada, as in Pascal, we must choose whether a particular data abstraction will have storage allocated for its objects on the stack or the heap. The
abstraction must use the heap if the size of its objects varies dynamically;
otherwise it can use the stack.
   Types using the heap differ substantially from those that use the stack.
A type using the heap has create operations, and its operations take their
arguments by value. It may also have a destroy operation if the language
in use does not have garbage collection (as in the case of Pascal); this
operation is not needed if the language has garbage collection (as in the case
of Ada). A type using the stack does not have these operations, but instead
has initialization operations, and its operations take their arguments by
reference (or by value/result in the case of Ada). (Initialization operations
are not needed if the language allows type definitions to have parameters
and allows these parameters to be used to initialize the newly allocated
storage. Ada allows this to a limited extent.) An additional point is that

{ Rep definitions for intset_100 and s100_elems. }

type
    intset_100 = **record**
        size: 0..100;
        els: **array** [1..100] **of integer**
        end;

    s100_elems = 0..100;

{ Operations of s100_elems }

procedure s100_elems_init (**var** s: intset_100; **var** g: s100_elems);
    begin
        g := 0
    end { s100_elems_init };

function s100_elems_done (**var** s: intset_100; **var** g: s100_elems): **boolean**;
    begin
        s100_elems_done := (g = s.size)
    end { s100_elems_done };

function s100_elems_next (**var** s: intset_100; **var** g: s100_elems): **integer**;
    begin
        g := g + 1;
        s100_elems_next := s.els[g]
    end { s100_elems_next };

Figure 15.4 Implementation of s100_elems.

the meaning of assignment is different in the two cases; assignment makes a copy for a type using the stack, but causes sharing if the heap is used. Also, the generators for heap-based types are different from those for stack-based types.

The decision whether a type uses the heap or the stack is made when it is designed, and is reflected in its specification, which should say explicitly which approach is intended. The implementation of a type must provide heap or stack allocation as indicated by the specification; it is an error not to provide the specified kind of allocation.

The difference between these two kinds of types also affects parameterized types. A parameterized collection can contain the objects passed to it as arguments if these objects are in the heap; however, in most languages it must contain copies of the objects if they are in the stack. This is because most languages do not support explicit pointers to locations in the stack. (Algol68 is an exception here.) If the collection contains the actual objects, then changes to those objects will be visible when they are accessed from the collection; if the collection contains copies, changes will not be visible.

queue = **data type** [etype: **type**] **is** queue[etype]_create, queue[etype]_enq,
     queue[etype]_deq, ...

Requires etype is a stack-based type.

Overview

   Queue[etype] is a heap-based type that provides access to its elements in
   first-in/first-out order.

Operations

   function queue[etype]_create ( ): queue[etype]
       effects Returns a new, empty queue.

   procedure queue[etype]_enq (q: queue[etype]; **var** e: etype)
       modifies s
       effects Enqueues a copy of e as the last element of q.

   procedure queue[etype]_deq (q: queue[etype]; **var** e: etype)
       requires q is not empty.
       modifies q
       effects Removes the first element of q and copies it into e.

   . . .

end queue[etype]

Figure 15.5 A parameterized type.

Since this is a substantial difference, it is not possible to define a parameterized type that can take either a stack-based type or a heap-based type as
its parameter. Instead each parameterized type must state in its requires
clause whether the parameter type is stack or heap based.

   An example of a specification of a parameterized type is given in figure 15.5. The naming conventions are those of chapter 7; instantiation is
to be done by text editing. Here the parameterized type is an unbounded
queue, and therefore its objects reside in the heap. However, it is intended
that the parameter type be stack based, as stated in the requires clause.
Note that since the parameter type is stack based, arguments of this type
are passed to queue operations by reference.

   Whether types are heap or stack based, in most languages it will be
necessary to design conventions for implementing them. In designing these
conventions, there is one major requirement: *To reimplement a type it
should not be necessary to modify the source code of any modules that use
the type.* Both methods shown for Pascal support this requirement.

   There are several other criteria to consider. First, the chosen method
and conventions must be *easy to use*. Making a method easy to use not

only increases the probability that it will be followed, but also decreases the likelihood that errors will arise from its misuse. It may be necessary to provide some support—for example, macros or a preprocessor—to satisfy this goal.

A second important criterion is *modifiability*. If possible, reimplementation of a data abstraction should not require recompilation of any using modules. Although violation of this criterion is not disastrous, the amount of recompilation needed in its absence can be substantial and can make program maintenance difficult and costly. Moreover, if programmers must remember what modules to recompile, the method will be hard to use and error prone. Some sort of program development aid that automatically detects the need for recompilation would help here.

Two additional criteria, *type-safety* and *protection*, are desirable, but can be difficult to achieve in some languages. Type-safety means that objects belonging to different types are kept distinct: An object of one type is never used in a context where some other type is required. For example, it should not be possible to pass a polynomial to a set operation. CLU supports security by compile-time type checking, and so do the Pascal and Ada methods. In other languages, compile-time checking may not be possible. For example, in PL/I objects of a type using the heap can be represented by untyped pointers; that is, the type of storage pointed to is not reflected in the type of the pointer. For example, the type of an intset would simply be **ptr**. Since the type of every abstract object allocated on the heap appears to be the same, violations of type-safety cannot be recognized by the PL/I compiler. The best that can be done is to implement runtime checking, but this may be too expensive.

Protection means that access to an object's representation is possible only from inside an implementation of the object's type. CLU supports protection by limiting the use of **up**, **down**, and **cvt** to clusters, and by defining them as providing conversions between that cluster's data abstraction and its rep. Ada also provides protection if data abstractions are implemented using private or limited private types. Note that in the absence of protection it is not possible to guarantee local control for a type's implementation, and therefore reasoning about its correctness is no longer local. Unfortunately, protection is very hard to achieve unless the language provides an encapsulation mechanism. If it does not, it is better to use management techniques such as having people read one another's code to enforce protection.

15.3 Summary

When using our programming method with some language other than CLU, there will be areas in which the language does not fit the method very well. In such a case we offer two suggestions:

1. Adapt the methodology to the language—for example, by changing the form of procedural abstractions.

2. Invent conventions for implementing abstractions in the language.

By changing the methodology, we ensure that the programs we design are suitable for implementation in the language. The conventions then ensure that implementations proceed in a straightforward way.

Choosing the right language for a programming project is important. Some languages make it much easier to do modular design and implementation than others. The methodology is more important than the programming language, though, and it can be used with any language. It is the programming method, not the language, that provides the power of modularity and abstraction.

Further Reading

Shaw, Mary, 1984. Abstraction techniques in modern programming languages. *IEEE Software* 1(4): 10–26.

Exercises

15.1   Specify and implement in Pascal a bounded first-in/first-out queue with room for 100 elements. Be sure to provide a generator to iterate over the elements of the queue. Use a stack-based implementation.

15.2   Define a technique for using the methodology for some language other than Pascal or CLU. Do an example program for each part of the method. Be sure to include exceptions, an iterator, a heap data type, and a stack data type.

15.3   Discuss the relative advantages and disadvantages of stack-based and heap-based approaches to providing data abstraction.

Appendix A
CLU Reference Manual

For a nominal fee universities can obtain an implementation of CLU by writing to the authors. At present there are implementations for the DEC20 and DEC Vax and Motorola 68000–based Unix systems.

A.1 Syntax

We use an extended BNF grammar to define the syntax. The general form of a production is

nonterminal ::= alternative | alternative | . . . | alternative

The following extensions are used:

a, . . .   a list of one or more *a*s separated by commas: "a" or "a, a" or "a, a, a", etc.

{ a }   a sequence of zero or more *a*s: " " or "a" or "a a", etc.

[a]   an optional *a*: " " or "a".

Nonterminal symbols appear in lightface. Reserved words appear in boldface. All other terminal symbols are nonalphabetic and appear in lightface:

```
module     ::=  { equate } procedure
           |    { equate } iterator
           |    { equate } cluster

procedure  ::=  idn = proc [ parms ] args [ returns ] [ signals ] [ where ]
                    routine_body
                end idn

iterator   ::=  idn = iter [ parms ] args [ yields ] [ signals ] [ where ]
                    routine_body
                end idn
```

cluster   ::= idn = **cluster** [parms] **is** idn , . . . [where]
                      cluster_body
                      end idn

parms   ::= [parm , . . .]

parm   ::= idn , . . . : **type** | idn , . . . : type_spec

args   ::= ([decl , . . .])

decl   ::= idn , . . . : type_spec

returns   ::= **returns** (type_spec , . . .)

yields   ::= **yields** (type_spec , . . .)

signals   ::= **signals** (exception , . . .)

exception   ::= name [(type_spec , . . .)]

where   ::= **where** restriction , . . .

restriction   ::= idn **has** oper_decl , . . .
                | idn **in** type_set

type_set   ::= { idn | idn **has** oper_decl , . . . { equate } } | idn

oper_decl   ::= op_name , . . . : type_spec

op_name   ::= name [[constant , . . .]]

constant   ::= expression | type_spec

routine_body ::= { equate }
                  { own_var }
                  { statement }

cluster_body ::= { equate } **rep** = type_spec { equate }
                  { own_var }
                  routine { routine }

routine   ::= procedure | iterator

equate   **::=** idn = constant **|** idn = type_set

own_var   **::= own** decl
                | **own** idn : type_spec := expression
                | **own** decl , **. . .** := invocation

type_spec   **::=** null **|** bool **|** int **|** real **|** char **|** string **|** any **| rep | cvt**
                | array [type_spec] **|** sequence [type_spec]
                | record [field_spec , **. . .**] **|** struct [field_spec , **. . .**]
                | oneof [field_spec , **. . .**] **|** variant [field_spec , **. . .**]
                | proctype ([type_spec , **. . .**]) [returns] [signals]
                | itertype ([type_spec , **. . .**]) [yields] [signals]
                | idn [constant , **. . .**] **|** idn

field_spec   **::=** name , **. . .** : type_spec

statement   **::=** decl
                | idn : type_spec := expression
                | decl , **. . .** := invocation
                | idn , **. . .** := invocation
                | idn , **. . .** := expression , **. . .**
                | primary **.** name := expression
                | primary [expression] := expression
                | invocation
                | **while** expression **do** body **end**
                | **for** [decl , **. . .**] **in** invocation **do** body **end**
                | **for** [idn , **. . .**] **in** invocation **do** body **end**
                | **if** expression **then** body
                        { elseif expression **then** body **}**
                        [**else** body]
                        end
                | **tagcase** expression
                        tag_arm **{** tag_arm **}**
                        [**others** : body]
                        end
                | **return** [(expression , **. . .**)]
                | **yield** [(expression , **. . .**)]
                | **signal** name [(expression , **. . .**)]
                | **exit** name [(expression , **. . .**)]
                | **break**
                | **continue**
                | **begin** body **end**

| statement **resignal** name , . . .
| statement **except** { when_handler }
                                    [others_handler]
                              end

tag_arm   ::= **tag** name , . . . [(idn : type_spec)]: body

when_handler   ::= **when** name , . . . [(decl , . . .)]: body
                 | **when** name , . . . (*) : body

others_handler   ::= **others** [(idn : type_spec)]: body

body   ::= { equate }
                     { statement }

expression   ::= primary
                 | (expression)
                 | ~ expression   %6 (precedence)
                 | − expression   %6
                 | expression ** expression   % 5
                 | expression // expression   %   4
                 | expression / expression   %   4
                 | expression * expression   %   4
                 | expression ‖ expression   %   3
                 | expression + expression   %   3
                 | expression − expression   %   3
                 | expression < expression   %   2
                 | expression <= expression   %   2
                 | expression = expression   %   2
                 | expression >= expression   %   2
                 | expression > expression   %   2
                 | expression ~< expression   %   2
                 | expression ~<= expression   %   2
                 | expression ~= expression   %   2
                 | expression ~>= expression   %   2
                 | expression ~> expression   %   2
                 | expression & expression   %   1
                 | expression **cand** expression   %   1
                 | expression | expression   %   0
                 | expression **cor** expression   %   0

primary   : : = **nil** | **true** | **false**
              | int_literal | real_literal | char_literal | string_literal
              | idn
              | idn [constant , **...**]
              | primary **.** name
              | primary [expression]
              | invocation
              | type_spec${ field , **...** }
              | type_spec$[**[** expression : **]** **[** expression , **...** **]**]
              | type_spec$name **[** [constant , **...**] **]**
              | **force** [type_spec]
              | **up** (expression)
              | **down** (expression)

invocation   : : = primary (**[** expression , **...** **]**)

field   : :   name , **...** : expression

A.2 Lexical Considerations

A module is written as a sequence of tokens and separators. A *token* is
a sequence of "printing" ASCII characters (octal value 40 through 176)
representing a reserved word, an identifier, a literal, an operator, or a
punctuation symbol. A *separator* is a "blank" character (space, vertical
tab, horizontal tab, carriage return, newline, form feed) or a comment. In
general, any number of separators may appear between tokens. Tokens and
separators are described in more detail in the following sections.

A.2.1 Reserved Words

The following character sequences are reserved words:

any	array	begin	bool	break
cand	char	cluster	continue	cor
cvt	do	down	else	elseif
end	except	exit	false	for
force	has	if	in	int
is	iter	itertype	nil	null
oneof	others	own	proc	proctype

real	record	rep	resignal	return
returns	sequence	signal	signals	string
struct	tag	tagcase	then	true
type	up	variant	when	where
while	yield	yields		

Upper and lowercase letters are not distinguished in reserved words. For example, "end," "END," and "eNd" are all the same reserved word. Reserved words appear in boldface in this document, except for names of types.

A.2.2 Identifiers

An *identifier* is a sequence of letters, digits, and underscores that begins with a letter or underscore and that is not a reserved word. As in reserved words, upper- and lowercase letters are not distinguished in identifiers.

In the syntax there are two different nonterminals for identifiers. The nonterminal *idn* is used when the identifier has scope (see section A.4.1); idns are used for variables, parameters, and module names, and as abbreviations for constants. The nonterminal *name* is used when the identifier is not subject to scope rules; names are used for record and structure selectors, oneof and variant tags, operation names, and exceptional condition names.

A.2.3 Literals

There are literals for naming objects of the built-in types null, bool, int, real, char, and string. Their forms are discussed in section A.3.

A.2.4 Operators and Punctuation Symbols

The following character sequences are used as operators and punctuation symbols:

(	:	"	<	~<	+	‖
)	:=	'	<=	~<=	−	**
{	,	\	=	~=	*	//
}	.		>=	~>=	/	&
[	$		>	~>		|
]				~		

A.2.5 Comments and Other Separators

A *comment* is a sequence of characters that begins with a percent sign (%), ends with a newline character, and contains only printing ASCII characters and horizontal tabs in between.

A *separator* is a blank character (space, vertical tab, horizontal tab, carriage return, newline, form feed) or a comment. Zero or more separators may appear between any two tokens, except that at least one separator is required between any two adjacent non-self-terminating tokens: reserved words, identifiers, integer literals, and real literals.

A.3 Type Specifications

Within a program, a type is specified by a syntactic construct called a *type_ spec*. The type specification for a type with no parameters is just the identifier (or reserved word) naming the type. For parameterized types, the type specification consists of the identifier (or reserved word) naming the parameterized type, together with the parameter values.

In addition to the built-in types (null, bool, int, real, char, string, and any) and the built-in type classes (array, sequence, record, struct, variant, oneof, proctype, and itertype), there are also type_ specs for user-defined types. These have the form

idn **[** [constant , **. . .**] **]**

where *idn* names the user_defined type, each *constant* must be computable at compile time (see section A.4.3), and constants of the appropriate types and number must be supplied in the order specified by the type.

There are three special type specifications that are used for implementing new abstractions: **rep, cvt**, and **type**. These forms are discussed in sections A.8.3 and A.8.4. Within a module, formal parameters declared with **type** can be used as type specifications.

Finally, identifiers that have been equated to type specifications can also be used as type specifications. Equates are discussed in section A.4.3.

Specifications for the built-in types and type classes are given in section A.9.

A.4 Scopes, Declarations, and Equates

We now describe how to introduce and use constants and variables, and the scope of constant and variable names.

A.4.1 Scoping Units

Scoping units follow the nesting structure of statements. Generally, a scoping unit is a body and an associated "heading." The scoping units are

1. from the start of a module to its end,

2. from a **cluster**, **proc**, or **iter** to the matching **end**,

3. from a **for**, **do**, or **begin** to the matching **end**,

4. from a **then** or **else** in an **if** statement to the end of the corresponding body,

5. from a **tag** or **others** in a **tagcase** statement to the end of the corresponding body,

6. from a **when** or **others** in an **except** statement to the end of the corresponding body, and

7. from the start of a type_ set to its end.

In this section we discuss only cases 1–6; the scope in a *type_ set* is discussed in section A.8.4.

The structure of scoping units is such that if one scoping unit overlaps another scoping unit (textually), then one is fully contained in the other. The contained scope is called a *nested* scope, and the containing scope is called a *surrounding* scope.

New constant and variable names may be introduced in a scoping unit. Names for constants are introduced by equates, which are syntactically restricted to appear grouped together at or near the beginning of scoping units. For example, equates may appear at the beginning of a body, but not after any statements in the body. By contrast, declarations, which introduce new variables, are allowed wherever statements are allowed, and hence may appear throughout a scoping unit.

In the syntax there are two distinct nonterminals for identifiers: *idn* and *name*. Any identifier introduced by an equate or declaration is an idn, as is the name of the module being defined and any operations the module has. An idn refers to a specific type or object. A name generally refers to a piece of something and is always used in context; for example, names are used as record selectors. The scope rules apply only to idns.

The scope rules are as follows:

1. An idn may not be redefined in its scope.

2. Any idn that is used as an external reference in a module may not be used for any other purpose in that module.

Thus, unlike other "block-structured" languages, CLU prohibits the redefinition of an identifier in a nested scope. An identifier used as an external reference must name a module or constant in its program.

A.4.2 Declarations

Declarations introduce new variables. The scope of a variable is from its declaration to the end of the smallest scoping unit containing its declaration; hence variables must be declared before use.

There are two sorts of declarations: those with initialization and those without. Simple declarations (those without initialization) take the form

decl $::=$ idn, . . . : type_spec

A simple declaration introduces a list of variables, all having the type given by *type_ spec*. This type determines the types of objects that can be assigned to the variable. The variables introduced in a simple declaration initially denote no objects; that is, they are uninitialized. Attempts to use uninitialized variables (if not detected at compile time) cause the runtime exception

failure("uninitialized variable")

A declaration with initialization combines declarations and assignments into a single statement. It is equivalent to one or more simple declarations followed by an assignment statement. The two forms of declaration with initialization are

idn : type_ spec := expression

and

$decl_1$, . . . , $decl_n$:= invocation

These are equivalent to (respectively)

idn : type_ spec
idn := expression

and

$decl_1$ . . . $decl_n$ % declaring idn_1 . . . idn_m
idn_1, . . . , idn_m := invocation

In the second form, the order of the idns in the assignment statement is the same as in the original declaration with initialization. (The invocation must return m objects.)

A.4.3 Equates and Constants

An equate allows a single identifier to be used as an abbreviation for a constant that may have a lengthy textual representation. An equate also permits a mnemonic identifier to be used in place of a commonly used constant, such as a numerical value. We use the term "constant" in a very narrow sense here: Constants, in addition to being immutable, must be computable at compile time. Constants are either types (built-in or user-defined), or objects that are the results of evaluating constant expressions. (Constant expressions are defined later.)

The syntax of equates is

equate ::= idn = constant | idn = type_set
constant ::= type_spec | expression

This section describes only the first form of equate; discussion of *type_ sets* is deferred to section A.8.4.

An equated identifier may be used as an expression. The value of such an expression is the constant to which the identifier is equated. An equated identifier cannot be used on the left-hand side of an assignment statement.

The scope of an equated identifier is the smallest scoping unit surrounding the equate defining it; here we mean the entire scoping unit, not just the portion after the equate. All the equates in a scoping unit must appear grouped near the beginning of the scoping unit. The exact placement of equates depends on the containing syntactic construct; usually equates appear at the beginnings of bodies.

Equates may be in any order within the group. Forward references among equates in the same scoping unit are allowed, but cyclic dependencies are illegal. For example,

x = y
y = z
z = 3

is a legal sequence of equates, but

x = y
y = z
z = x

is not. Since equates introduce idns, the scoping restrictions on idns apply (that is, the idns may not be defined more than once).

Identifiers may be equated to type specifications, thus giving abbreviations for type names. Since equates cannot have cyclic dependencies, though, directly recursive type specifications cannot be written. This does not prevent the definition of recursive types, since we can use clusters to write them.

A *constant expression* is an expression that can be evaluated at compile time to produce an immutable object of a built-in type. Specifically this includes

1. literals,

2. identifiers equated to constants,

3. formal parameters (see section A.8.4),

4. procedure and iterator names, including **force**[*t*] for any type *t*, and

5. invocations of procedure operations of the built-in constant types, provided that all operands and all results are constant expressions; however, we explicitly forbid the use of formal parameters as operands to invocations in constant expressions, since the values of formal parameters are not known at compile time.

The built-in constant types are null, int, real, bool, char, string, sequence types, oneof types, structure types, procedure types, and iterator types. Any invocation in a constant expression must terminate normally; a program is illegal if evaluation of any constant expression would signal an exception. Illegal programs will not be executed.

A.5 Assignment and Invocation

Two fundamental actions of CLU are assignment of computed objects to variables and invocation of procedures (and iterators) to compute objects. Other actions are composed from these two by using various control flow mechanisms. Since the correctness of assignments and invocations depends on a type-checking rule, we describe that rule first, then assignment, and finally invocation.

A.5.1 Type Inclusion

CLU is designed to allow complete compile-time type checking. The type of each variable is known by the compiler. Furthermore, the type of object that could result from the evaluation of any expression (invocation) is known at compile time. Hence every assignment can be checked at compile time to make sure that the variable is only assigned objects of its declared type. The rule is that an assignment $v := E$ is legal only if the set of objects defined by the type of E (loosely, the set of all objects that could possibly result from evaluating the expression) is included in the set of all objects that could be denoted by v.

Instead of speaking of the set of objects defined by a type, we generally speak of the type and say that the type of the expression must be *included in* the type of the variable. If it were not for the type any, the inclusion rule would be an equality rule. This leads to a simple interpretation of the type inclusion rule: *The type of a variable being assigned an expression must be either the type of the expression or the type any.*

A.5.2 Assignment

Assignment is the means of causing a variable to denote an object. The simplest form of assignment is

   idn := expression

In this case the expression is evaluated and the resulting object is assigned to the variable. The expression must return a single object (whose type must be included in that of the variable).

There are two forms of assignment that assign to more than one variable at the same time:

   idn , . . . := expression , . . .

and

   idn , . . . := invocation

The first form of multiple assignment is a generalization of simple assignment. The first variable is assigned the first expression, the second variable the second expression, and so on. The expressions are all evaluated (from left to right) before any assignments are performed. The number of variables in the list must equal the number of expressions, no variable may occur more than once, and the type of each variable must include the type of the corresponding expression. There is no form of this statement with declarations.

The second form of multiple assignment allows retention of the objects resulting from an invocation returning two or more objects. The first variable is assigned the first object, the second variable the second object, and so on. The order of the objects is the same as in the **return** statement of the invoked routine. The number of variables must equal the number of objects returned, no variable may occur more than once, and the type of each variable must include the corresponding return type of the invoked procedure.

The assignment symbol := is used in two other syntactic forms that are not true assignments, but rather abbreviations for certain invocations. These forms are used for updating collections such as records and arrays (see section A.7.2).

A.5.3 Invocation

Invocation is the other fundamental action of CLU. In this section we discuss procedure invocation; iterator invocation is discussed in section A.7.6. However, up to and including passing of arguments, the two are the same.

Invocations take the form

primary ([expression , . . .   ])

A *primary* is a slightly restricted form of expression, which includes variables and routine names, among other things (see the next section).

The sequence of activities in performing an invocation is as follows:

1. The primary is evaluated. It must evaluate to a procedure or iterator.

2. The expressions are evaluated, from left to right.

3. New variables are introduced corresponding to the formal arguments of the routine being invoked (that is, a new environment is created for the invoked routine to execute in).

4. The objects resulting from evaluating the expressions (the actual arguments) are assigned to the corresponding new variables (the formal arguments). The first formal is assigned the first actual, the second formal the second actual, and so on. The type of each expression must be included in the type of the corresponding formal argument.

5. Control is transferred to the routine at the start of its body.

An invocation is considered legal in exactly those situations where all the (implicit) assignments involved in its execution are legal.

It is permissible for a routine to assign an object to a formal argument variable; the effect is just as if that object were assigned to any other variable. From the point of view of the invoked routine, the only difference between its formal argument variables and its other local variables is that the formals are initialized by its caller.

Procedures can terminate in two ways: *normally*, returning zero or more objects, or *exceptionally*, signaling an exception. When a procedure terminates normally, the result objects become available to the caller and will (usually) be assigned to variables or passed as arguments to other routines. When a procedure terminates exceptionally, the flow of control will not go to the point of return of the invocation, but rather will go to an exception handler, as described in section A.7.13.

A.6 Expressions

An expression evaluates to an object in the CLU universe. This object is said to be the *result* or *value* of the expression. Expressions are used to name the object to which they evaluate. The simplest expressions are literals, variables, parameters, and routine names. These forms directly name their result object. More complex expressions are generally built up out of nested procedure invocations. The result of such an expression is the value returned by the outermost invocation.

Like many other languages, CLU has prefix and infix operators for the common arithmetic and comparison operations and uses the familiar syntax for array indexing and record component selection (for example, $a[i]$ and $r.s$). However, in CLU these notations are considered to be abbreviations for procedure invocations. This allows built-in types and user-defined types to be treated as uniformly as possible, and also allows the programmer to use familiar notation when appropriate.

In addition to invocation, four other forms are used to build complex expressions out of simpler ones. These are the conditional operators **cand** and **cor** (see section A.6.10), and the type conversion operations **up** and **down** (see section A.6.12).

There is a syntactically restricted form of expression called a *primary*. A primary is any expression that does not have a prefix or infix operator, or parentheses, at the top level. In certain places the syntax requires a primary rather than a general expression. This has been done to increase the readability of the resulting programs.

As a general rule, procedures with side effects should not be used in expressions, and programs should not depend on the order in which expressions are evaluated. However, to avoid surprises, the subexpressions of any expression are evaluated from left to right.

A.6.1 Literals

Integer, real, character, string, boolean, and null literals are expressions. The syntax for literals is given in the sections describing these types. The type of a literal expression is the type of the object named by the literal.

A.6.2 Variables

Variables are identifiers that denote objects of a given type. The type of a variable is the type given in the declaration of that variable and determines which objects may be denoted by the variable.

A.6.3 Parameters

Parameters are identifiers that denote constants supplied when a parameterized module is instantiated (see section A.8.4). The type of a parameter is the type given in the declaration of that parameter. Parameters of type **type** cannot be used as expressions.

A.6.4 Equated Identifiers

Equated identifiers denote constants. The type of an equated identifier is the type of the constant that it denotes. Identifiers equated to types and type_sets cannot be used as expressions.

A.6.5 Procedure and Iterator Names

Procedures and iterators may be defined either as separate modules or within a cluster. Those defined as separate modules are named by expressions of the form

idn **[** [constant , **. . .**] **]**

The optional constants are the parameters of the procedure or iterator abstraction. (Constants are discussed in section A.4.3.)

When a procedure or iterator is defined as an operation of a type, the type name must be part of the name of the routine. The form for naming an operation of a type is

type_spec$name **[** [constant , **. . .**] **]**

The type of a procedure or iterator name is just the type of the named routine.

A.6.6 Procedure Invocations

Procedure invocations have the form

primary (**[** expression , **. . .**])

The primary is evaluated to obtain a procedure object, and then the expressions are evaluated left to right to obtain the argument objects. The procedure is invoked with these arguments, and the object returned is the result of the entire expression. For a more detailed discussion see section A.5.3.

Any procedure invocation $P(E_1, \ldots, E_n)$ must satisfy two constraints: The type of P must be of the form

proctype $(T_1, \ldots, T_n)$ **returns** (R) **signals** $(\ldots)$

and the type of each expression E_i must be included in the corresponding type T_i. The type of the entire invocation expression is given by R.

Procedures can also be invoked as statements (see section A.7.1).

A.6.7 Selection Operations

Arrays, sequences, records, and structures are collections of objects. Selection operations provide access to the individual elements or components of the collection. Simple notations are provided for invoking the *fetch* and *store* operations of array types, the *fetch* operation of sequence types, the *get* and *set* operations of record types, and the *get* operations of structure types. In addition, these short forms may be used for user-defined types with the appropriate properties.

An element selection expression has the form

primary [expression]

This form is a short form for an invocation of a *fetch* operation and is completely equivalent to

T$fetch(primary, expression)

where T is the type of *primary*. For example, if *a* is an array of integers, then

a[27]

is completely equivalent to the invocation

array[int]$fetch(a, 27)

The expression is legal whenever the corresponding invocation is legal. In other words, T (the type of *primary*) must provide a procedure operation named *fetch*, which takes two arguments whose types include the types of *primary* and *expression*, and which returns a single result.

The use of *fetch* for user-defined types should be restricted to types with arraylike behavior. Objects of such types will contain (along with other information) a collection of objects, where the collection can be indexed in some way. For example, it might make sense for an associative memory type to provide a *fetch* operation to access the value associated with a key. *Fetch* operations are intended for use in expressions; thus they should never have side effects.

The component selection expression has the form

   primary . name

This form is short for an invocation of a *get_name* operation and is completely equivalent to

   T$get_*name*(primary)

where T is the type of *primary*. For example, if x has type RT = record[first: int, second: real], then

   x.first

is completely equivalent to

   RT$get_first(x)

The expression is legal whenever the corresponding invocation is legal. In other words, T (the type of *primary*) must provide a procedure operation named *get_ name*, which takes one argument whose type includes the type of *primary* and returns a single result.

The use of *get* operations for user-defined types should be restricted to types with recordlike behavior. Objects of such types will contain (along with other information) one or more named objects. For example, it might make sense for a type that implements channels to files to provide a get_author operation, which returns the name of the file's creator. *Get* operations are intended for use in expressions; thus they should never have side effects.

A.6.8 Constructors

Constructors are expressions that enable users to create and initialize arrays, sequences, records, and structures. Constructors are not provided for user-defined types.

An array constructor has the form

   type_ spec$[[expression:] [expression , . . .]]

The type specification must name an array type: array[T]. This is the type of the constructed array. The expression preceding the colon must evaluate to an integer and becomes the low bound of the constructed array. If this expression is omitted, the low bound is 1. The expressions following the colon are evaluated to obtain the elements of the array. They correspond (from left to right) to the indexes *low_bound, low_bound*+1, *low_bound*+2, ... For an array of type array[T], the type of each element expression in the constructor must be included in T.

An array constructor is equivalent to an array *create* operation, followed by a number of array *addh* operations. However, such a sequence of operations cannot be written as an expression.

A sequence constructor has the form

   type_spec$[[expression , . . .]]

The type specification must name a sequence type: sequence[T]. This is the type of the constructed sequence. The expressions are evaluated to obtain the elements of the sequence. They correspond (from left to right) to the indexes 1, 2, 3, ... For a sequence of type sequence[T], the type of each element expression in the constructor must be included in T.

A sequence constructor is equivalent to a sequence *new* operation, followed by a number of sequence *addh* operations.

A record constructor has the form

   type_spec${ field , . . . }

where

   field ::= name , . . . : expression

Whenever a field has more than one name, it is equivalent to a sequence of fields, one for each name. Thus the following two constructors are equivalent:

   R${a, b: 7,   c: 9}
   R${a: 7,   b: 7,   c: 9}

where

   R = record[a: int, b: int, c: int]

In a record constructor, the type specification must name a record type: record[S_1: T_1, ..., S_n: T_n]. This will be the type of the constructed record. The component names in the field list must be exactly the names

$S_1, \ldots, S_n$, although these names may appear in any order. The expressions are evaluated from left to right, and there is one evaluation per component name even if several component names are grouped with the same expression. The type of the expression for component S_i must be included in T_i. The results of these evaluations form the components of a newly constructed record. This record is the value of the entire constructor expression.

A structure constructor has the form

type_spec${ field , . . . }

where (as for records)

field ::= name , . . . : expression

Whenever a field has more than one name, it is equivalent to a sequence of fields, one for each name.

In a structure constructor, the type specification must name a structure type: struct[S_1: T_1, ..., S_n: T_n]. This will be the type of the constructed structure. The component names in the field list must be exactly the names $S_1, \ldots, S_n$, although these names may appear in any order. The expressions are evaluated from left to right, and there is one evaluation per component name even if several component names are grouped with the same expression. The type of the expression for component S_i must be included in T_i. The results of these evaluations form the components of a newly constructed structure. This structure is the value of the entire constructor expression.

A.6.9 Prefix and Infix Operators

CLU allows prefix and infix notation to be used as a shorthand for several operations. Table A.1 shows the short form and the equivalent expanded form for each such operation. Here T is the type of the first operand.

Operator notation is used most heavily for the built-in types, but may be used for user-defined types as well. When these operations are provided for user-defined types, they should always be side-effect-free, and they should mean roughly the same thing as they do for the built-in types. For example, the comparison operations should only be used for types that have a natural partial or total order. Usually the comparison operations (*lt, le, equal, ge, gt*) will be of type

proctype (T, T) **returns** (**bool**)

Table A.1 Short forms for operations.

Shorthand Form	Expansion
$expr_1$ ** $expr_2$	T\$power($expr_1$, $expr_2$)
$expr_1$ // $expr_2$	T\$mod($expr_1$, $expr_2$)
$expr_1$ / $expr_2$	T\$div($expr_1$, $expr_2$)
$expr_1$ * $expr_2$	T\$mul($expr_1$, $expr_2$)
$expr_1$ || $expr_2$	T\$concat($expr_1$, $expr_2$)
$expr_1$ + $expr_2$	T\$add($expr_1$, $expr_2$)
$expr_1$ − $expr_2$	T\$sub($expr_1$, $expr_2$
$expr_1$ < $expr_2$	T\$lt($expr_1$, $expr_2$)
$expr_1$ <= $expr_2$	T\$le($expr_1$, $expr_2$)
$expr_1$ = $expr_2$	T\$equal($expr_1$, $expr_2$)
$expr_1$ >= $expr_2$	T\$ge($expr_1$, $expr_2$)
$expr_1$ > $expr_2$	T\$gt($expr_1$, $expr_2$)
$expr_1$ ~< $expr_2$	~($expr_1$ < $expr_2$)
$expr_1$ ~<= $expr_2$	~($expr_1$ <= $expr_2$)
$expr_1$ ~= $expr_2$	~($expr_1$ = $expr_2$)
$expr_1$ ~>= $expr_2$	~($expr_1$ >= $expr_2$)
$expr_1$ ~> $expr_2$	~($expr_1$ > $expr_2$)
$expr_1$ & $expr_2$	T\$and($expr_1$, $expr_2$)
$expr_1$ | $expr_2$	T\$or($expr_1$, $expr_2$)
−expr	T\$minus(expr)
~expr	T\$not(expr)

The other binary operations (for example, *add* and *sub*) will be of type

proctype (T, T) **returns** (T) **signals** (...)

and the unary operations will be of type

proctype (T) **returns** (T) **signals** (...)

A.6.10 Cand and Cor

Two additional binary operators are provided. These are the conditional *and* operator, **cand**, and the conditional *or* operator, **cor**.

expression₁ **cand** expression₂

is the boolean *and* of *expression₁* and *expression₂*. However, if *expression₁* is false, *expression₂* is never evaluated.

expression₁ **cor** expression₂

is the boolean *or* of *expression₁* and *expression₂*, but *expression₂* is not

evaluated unless *expression*$_1$ is false. For both **cand** and **cor**, *expression*$_1$
and *expression*$_2$ must have type bool.

Because of the conditional expression evaluation involved, uses of **cand**
and **cor** are not equivalent to any procedure invocation.

A.6.11 Precedence

When an expression is not fully parenthesized, the proper nesting of subexpressions may be ambiguous. Precedence rules are used to resolve such
ambiguity. The precedence of each infix operator is given below—note that
higher precedence operations are performed first and that prefix operators
always have precedence over infix operators:

Precedence	Operators
5	**
4	* / //
3	+ − ||
2	< <= = >= > ~< ~<= ~= ~>= ~>
1	& **cand**
0	| **cor**

The order of evaluation for operators of the same precedence is from left
to right, except for **, which is from right to left. The following examples
illustrate the precedence rules:

Expression	Equivalent Form
a + b // c	a + (b // c)
a + b − c	(a + b) − c
a + b ** c ** d	a + (b ** (c ** d))
a = b | c = d	(a = b) | (c = d)
−a * b	(−a) * b

A.6.12 Up and Down

There are no implicit type conversions in CLU. Two forms of expression
exist for explicit conversions. These are

up (expression)
down (expression)

Up and **down** may be used only within the body of a cluster operation.
Up changes the type of the expression from the rep type of the cluster
to the abstract type. **Down** converts the type of the expression from the

abstract type to the rep type. These conversions are explained further in
section A.8.3.

A.6.13 Force

CLU has a single built-in procedure generator called **force**. **Force** takes
one type parameter and is written

force [type_ spec]

The procedure **force**[T] has type

proctype (any) **returns** (T) **signals** (wrong_type)

If **force**[T] is applied to an object that is included in type T, it returns
that object. If **force**[T] is applied to an object that is not included in type
T, it signals "wrong_type."

Force is a necessary companion to the type any. The type any allows
programs to pass around objects of arbitrary type. However, to do anything
substantive with an object, one must use the primitive operations of that
object's type. This raises a conflict with compile-time type checking, since
an operation can be applied only when the arguments are known to be of
the correct types. This conflict is resolved by using **force**. **Force**[T] allows
a program to check, at runtime, that a particular object is actually of type
T. If this check succeeds, the object can be used in all the ways appropriate
for objects of type T.

For example, the procedure **force**[T] allows us to write the following
code:

x: any := 3
y: int := **force**[int](x)

while the following is illegal because the type of y (int) does not include
the type of the expression x (any):

x: any := 3
y: int := x

A.7 Statements

CLU is a statement-oriented language; that is, statements are executed for
their side effects and do not return any values. Most statements are *control*
statements that permit the programmer to define how control flows through

the program. The real work is done by the *simple* statements: assignment
and invocation. Assignment has already been discussed in section A.5.2;
the invocation statement is discussed in the following. Two special statements that look like assignments but are really invocations are discussed
in section A.7.2.

The syntax of CLU is defined to permit a control statement to control
a group of equates, declarations, and statements rather than just a single
statement. Such a group is called a *body* and has the form

body ::= { equate }
{ statement }   % statements include declarations

Scope rules for bodies are discussed in section A.4.1. No special termi-
nator is needed to signify the end of a body; reserved words used in the
various compound statements serve to delimit the bodies. Occasionally it is
necessary to indicate explicitly that a group of statements should be treated
like a single statement; this is done by the **begin** statement, discussed in
section A.7.3.

A.7.1 Invocation Statement

An invocation statement invokes a procedure. Its form is the same as an
invocation expression:

primary ([expression , . . .])

The primary must evaluate to a procedure object, and the type of each
expression must be included in the type of the corresponding formal argument for that procedure. The procedure may or may not return results; if
it does return results, they are discarded.

A.7.2 Update Statements

Two special statements are provided for updating components of records
and arrays. They may also be used with user-defined types with the appropriate properties. These statements resemble assignments syntactically,
but they are really invocations.

The *element update* statement has the form

primary [$expression_1$] := $expression_2$

This form is merely a shorthand for an invocation of a *store* operation and is completely equivalent to the invocation statement

T\$store(primary, $expression_1$, $expression_2$)

where T is the type of *primary*.

The element update statement is not restricted to arrays. The statement is legal if the corresponding invocation statement is legal. In other words, T (the type of *primary*) must provide a procedure operation named *store*, which takes three arguments whose types include those of *primary*, $expression_1$, and $expression_2$, respectively.

We recommend that the use of *store* for user-defined types be restricted to types with arraylike behavior, that is, types whose objects contain mutable collections of indexable elements. For example, it might make sense for an associative memory type to provide a *store* operation for changing the value associated with a key.

The *component update* statement has the form

primary . name := expression

This form is merely a shorthand for an invocation of a *set_ name* operation and is completely equivalent to the invocation statement

T\$set_ name(primary, expression)

where T is the type of *primary*. For example, if x has type RT = record[first: int, second: real], then

x.first := 6

is completely equivalent to

RT\$set_ first(x, 6)

The component update statement is not restricted to records. The statement is legal if the corresponding invocation statement is legal. In other words, T (the type of *primary*) must provide a procedure operation called *set_ name*, which takes two arguments whose types include the types of *primary* and *expression*, respectively.

We recommend that *set_* operations be provided for user-defined types only if recordlike behavior is desired, that is, only if it is meaningful to permit selected parts of the abstract object to be modified. In general, *set_* operations should not perform any substantial computation, except possibly checking that the arguments satisfy certain constraints.

A.7.3 Begin Statement

The **begin** statement permits a sequence of statements to be grouped together into a single statement. Its form is

begin body **end**

Since the syntax already permits bodies inside control statements, the main use of the **begin** statement is to group statements together for use with the **except** statement (see section A.7.13).

A.7.4 Conditional Statement

The form of the conditional statement is

if expression **then** body
   { **elseif** expression **then** body }
   [**else** body]
   end

The expressions must be of type bool. They are evaluated successively until one is found to be true. The body corresponding to the first true expression is executed, and the execution of the **if** statement then terminates. If none of the expressions is true, the body in the **else** clause is executed (if an **else** clause exists). The **elseif** form provides a convenient way to write a multipath branch.

A.7.5 While Statement

The **while** statement has the form

while expression **do** body **end**

Its effect is to execute the body repeatedly as long as the expression remains true. The expression must be of type bool. If the value of the expression is true, the body is executed, and then the entire **while** statement is executed again. When the expression evaluates to false, execution of the **while** statement terminates.

A.7.6 For Statement

The only way an iterator can be invoked is by use of a **for** statement. The iterator produces a sequence of *items* (where an item is a group of zero or more objects) one item at a time; the body of the **for** statement is executed for each item in the sequence.

The **for** statement has the form

> **for** [idn , **. . .**] **in** invocation **do** body **end**

or

> **for** [decl , **. . .**] **in** invocation **do** body **end**

The invocation must be an iterator invocation. The *idn* form uses previously declared variables to serve as the loop variables, while the *decl* form introduces new variables, local to the **for** statement, for this purpose. In either case, the type of each variable must include the corresponding yield type of the invoked iterator.

Execution of the **for** statement proceeds as follows. First the iterator is invoked, and it either yields an item or terminates. If the iterator yields an item, its execution is temporarily suspended, the objects in the item are assigned to the loop variables, and the body of the **for** statement is executed. The next cycle of the loop is begun by resuming execution of the iterator from its point of suspension. Whenever the iterator terminates, the entire **for** statement terminates. If the **for** statement terminates, this also terminates the iterator.

A.7.7 Continue Statement

The **continue** statement has the form

> **continue**

Its effect is to terminate execution of the body of the smallest loop statement in which it appears and to start the next cycle of that loop (if any).

A.7.8 Break Statement

The **break** statement has the form

> **break**

Its effect is to terminate execution of the smallest loop statement in which it appears.

A.7.9 Tagcase Statement

The **tagcase** statement is a special statement provided for decomposing oneof and variant objects; it permits the selection of a body based on the tag of the object.

The form of the **tagcase** statement is

tagcase expression
    tag_arm **{** tag_arm **}**
    [others : body **]**
    end

where

    tag_arm **::= tag** name , ... **[** (idn: type_spec) **]** : body

The expression must evaluate to a oneof or variant object. The tag of this
object is then matched against the names on the tag_arms. When a match
is found, if a declaration (*idn: type_spec*) exists, the value component of
the object is assigned to the local variable *idn*. The matching body is then
executed; *idn* is defined only in that body. If no match is found, the body
in the **others** arm is executed.

In a syntactically correct **tagcase** statement, the following constraints
are satisfied. The type of the expression must be some oneof or variant
type T. The tags named in the tag_arms must be a subset of the tags of T,
and no tag may occur more than once. If all tags of T are present, there is
no **others** arm; otherwise an **others** arm must be present. Finally, on any
tag_arm containing a declaration (*idn: type_spec*), *type_spec* must equal
(not include) the type specified as corresponding in T to the tag or tags
named in the tag_arm.

A.7.10 Return Statement

The form of the **return** statement is

    return [(expression , ...) **]**

The **return** statement terminates execution of the containing procedure or
iterator. If the **return** statement is in a procedure, the type of each expression must be included in the corresponding return type of the procedure.
The expressions (if any) are evaluated from left to right, and the objects
obtained become the results of the procedure. If the **return** statement
occurs in an iterator, no results can be returned.

A.7.11 Yield Statement

Yield statements may occur only in the body of an iterator. The form of a **yield** statement is

   yield [(expression , . . .)]

It has the effect of suspending operation of the iterator and returning control to the invoking **for** statement. The values obtained by evaluating the expressions (from left to right) are passed to the **for** statement to be assigned to the corresponding list of identifiers. The type of each expression must be included in the corresponding yield type of the iterator. After the body of the **for** loop has been executed, execution of the iterator is resumed at the statement following the **yield** statement.

A.7.12 Signal Statement

An exception is signaled with a **signal** statement, which has the form

   signal name [(expression , . . .)]

A **signal** statement may appear anywhere in the body of a routine. The execution of a **signal** statement begins with evaluation of the expressions (if any), from left to right, to produce a list of *exception results*. The activation of the routine is then terminated. Execution continues in the caller, as described in section A.7.13.

   The exception name must be either one of the exception names listed in the routine heading or *failure*. If the corresponding exception specification in the heading has the form

   name(T_1, . . . , T_n)

there must be exactly n expressions in the **signal** statement, and the type of the ith expression must be included in T_i. If the name is *failure*, there must be exactly one expression present, of type string.

A.7.13 Except Statement

When a routine activation terminates by signaling an exception, the corresponding invocation (the text of the call) is said to *raise* that exception. By attaching *handlers* to statements, the caller can specify the action to be taken when an exception is raised.

A statement with handlers attached is called an **except** statement and
has the form

statement **except** **{** when_handler **}**
   [others_handler **]**
    end

where

when_handler **::=** **when** name , . . . **[** (decl , . . .) **]** : body
              | **when** name , . . . (*) : body

others_handler **::=** **others** **[** (idn : type_spec) **]** : body

Let S be the statement to which the handlers are attached, and let X
be the entire **except** statement. Each when_handler specifies one or more
exception names and a body. The body is executed if an exception with
one of those names is raised by an invocation in S. All the names listed in
the when_handlers must be distinct. The optional others_handler is used
to handle all exceptions not explicitly named in the when_handlers. Here
S can be any form of statement, even another **except** statement.

If, during the execution of S, some invocation in S raises an exception
E, control immediately transfers to the closest applicable handler—that is,
the closest handler for E that is attached to a statement containing the
invocation. When execution of the handler is complete, control passes to
the statement following the one to which the handler is attached. Thus if
the closest handler is attached to S, the statement following X is executed
next. If execution of S completes without raising an exception, the attached
handlers are not executed.

An exception raised inside a handler is treated the same as any other exception: Control passes to the closest handler for that exception. Note that
an exception raised in some handler attached to S cannot be handled by any
handler attached to S; either the exception is handled within the handler
or it is handled by some handler attached to a statement containing X.

We now consider the forms of handlers in more detail. The form

when name , . . . **[** (decl , . . .) **]** : body

is used to handle exceptions with the given names when the exception
results are of interest. The optional declared variables, which are local to
the handler, are assigned the exception results before the body is executed.
Every exception potentially handled by this form must have the same number of results as there are declared variables, and the types of the results

must equal (rather than include) the types of the variables. The form

when name , . . . (*) : body

handles all exceptions with the given names, regardless of the existence of exception results; any actual results are discarded. Hence exceptions with differing numbers and types of results can be handled together.

The form

others [(idn : type_ spec)] : body

is optional and must appear last in a handler list. This form handles any exception not handled by other handlers in the list. If a variable is declared, it must be of type string. The variable, which is local to the handler, is assigned a lowercase string representing the actual exception name; any results are discarded.

Note that exception results are ignored when matching exceptions to handlers; only the names of exceptions are used. Thus the following is illegal, in that int$*div* signals *zero_divide* without any results, but the closest handler has a declared variable:

begin
    y: int := 0
    x: int := 3 / y **except when** zero_divide (z: int): **return end**
    end except when zero_divide: **return end**

An invocation need not be surrounded by **except** statements that handle all potential exceptions. This policy was adopted because in many cases the programmer can prove that a particular exception will not arise. For example, the invocation int$*div*(x, 7) will never signal *zero_divide*. However, this policy does lead to the possibility that some invocation may raise an exception for which there is no handler. To avoid this situation, every routine body is contained in an implicit **except** statement of the form

    begin *routine_body* **end**
        except when failure (s: **string**): **signal** failure(s)
            others (s: **string**): **signal** failure("unhandled exception: " || s)
        end

Failure exceptions are propagated unchanged; an exception named *name* becomes

    failure("unhandled exception: *name*")

A.7.14 Resignal Statement

A **resignal** statement is a syntactically abbreviated form of exception handling:

statement **resignal** name , . . .

Each name listed must be distinct, and each must be either one of the condition names listed in the routine heading or *failure*. The **resignal** statement acts like an **except** statement containing a handler for each condition named, where each handler simply signals that exception with exactly the same results. Thus if the **resignal** clause names an exception with a specification in the routine heading of the form

name(T_1, . . . , T_n)

then effectively there is a handler of the form

when name (x_1: T_1, . . . , x_n: T_n): **signal** name(x_1, . . . , x_n)

As in an explicit handler of this form, every exception potentially handled by this implicit handler must have the same number of results as are declared in the exception specification, and the types of the results must equal the types listed in the exception specification.

A.7.15 Exit Statement

A *local* transfer of control can be effected by using an **exit** statement, which has the form

exit name [(expression , . . .)]

An **exit** statement is similar to a **signal** statement, except that where the **signal** statement *signals* an exception to the *calling* routine, the **exit** statement *raises* the exception directly in the *current* routine. An exception raised by an **exit** statement must be handled explicitly by a containing **except** statement with a handler of the form

when name , . . . [(decl , . . .)] : body

As usual, the types of the expressions in the **exit** statement must equal the types of the variables declared in the handler. The handler must be an explicit one; that is, **exit**s to the implicit handlers of **resignal** statements or to the implicit *failure* handler enclosing a routine body are illegal.

A.8 Modules

A CLU program consists of a group of modules. Three kinds of modules are provided, each representing a kind of abstraction that we have found to be useful in program construction:

module ::= { equate } procedure
       | { equate } iterator
       | { equate } cluster

Procedures support procedural abstraction, iterators support control abstraction, and clusters support data abstraction.

A module defines a new *scope*. The identifiers introduced in the equates (if any) and the identifier naming the abstraction (the *module name*) are local to that scope (and therefore may not be redefined in an inner scope). Abstractions implemented by other modules are referred to by using nonlocal identifiers.

The existence of an externally established meaning for an identifier does not preclude a local definition for that identifier. Within a module, any identifier may be used in a purely local fashion or in a purely nonlocal fashion, but no identifier may be used in both ways.

A.8.1 Procedures

A procedure performs an action on zero or more *arguments*, and terminates by returning zero or more *results*. It may terminate in one of a number of conditions; one of these is the *normal* condition, while the others are *exceptional* conditions. Differing numbers and types of results may be returned in the different conditions.

The form of a procedure is

idn = **proc** [parms] args [returns] [signals] [where]
        routine_body
        end idn

where

args   ::= ([decl , ...])
returns   ::= **returns** (type_spec , ...)
signals   ::= **signals** (exception , ...)
exception   ::= name [(type_spec , ...)]
routine_body ::= { equate } { own_var } { statement }

The *idn* following the **end** of the procedure must be the same as the *idn* naming the procedure. In this section we discuss nonparameterized procedures, in which the *parms* and **where** clauses are missing. Parameterized modules are discussed in section A.8.4; own variables are discussed in section A.8.5.

The header of a procedure describes the way in which the procedure communicates with its caller. The *args* clause specifies the number, order, and types of arguments required to invoke the procedure, while the **returns** clause specifies the number, order, and types of results returned when the procedure terminates normally (by executing a **return** statement or reaching the end of its body). A missing **returns** clause indicates that no results are returned.

The **signals** clause names the exceptional conditions in which the procedure can terminate and specifies the number, order, and types of result objects returned in each exception. In addition to the exceptions explicitly named in the **signals** clause, any procedure can terminate in the *failure* exception. The *failure* exception returns with one result, a string object. All names of exceptions in the **signals** clause must be distinct, and none can be *failure*.

A procedure is an object of some procedure type. For a nonparameterized procedure, this type is derived from the procedure heading by removing the procedure name, rewriting the formal argument declarations with one *idn* per *decl*, deleting the names of formal arguments, and, finally, replacing **proc** by **proctype**.

As was discussed in section A.5.3, the invocation of a procedure causes the introduction of the formal arguments, and the actual arguments are assigned to these variables. Then the procedure body is executed. Execution terminates when a **return** statement or a **signal** statement is executed or when the textual end of the body is reached. If a procedure that should return results reaches the textual end of the body, the procedure terminates in the condition

failure("no return values")

At termination the result objects, if any, are passed back to the invoker of the procedure.

A.8.2 Iterators

An iterator computes a sequence of items, one at a time, where each item is a group of zero or more objects. It has the form

   idn = **iter** [parms] args [yields] [signals] [where]
      routine_body
      end idn

where

   yields **: : = yields** (type_spec , . . .)

The *idn* following the **end** of the iterator must be the same as the *idn* naming the iterator. In this section we discuss nonparameterized iterators, in which the *parms* and **where** clauses are missing. Parameterized modules are discussed in section A.8.4; own variables are discussed in section A.8.5.

   The form of an iterator is very similar to the form of a procedure. There are only two differences:

1. An iterator has a **yields** clause in its header in place of the **returns** clause of a procedure. The **yields** clause specifies the number, order, and types of objects yielded each time the iterator produces the next item in the sequence. If zero objects are yielded, the **yields** clause is omitted.

2. Within the iterator body, the **yield** statement is used to present the caller with the next item in the sequence. An iterator terminates in the same manner as a procedure, but it may not return any results.

   An iterator is an object of some iterator type. For a nonparameterized iterator, this type is derived from the iterator heading by removing the iterator name, rewriting the formal argument declarations with one *idn* per *decl*, deleting the formal argument names, and, finally, replacing **iter** by **itertype**.

   An iterator can be invoked only by a **for** statement. The execution of iterators is described in section A.7.6.

A.8.3 Clusters

A cluster is used to implement a new data type, distinct from any other built-in or user-defined data type. The form is

   idn = **cluster** [parms] **is** idn , . . . [where]
      cluster_body
      end idn

where

> cluster_body $::= \{$ equate $\}$ **rep** $=$ type_spec $\{$ equate $\}$
> $\qquad\qquad\qquad \{$ own_var $\}$
> $\qquad\qquad\qquad$ routine $\{$ routine $\}$
> routine $\qquad ::=$ procedure $|$ iterator

The *idn* following the **end** of the cluster must be the same as the *idn* naming the cluster. In this section we discuss nonparameterized clusters, in which the *parms* and **where** clauses are missing. Parameterized modules are discussed in section A.8.4; own variables are discussed in section A.8.5.

The primitive operations are named by the list of *idns* following the reserved word **is**. All of the *idns* in this list must be distinct.

To define a new data type, it is necessary to choose a *representation* for the objects of the type. The special equate

rep $=$ type_spec

within the cluster body identifies *type_spec* as the representation. Within the cluster, **rep** may be used as an abbreviation for *type_spec*.

The identifier naming the cluster is available for use in the cluster body. Use of this identifier within the cluster body permits the definition of recursive types.

In addition to specifying the rep of objects, the cluster must implement the primitive operations of the type. The operations may be either procedures or iterators. Most of the routines in the cluster body define the primitive operations (those whose names are listed in the cluster header). Any additional routines are *hidden*; they are private to the cluster and may not be named directly by users of the abstract type. All the routines must be named by distinct identifiers; the scope of these identifiers is the entire cluster.

Outside the cluster, the type's objects may only be treated abstractly (that is, manipulated by using the primitive operations). To implement the operations, however, it is usually necessary to manipulate the objects in terms of their rep. It is also sometimes convenient to manipulate the objects abstractly. Therefore, inside the cluster it is possible to view the type's objects either abstractly or in terms of their rep. The syntax is defined to specify unambiguously, for each variable that refers to one of the type's objects, which view is being taken. Thus inside a cluster named T, a declaration v: T

indicates that the object referred to by v is to be treated abstractly, while a declaration

w: **rep**

indicates that the object referred to by w is to be treated concretely. Two primitives, **up** and **down**, are available for converting between these two points of view. The use of **up** permits an object of type **rep** to be viewed abstractly, while **down** permits an abstract object to be viewed concretely. For example, the following two assignments are legal for the declarations just given:

v := **up**(w)
w := **down**(v)

Only routines inside a cluster may use **up** and **down**. Note that **up** and **down** are used merely to inform the compiler that the object is going to be viewed abstractly or concretely, respectively.

A common place where the view of an object changes is at the interface to one of the type's operations. The user, of course, views the object abstractly, while inside the operation the object is viewed concretely. To facilitate this use, a special type specification, **cvt**, is provided. The use of **cvt** is restricted to the *args*, **returns**, **yields**, and **signals** clauses of routines inside a cluster, and it may be used at the top level only (for example, array[**cvt**] is illegal). When used inside the *args* clause, it means that **down** is applied implicitly to the argument object when it is assigned to the formal argument variable. When **cvt** is used in the **returns**, **yields**, or **signals** clause, it means that **up** is applied implicitly to the result object as it is returned (or yielded) to the caller. Thus **cvt** means abstract outside, concrete inside; when constructing the type of a routine, **cvt** is equivalent to the abstract type, but when type checking the body of a routine, **cvt** is equivalent to the representation type.

The **cvt** form does not introduce any new ability over what is provided by **up** and **down**. It is merely a shorthand for a common case.

The type of each routine is derived from its heading in the usual manner, except that each occurrence of **cvt** is replaced by the abstract type.

Inside the cluster, it is not necessary to use the compound form (*type_ spec*$*op_name*) for naming locally defined routines. Furthermore, the compound form cannot be used for invoking hidden routines.

A.8.4 Parameters

Procedures, iterators, and clusters can all be parameterized. Parameterization permits a set of related abstractions to be defined by a single module. Recall that in each module heading there is an optional *parms* clause and an optional **where** clause. The presence of the *parms* clause indicates that the module is parameterized; the **where** clause states certain constraints on permissible actual values for the parameters.

The form of the *parms* clause is

[parm , . . .]

where

    parm **: :** **=** idn , . . . : type_spec
         | idn , . . . : **type**

Each parameter is declared like an argument. However, only the following types of parameters are legal: int, real, bool, char, string, null, and **type**. The actual values for parameters are required to be constants that can be computed at compile time. This requirement ensures that all types are known at compile time and permits complete compile-time type checking.

In a parameterized module, the scope rules permit the parameters to be used throughout the remainder of the module. Type parameters can be used freely as type specifications, and all other parameters can be used freely as expressions. For example, type parameters can be used in defining the types of arguments and results:

p = **proc** [t: **type**] (x: t) **returns** (t)

To use a parameterized module, we must first *instantiate* it, that is, provide actual, constant values for the parameters. (The exact forms of such constants are discussed in section A.4.3.) The result of instantiation is a procedure, iterator, or type (where the parameterized module was a procedure, iterator, or cluster, respectively) that may be used just like a nonparameterized module of the same kind. For each distinct instantiation (that is, for each distinct list of actual parameters), a distinct procedure, iterator, or type is produced.

The meaning of a parameterized module is most easily understood in terms of rewriting. When the module is instantiated, the actual parameter values are substituted for the formal parameters throughout the module, and the *parms* clause and **where** clause are deleted. The resulting module is a regular (nonparameterized) module. In the case of a cluster, some of the operations may have additional parameters; further rewriting will be performed when these operations are used.

In the case of a type parameter, constraints on permissible actual types can be given in the **where** clause. The **where** clause lists a set of operations that the actual type is required to have and also specifies the type of each required operation. The **where** clause constrains the parameterized module as well; the only primitive operations of the type parameter that can be used are those listed in the **where** clause.

The form of the **where** clause is

where **::= where** restriction , . . .

where

    restriction   **::=** idn **has** oper_decl , . . .
                  | idn **in** type_set
    oper_decl   **::=** op_name , . . . : type_spec
    op_name   **::=** name [[constant , . . .]]
    type_set   **::=** { idn | idn **has** oper_decl , . . . { equate } }
                  | idn

There are two forms of restrictions. In both forms, the initial *idn* must be a type parameter. The **has** form lists the set of required operations directly, by means of *oper_decls*. The *type_spec* in each *oper_decl* must name a routine type. Note that if some of the type's operations are parameterized, particular instantiations of those operations must be given. The **in** form requires that the actual type be a member of a *type_set*, a set of types having the required operations. The two identifiers in the type_set must match, and the notation is read like set notation; an example is

{t | t **has** f: ... }

which means "the set of all types *t* such that *t* **has** *f* ...." The scope of the identifier is the type_set.

The **in** form is useful because an abbreviation can be given for a type_set by means of an equate. If it is helpful to introduce some abbreviations in defining the type_set, these are given in the optional equates within the type_set. The scope of these equates is the type_set.

A routine in a parameterized cluster may have a **where** clause in its heading and can place further constraints on the cluster parameters. For example, any type is permissible for the array element type, but the array *similar* operation requires that the element type have a *similar* operation.

This means that array[T] exists for any type T, but array[T]$*similar* exists only when T$*similar* exists. Note that a routine need not include in its **where** clause any of the restrictions included in the cluster **where** clause.

A.8.5 Own Variables

Occasionally it is desirable to have a module that retains information internally between invocations. Without such an ability, the information would have to be either reconstructed at every invocation, which can be expensive (and may even be impossible if the information depends on previous invocations), or passed in through arguments, which is undesirable because the information is then subject to uncontrolled modification in other modules.

Procedures, iterators, and clusters can all retain information through the use of own variables. An own variable is similar to a normal variable, except that it exists for the life of the program, rather than being bound to the life of any particular routine activation. Syntactically, own variable declarations must appear immediately after the equates in a routine or cluster body; they cannot appear in bodies nested within statements. Own variable declarations have the form

```
own_var  ::= own decl
         |   own idn : type_spec := expression
         |   own decl , . . . := invocation
```

Note that initialization is optional.

Own variables are created when a program begins execution, and they always start out uninitialized. The own variables of a routine (including cluster operations) are initialized in textual order as part of the first invocation of that routine, before any statements in the body of the routine are executed. Cluster own variables are initialized in textual order as part of the first invocation of the first cluster operation to be invoked (even if the operation does not use the own variables). Cluster own variables are initialized before any operation own variables are initialized.

Aside from the placement of their declarations, the time of their initialization, and their lifetime, own variables act just like normal variables and can be used in all the same places. As for normal variables, attempts to use uninitialized own variables (if not detected at compile time) cause the runtime exception

failure("uninitialized variable")

Own variable declarations in different modules always refer to distinct

C = **cluster** [t: **type**] **is** . . .
   • • •
   own x: **int** := init(. . .)

   P = **proc** (. . .)
      own y: • • •
      • • •
      end P

   Q = **proc** [i: **int**] (. . .)
      own z: • • •
      • • •
      end Q

   end C

Figure A.1 Own variables.

own variables, and distinct executions of programs never share own variables (even if the same module is used in several programs). Furthermore, own variable declarations within a parameterized module produce distinct own variables for each instantiation of the module. For a given instantiation of a parameterized cluster, all instantiations of the type's operations share the same set of cluster own variables, but distinct instantiations of parameterized operations have distinct routine own variables. For example, in the cluster in figure A.1 there is a distinct x and y for every type t, and a distinct z for every type-integer pair (t, i).

Own variable declarations cannot be enclosed by an **except** statement, so care must be exercised when writing initialization expressions. If an exception is raised by an initialization expression, it will be treated as an exception raised, but not handled, in the body of the routine whose invocation caused the initialization to be attempted. This routine will then signal *failure* to its caller. In the example cluster just given, if procedure P were the first operation of C[string] to be invoked, causing initialization of x to be attempted, then an *overflow* exception raised in the initialization of x would result in P signaling

   failure("unhandled exception: overflow")

to its caller.

With the introduction of own variables, procedures and iterators become potentially mutable objects. If the abstract behavior of a routine depends on its history, this should be stated in its specification. In general, own variables should not be used to modify the abstract behavior of a module.

A.9 Built-in Types

In this section we describe the built-in types and type classes. Familiarity with chapter 2 is assumed in this discussion. In defining the built-in type classes, we do not depend on users satisfying any constraints beyond those that can be checked at compile time (that is, that a type parameter has an operation with a particular name and signature). This decision leads to more complicated specifications. For example, for the *elements* iterator for arrays, we must define the behavior when the loop modifies the array.

A.9.1 Null

null = **data type is** copy, equal, similar

Overview

> The type null has exactly one, immutable object, represented by the literal **nil**. **Nil** is generally used as a place holder in type definitions using oneofs or variants.

Operations

> copy = **proc** (n: null) **returns** (null)
>   **effects** Returns nil.
>
> equal = **proc** (n1, n2: null) **returns** (bool)
>   **effects** Returns true.
>
> similar = **proc** (n1, n2: null) **returns** (bool)
>   **effects** Returns true.

end null

A.9.2 Booleans

bool = **data type is** and, or, not, equal, similar, copy

Overview

> The two immutable objects of type bool, with literals **true** and **false**, represent logical truth values.

Operations

> and = **proc** (b1, b2: bool) **returns** (bool)
>   **effects** Returns the logical and of *b1* and *b2*.

or = **proc** (b1, b2: bool) **returns** (bool)
>   **effects** Returns the logical or of *b1* and *b2*.

not = **proc** (b: bool) **returns** (bool)
>   **effects** Returns the logical negation of *b*.

equal = **proc** (b1, b2: bool) **returns** (bool)
similar = **proc** (b1, b2: bool) **returns** (bool)
>   **effects** Returns true if *b1* and *b2* are both true or both false;
>   otherwise returns false.

copy = **proc** (b: bool) **returns** (bool)
>   **effects** If *b* is true returns true; otherwise returns false.

end bool

A.9.3 Integers

int = **data type is** add, sub, mul, minus, div, mod, power, abs, max,
>   min, lt, le, ge, gt, equal, similar, copy, from_to_by, from_to,
>   parse, unparse

Overview

Objects of type int are immutable and are intended to model a
subrange of the mathematical integers. The exact range is not part of
the language definition and can vary somewhat from implementation
to implementation. Each implementation is constrained to provide a
closed interval [*int_min*, *int_max*], with *int_min* < 0 and
int_max ≥ *char_top* (the number of characters—see section A.9.5).
An *overflow* exception is signaled by an operation if the result would
lie outside this interval.

Integer literals are written as a sequence of one or more decimal
digits.

Operations

add = **proc** (x, y: int) **returns** (int) **signals** (overflow)
sub = **proc** (x, y: int) **returns** (int) **signals** (overflow)
mul = **proc** (x, y: int) **returns** (int) **signals** (overflow)
>   **effects** These are the standard integer addition, subtraction,
>   and multiplication operations. They signal *overflow* if the
>   result would lie outside the represented interval.

minus = **proc** (x: int) **returns** (int) **signals** (overflow)
    effects Returns the negative of x; signals *overflow* if the result
    would lie outside the represented interval.

div = **proc** (x, y: int) **returns** (int) **signals** (zero_divide, overflow)
    effects Signals *zero_divide* if $y = 0$. Otherwise returns the
    integer quotient of dividing x by y; signals *overflow* if the
    result would lie outside the represented interval.

mod = **proc** (x, y: int) **returns** (int) **signals** (zero_divide, overflow)
    effects Signals *zero_divide* if $y = 0$. Otherwise returns the
    integer remainder of dividing x by y; signals *overflow* if the
    quotient would lie outside the represented interval.

power = **proc** (x, y: int) **returns** (int)
    signals (negative_exponent, overflow)
    effects Signals *negative_exponent* if $y < 0$. Otherwise returns
    x^y; signals *overflow* if the result would lie outside the repre-
    sented interval.

abs = **proc** (x: int) **returns** (int) **signals** (overflow)
    effects Returns the absolute value of x; signals *overflow* if the
    result would lie outside the represented interval.

max = **proc** (x, y: int) **returns** (int)
    effects Returns the larger of x and y.

min = **proc** (x, y: int) **returns** (int)
    effects Returns the smaller of x and y.

lt = **proc** (x, y: int) **returns** (bool)
gt = **proc** (x, y: int) **returns** (bool)
le = **proc** (x, y: int) **returns** (bool)
ge = **proc** (x, y: int) **returns** (bool)
equal = **proc** (x, y: int) **returns** (bool)
    effects These are the standard ordering relations.

similar = **proc** (x, y: int) **returns** (bool)
    effects Returns ($x = y$).

copy = **proc** (x: int) **returns** (y: int)
    effects Returns y such that $x = y$.

from_to_by = **iter** (from, to, by: int) **yields** (int)

    effects Yields the integers from *from* to *to*, incrementing
by *by* each time, that is, yields *from*, *from* + *by*, ... ,
from + $n * by$, where n is the largest positive integer such
that *from* + $n * by \leq to$. If *by* = 0, then yields *from* indefi-
nitely. Yields nothing if *from* > *to* and *by* > 0, or if
from < *to* and *by* < 0.

from_to = **iter** (from, to: int) **yields** (int)

    effects Identical to *from_to_by(from, to, 1)*.

parse = **proc** (s: string) **returns** (int)

    signals (bad_format, overflow)

    effects s must be an integer literal, with an optional leading
plus or minus sign; if s is not of this form, signals
bad_format. Otherwise returns the integer corresponding
to s; signals *overflow* if the result would be outside of the
represented interval.

unparse = **proc** (x: int) **returns** (string)

    effects Produces the string corresponding to the integer value
of x, preceded by a minus sign if $x < 0$. Leading zeros are
suppressed, and there is no leading plus sign for positive
integers.

end int

A.9.4 Reals

real = **data type is** add, sub, mul, minus, div, power, abs, max, min,
exponent, mantissa, i2r, r2i, trunc, parse, unparse, lt, le, ge,
gt, equal, similar, copy

Overview

The type real models a subset of the mathematical real numbers.
Reals are immutable and are written as a *mantissa* with an optional
exponent. A mantissa is either a sequence of one or more decimal digits or two sequences (one of which may be empty) joined by a period.
The mantissa must contain at least one digit. An exponent is E or e,
optionally followed by + or −, followed by one or more decimal digits.
An exponent is required if the mantissa does not contain a period. As
is usual, $mEx = m * 10^x$. Examples of real literals are:

    3.14   3.14E0   314e−2   .0314E+2   3.   .14

Each implementation represents numbers in

$$D = \{-\text{real\_max}, -\text{real\_min}\} \cup \{0\} \cup \{\text{real\_min}, \text{real\_max}\}$$

where

$$0 < \text{real\_min} < 1 < \text{real\_max}$$

Numbers in D are approximated by the implementation with precision p such that

$\forall r \in D$	$\text{Approx}(r) \in \text{Real}$
$\forall r \in \text{Real}$	$\text{Approx}(r) = r$
$\forall r \in D -\{0\}$	$\mid (\text{Approx}(r) - r)/r\mid < 10^{1-p}$
$\forall r, s \in D$	$r \leq s \Rightarrow \text{Approx}(r) \leq \text{Approx}(s)$
$\forall r \in D$	$\text{Approx}(-r) = -\text{Approx}(r)$

We define *Max_width* and *Exp_width* to be the smallest integers such that every nonzero element of real can be represented in "standard" form (exactly one digit, not zero, before the decimal point) with no more than *Max_width* digits of mantissa and no more than *Exp_width* digits of exponent.

Real operations signal an exception if the result of a computation lies outside of D; *overflow* occurs if the magnitude exceeds *real_max*, and *underflow* occurs if the magnitude is less than *real_min*.

Operations

add = **proc** (x, y: real) **returns** (real) **signals** (overflow, underflow)
    effects Computes the sum z of x and y; signals *overflow* or *underflow* if z is outside of D, as explained earlier. Otherwise returns $Approx(z)$ such that

$$(\text{x, y} \geq 0 \vee \text{x, y} \leq 0) \Rightarrow \text{add(x, y)} = \text{Approx(x + y)}$$
$$\text{add(x, y)} = (1 + \epsilon)(\text{x + y}) \quad |\epsilon| < 10^{1-p}$$
$$\text{add(x, 0)} = \text{x}$$
$$\text{add(x, y)} = \text{add(y, x)}$$
$$\text{x} \leq \text{x}' \Rightarrow \text{add(x, y)} \leq \text{add(x', y)}$$

sub = **proc** (x, y: real) **returns** (real) **signals** (overflow, underflow)
    effects Computes $x - y$; the result is identical to $add(x, -y)$.

minus = **proc** (x: real) **returns** (real)
    effects Returns $-x$.

mul = **proc** (x, y: real) **returns** (real) **signals** (overflow, underflow)
    effects Returns $Approx(x * y)$; signals *overflow* or *underflow* if $x * y$ is outside of D.

div = **proc** (x, y: real) **returns** (real)
      signals (zero_divide, overflow, underflow)
      effects If $y = 0$, signals *zero_divide*. Otherwise returns
      Approx(x/y); signals *overflow* or *underflow* if x/y is outside
      of D.

power = **proc** (x, y: real) **returns** (real)
      signals (zero_divide, complex_result, overflow, underflow)
      effects If $x = 0$ and $y < 0$, signals *zero_divide*. If $x < 0$ and y
      is nonintegral, signals *complex_result*. Otherwise returns an
      approximation to x^y; signals *overflow* or *underflow* if x^y is
      outside of D.

abs = **proc** (x: real) **returns** (real)
      effects Returns the absolute value of x.

max = **proc** (x, y: real) **returns** (real)
      effects Returns the larger of x and y.

min = **proc** (x, y: real) **returns** (real)
      effects Returns the smaller of x and y.

exponent = **proc** (x: real) **returns** (int) **signals** (undefined)
      effects If $x = 0$, signals *undefined*. Otherwise returns the
      exponent that would be used in representing x as a literal
      in standard form, that is, returns
          max $(\{i \mid abs(x) \geq 10^i\})$

mantissa = **proc** (x: real) **returns** (real)
      effects Returns the mantissa of x when represented in stan-
      dard form, that is, returns *Approx*$(x/10^e)$, where
      $e = exponent(x)$. If $x = 0.0$, returns 0.0.

i2r = **proc** (i: int) **returns** (real) **signals** (overflow)
      effects Returns *Approx*(i); signals overflow if i is not in D.

r2i = **proc** (x: real) **returns** (int) **signals** (overflow)
      effects Rounds x to the nearest integer and toward zero in
      case of a tie. Signals *overflow* if the result lies outside the
      represented range of integers.

trunc = **proc** (x: real) **returns** (int) **signals** (overflow)
      effects Truncates x toward zero; signals *overflow* if the result
      would be outside the represented range of integers.

parse = **proc** (s: string) **returns** (real)
      signals (bad_ format, overflow, underflow)
      effects Computes the exact value z corresponding to a real or integer literal and returns $Approx(z)$. s must be a real or integer literal with an optional leading plus or minus sign; otherwise signals *bad_ format*. Signals *underflow* or *overflow* if z is not in D.

unparse = **proc** (x: real) **returns** (string)
      effects Returns a real literal such that *parse*(*unparse*(x)) = x. The general form of the literal is
$$[-] i\_field.f\_field [e \pm x\_field]$$
      Leading zeros in *i_ field* and trailing zeros in *f_ field* are suppressed. If x is integral and within the range of CLU integers, then *f_ field* and the exponent are not present. If x can be represented by a mantissa of no more than *Max_width* digits and no exponent (that is, if $-1 \leq exponent(arg1) < Max\_width$), then the exponent is not present. Otherwise the literal is in standard form, with *Exp_width* digits of exponent.

lt = **proc** (x, y: real) **returns** (bool)
le = **proc** (x, y: real) **returns** (bool)
ge = **proc** (x, y: real) **returns** (bool)
gt = **proc** (x, y: real) **returns** (bool)
equal = **proc** (x, y: real) **returns** (bool)
      effects The standard ordering relations.

similar = **proc** (x, y: real) **returns** (bool)
      effects Returns $(x = y)$.

copy = **proc** (x: real) **returns** (real)
      effects Returns y such that $x = y$.

end real

A.9.5 Characters

char = **data type is** i2c, c2i, lt, le, ge, gt, equal, similar, copy

Overview

Type char provides the alphabet for text manipulation. Characters are immutable and form an ordered set. Every implementation must provide at least 128, but no more than 512, characters; the first 128 characters are the ASCII characters in their standard order.

Operations *i2c* and *c2i* convert between ints and chars. The smallest character corresponds to zero, and characters are numbered sequentially up to *char_top*, the integer corresponding to the largest character. This numbering determines the ordering of the characters.

Printing ASCII characters (octal 40 through octal 176), other than single quote or backslash, can be written as that character enclosed in single quotes. Any character can be written by enclosing one of the following escape sequences in single quotes:

escape sequence	character
\'	' (single quote)
\"	" (double quote)
\\	\ (backslash)
\n	NL (newline)
\t	HT (horizontal tab)
\p	FF (form feed, newpage)
\b	BS (backspace)
\r	CR (carriage return)
\v	VT (vertical tab)
\ * **	specified by octal value
	(exactly three octal digits)

The escape sequences may also be written using upper case letters. Examples of character literals are

'7'   'a'   " "   '\"'   '\"   '\B'   '\177'

Operations

i2c = **proc** (x: int) **returns** (char) **signals** (illegal_char)
       effects Returns the character corresponding to x; signals *illegal_argument* if x is not in the range $[0,\ char\_top]$.

c2i = **proc** (c: char) **returns** (int)
       effects Returns the integer corresponding to c.

lt = **proc** (c1, c2: char) **returns** (bool)
le = **proc** (c1, c2: char) **returns** (bool)
ge = **proc** (c1, c2: char) **returns** (bool)
gt = **proc** (c1, c2: char) **returns** (bool)
equal = **proc** (c1, c2: char) **returns** (bool)
       effects These are the standard ordering relations, where the order is consistent with the numbering of characters.

similar = **proc** (c1, c2: char) **returns** (bool)
    effects Returns ($c1 = c2$).

copy = **proc** (c1: char) **returns** (c2: char)
    effects Returns $c2$ such that $c1 = c2$.

end char

A.9.6 Strings

string = **data type is** size, empty, concat, append, fetch, rest, indexs,
                indexc, substr, lt, le, ge, gt, equal, similar, copy, c2s, s2ac,
                ac2s, s2sc, sc2s, chars

Overview

Type string is used for representing text. A string is an immutable
sequence of zero or more characters. The characters of a string are
indexed sequentially starting from one. Strings are alphanumerically
ordered based on the ordering for characters.

A string is written as a sequence of zero or more character representations enclosed in double quotes. Within a string literal, a printing
ASCII character other than double quote or backslash is represented
by itself. Any character can be represented by using the escape sequences listed for characters. Examples of string literals are
    "Item\tCost"   "altmode (\033) = \\033" " " " "
If the result of a string operation would be a string containing more
than int_max characters, the operation signals *failure*.

Operations

size = **proc** (s: string) **returns** (int)
    effects Returns the number of characters in s.

empty = **proc** (s: string) **returns** (bool)
    effects Returns true if s is empty (contains no characters);
        otherwise returns false.

concat = **proc** (s1, s2: string) **returns** (string)
    effects Returns a new string containing the characters of $s1$
        followed by the characters of $s2$. Signals *failure* if the new
        string would contain more than int_max characters.

append = **proc** (s: string, c: char) **returns** (string)
  effects Returns a new string containing the characters of *s* fol-
    lowed by *c*. Signals *failure* if the new string would contain
    more than *int_max* characters.

fetch = **proc** (s: string, i: int) **returns** (char) **signals** (bounds)
  effects Signals *bounds* if $i < 0$ or $i > size(s)$; otherwise returns
    the *i*th character of *s*.

rest = **proc** (s: string, i: int) **returns** (string) **signals** (bounds)
  effects Signals *bounds* if $i < 0$ or $i > size(s) + 1$; otherwise
    returns a new string containing the characters $s[i]$, $s[i + 1]$,
    ..., $s[size(s)]$. Note that if $i = size(s) + 1$, *rest* returns the
    empty string.

indexs = **proc** (s1, s2: string) **returns** (int)
  effects If *s1* occurs as a substring in *s2*, returns the least index
    at which *s1* occurs. Returns 0 if *s1* does not occur in *s2*,
    and 1 if *s1* is the empty string. For example,
      indexs("bc", "abcbc") = 2
      indexs(" ", "abcde") = 1

indexc = **proc** (c: char, s: string) **returns** (int)
  effects If *c* occurs in *s*, returns the least index at which *c*
    occurs; returns 0 if *c* does not occur in *s*.

substr = **proc** (s: string, at, cnt: int) **returns** (string)
        signals (bounds, negative_size)
  effects If $cnt < 0$, signals *negative_size*. If $at < 1$ or
    $at > size(s) + 1$, signals *bounds*. Otherwise returns a new
    string containing the characters $s[at]$, $s[at + 1]$, ...; the
    new string contains $min(cnt, size - at + 1)$ characters. For
    example,
      substr ("abcdef", 2, 3) = "bcd"
      substr ("abcdef", 2, 7) = "bcdef"
      substr ("abcdef", 7, 1) = " "
    Note that if $min(cnt, size - at + 1) = 0$, *substr* returns the
    empty string.

lt = **proc** (s1, s2: string) **returns** (bool)
le = **proc** (s1, s2: string) **returns** (bool)
ge = **proc** (s1, s2: string) **returns** (bool)
gt = **proc** (s1, s2: string) **returns** (bool)
equal = **proc** (s1, s2: string) **returns** (bool)
> **effects** These are the usual lexicographic ordering relations on strings, based on the ordering of characters. For example,
>> "abc" < "aca"
>> "abc" < "abca"

similar = **proc** (s1, s2: string) **returns** (bool)
> **effects** Returns true if $s1 = s2$; otherwise returns false.

copy = **proc** (s1: string) **returns** (s2: string)
> **effects** Returns $s2$ such that $s1 = s2$.

c2s = **proc** (c: char) **returns** (string)
> **effects** Returns a string containing c as its only character.

s2ac = **proc** (s: string) **returns** (array[char])
> **effects** Stores the characters of s as elements of a new array of characters, a. The low bound of the array is 1, the size is $size(s)$, and the ith element of the array is the ith character of s, for $1 \leq i \leq size(s)$.

ac2s = **proc** (a: array[char]) **returns** (string)
> **effects** Does the inverse of $s2ac$. The result is a string with characters in the same order as in a. That is, the ith character of the string is the $(i + low(a) - 1)$th element of a.

s2sc = **proc** (s: string) **returns** (sequence[char])
> **effects** Transforms a string into a sequence of characters. The size of the sequence is $size(s)$. The ith element of the sequence is the ith character of s.

sc2s = **proc** (s: sequence[char]) **returns** (string)
> **effects** Does the inverse of $s2sc$. The result is a string with characters in the same order as in s. That is, the ith character of the string is the ith element of s.

chars = **iter** (s: string) **yields** (char)
> **effects** Yields, in order, each character of s.

end string

A.9.7 Any

A type specification is used to restrict the class of objects that a variable can denote, a procedure or iterator can take as arguments, a procedure can return, and so forth. There are times when no restrictions are desired, when any object is acceptable. At such times the type specification *any* is used. For example, one might wish to implement a table mapping strings to arbitrary objects, with the intention that different strings should map to objects of different types. The lookup operation, used to get the object corresponding to a string, would have its result declared to be of type any.

The type any is the union of all possible types. Every object is of type any, as well as being of some base type. The type any has no operations; however, the base type of an object can be tested at runtime (see section A.6.13).

A.9.8 Arrays

array = **data type** [t: type] **is** create, new, predict, fill, low, high, size,
                         empty, set_low, trim, fetch, bottom, top, store, addh, addl,
                         remh, reml, elements, indexes, equal, similar, similar1, copy,
                         copy1, fill_copy

Overview

Arrays are one-dimensional mutable objects that can grow and shrink dynamically. Each array has a low bound and a sequence of elements that are indexed sequentially, starting from the low bound. All elements of an array are of the same type.

Arrays can be created by calling array operations *create, new, fill, fill_copy*, and *predict*. They can also be created by means of a special constructor form that specifies the array low bound, and an arbitrary number of initial elements. For example,

array[int]$[5: 1, 2, 3, 4]

creates an integer array with low bound 5 and four elements. The low bound can be omitted if it is 1; for example,

array[string]$["a", "b", "c"]

creates a three-element array with low bound 1. Array constructors are discussed further in section A.6.8.

Operations *low, high,* and *size* return the current low and high bounds and size of the array. For array a, $size(a)$ is the number of elements in a, and $high(a) = low(a) + size(a) - 1$.

For any index i between the low and high bound of an array, there is a defined element $a[i]$. Any operation call that receives as an index an integer outside the defined range terminates with a *bounds* exception. Any call that would lead to an array whose low or high bound or size is outside the defined range of integers terminates with the *failure* exception.

Operations *similar*, *similar1*, *copy*, and *fill_copy* require that the element type t provide certain operations. No constraints are assumed on the behavior of these t operations; the behavior of the array operations is defined even if the t operation behaves strangely (for example, modifies the array being copied).

Operations

create = **proc** (lb: int) **returns** (array[t])
    effects Returns a new, empty array with low bound *lb*.

new = **proc** ( ) **returns** (array[t])
    effects Returns a new, empty array with low bound 1.

predict = **proc** (lb, cnt: int) **returns** (array[t])
    effects Returns a new, empty array with low bound *lb*. Argument *cnt* is a prediction of how many *addh*s or *addl*s are likely to be performed on this new array. If $cnt > 0$, *addh*s are expected; otherwise *addl*s are expected. These operations may execute faster than if the array had been produced by calling *create*.

fill = **proc** (lb, cnt: int, elem: t) **returns** (array[t])
    signals (negative_size)
    effects If $cnt < 0$, signals *negative_size*. Returns a new array with low bound *lb* and size *cnt*, and with *elem* as each element; if this new array would have high bound $< int\_min$ or $> int\_max$, signals failure.

low = **proc** (a: array[t]) **returns** (int)
    effects Returns the low bound of *a*.

high = **proc** (a: array[t]) **returns** (int)
    effects Returns the high bound of *a*.

size = **proc** (a: array[t]) **returns** (int)
    effects Returns a count of the number of elements of *a*.

empty = **proc** (a: array[t]) **returns** (bool)
    effects Returns true if a contains no elements; otherwise
    returns false.

set_low = **proc** (a: array[t], lb: int)
    modifies a
    effects Modifies the low and high bounds of a; the new
    low bound of a is lb and the new high bound is
    $high(a_{pre}) - lb + low(a_{pre})$. (Here a_{pre} is the value of a at
    the time of the call.) If the new high bound is outside the
    represented range of integers, signals *failure* and does not
    modify a.

trim = **proc** (a: array[t], lb, cnt: int)
        signals (bounds, negative_size)
    modifies a
    effects If $lb < low(a)$ or $lb > high(a) + 1$, signals *bounds*.
    If $cnt < 0$, signals *negative_size*. Otherwise modifies a by
    removing all elements with index $< lb$ or greater than or
    equal to $lb + cnt$; the new low bound is lb. For example, if
    a =array[int]$[1,2,3,4,5], then
        trim(a, 2, 2) results in a having value [2: 2, 3]
        trim(a, 4, 3) results in a having value [4: 4, 5]

fetch = **proc** (a: array[t], i: int) **returns** (t) **signals** (bounds)
    effects If $i < low(a)$ or $i > high(a)$, signals *bounds*; otherwise
    returns the element of a with index i.

store = **proc** (a: array[t], i: int, elem: t) **signals** (bounds)
    modifies a
    effects If $i < low(a)$ or $i > high(a)$, signals *bounds*; otherwise
    makes *elem* the element of a with index i.

bottom = **proc** (a: array[t]) **returns** (t) **signals** (bounds)
    effects If a is empty, signals *bounds*; otherwise returns
    $a[low(a)]$.

top = **proc** (a: array[t]) **returns** (t) **signals** (bounds)
    effects If a is empty, signals *bounds*; otherwise returns
    $a[high(a)]$.

addh = **proc** (a: array[t], elem: t)
    modifies a
    effects If extending a on the high end causes the high bound
    or size of a to be outside the defined range of integers, sig-
    nals *failure*. Otherwise extends a by 1 in the high direction
    and stores *elem* as the new element.

addl = **proc** (a: array[t], elem: t)
    modifies a
    effects If extending a on the low end causes the low bound or
    size of a to be outside the defined range of integers, signals
    failure. Otherwise extends a by 1 in the low direction and
    stores *elem* as the new element.

remh = **proc** (a: array[t]) **returns** (t) **signals** (bounds)
    modifies a
    effects If a is empty, signals *bounds*. Otherwise shrinks a by
    removing its high element and returning that element.

reml = **proc** (a: array[t]) **returns** (t) **signals** (bounds)
    modifies a
    effects If a is empty, signals *bounds*. Otherwise shrinks a by
    removing its low element and returning that element.

elements = **iter** (a: array[t]) **yields** (t)
    effects The effect is equivalent to the following body:
        for i: int **in**
          int\$from_to(array[t]\$low(a), array[t]\$high(a)) **do**
          yield (a[i])
        end
    Note that if a is not modified by the loop body, the effect is
    to yield the elements of a, each exactly once, from the low
    bound to the high bound.

indexes = **iter** (a: array[t]) **yields** (int)
    effects Yields the indexes of a from the low bound of a_{pre} to
    the high bound of a_{pre}, where a_{pre} is the value of a at the
    time of the call. Note that *indexes* is unaffected by any
    modifications done by the loop body.

equal = **proc** (a1, a2: array[t]) **returns** (bool)
    effects Returns true if *a1* and *a2* refer to the same array
    object; otherwise returns false.

similar = **proc** (a1, a2: array[t]) **returns** (bool)
  requires *t* has operation
      similar: **proctype** (t, t) **returns** (bool)
  effects Returns true if *a1* and *a2* have the same low and high
      bounds and if their elements are pairwise similar as de-
      termined by *t$similar*. This operation is equivalent to the
      following body:
      at = array[t]
      if at$low(a) $\sim=$ at$low(a2) **cor**
      at$size(a1) $\sim=$ at$size(a2)
          then return (false)
          end
      for i: int **in** at$indexes(a1) **do**
          if $\sim$t$similar(a1[i], a2[i]) **then return (false) end**
          end
      return (true)

similar1 = **proc** (a1, a2: array[t]) **returns** (bool)
  requires *t* has operation
      equal: **proctype** (t, t) **returns** (bool)
  effects Returns true if *a1* and *a2* have the same low and high
      bounds and if their elements are pairwise equal as deter-
      mined by *t$equal*. This operation works the same way as
      similar, except that *t$equal* is used instead of *t$similar*.

copy = **proc** (a: array[t]) **returns** (b: array[t])
  requires *t* has operation
      copy: **proctype** (t) **returns** (t)
  effects Returns a new array *b* with the same low and high
      bounds as *a* and such that each element *b[i]* contains
      t$copy(a[i]). This operation is equivalent to the following
      body:
      b := array[t]$copy1(a)
      for i: int **in** array[t]$indexes(a) **do**
          b[i] := t$copy(a[i])
          end
      return (b)

copy1 = **proc** (a: array[t]) **returns** (b: array[t])
  effects Returns a new array *b* with the same low and high
      bounds as *a* and such that each element *b[i]* contains the
      same element as *a[i]*.

fill_copy = **proc** (lb, cnt: int, elem: t) **returns** (array[t])
        signals (negative_size)
     requires t has operation
          copy: **proctype** (t) **returns** (t)
    effects The effect is like *fill* except that *elem* is copied. If
        cnt < 0, signals *negative_size*. Otherwise returns a new
        array x with low bound *lb* and size *cnt* and with each ele-
        ment a distinct copy of *elem*, as produced by *t$copy*; if the
        new array has high bound < *int_min* or > *int_max*, signals
        failure. This operation is equivalent to the following body:
           if cnt < 0 **then signal** negative_size **end**
           x: array[t] := array[t]$predict(lb, cnt)
           for j: int **in** int$from_to(1, cnt) **do**
              array[t]$addh(x, t$copy(elem))
           end
           return (x)

end array

A.9.9 Sequences

sequence = **data type** [t: type] **is** new, fill, size, empty, fetch, bottom,
              top, replace, addh, addl, remh, reml, concat, subseq, e2s,
              a2s, s2a, elements, indexes, equal, similar, copy, fill_copy

Overview

Sequences are immutable sequences of elements; they always have low
bound 1. The elements of the sequence can be indexed sequentially
from 1 up to the size of the sequence.

Sequences can be created by calling sequence operations and by
means of a special constructor; arguments of the constructor are simply the elements of the new sequence; for example,
     sequence[int]$[3, 7]
creates a two-element sequence containing 3 at index 1 and 7 at index
2. The special constructor is discussed further in section A.6.8.

Any operation call that attempts to access a sequence with an index
that is not within the defined range terminates with the *bounds* exception. Any operation call that would give rise to a sequence whose
size is greater than *int_max* terminates with the *failure* exception.

Operations

new = **proc** ( ) **returns** (sequence[t])
    effects Returns the empty sequence.

fill = **proc** (cnt: int, elem: t) **returns** (sequence[t])
    signals (negative_ size)
    effects If *cnt* < 0, signals *negative_ size*. Otherwise returns a
    sequence containing *cnt* elements each of which is *elem*.

size = **proc** (s: sequence[t]) **returns** (int)
    effects Returns a count of the number of elements in *s*.

empty = **proc** (s: sequence[t]) **returns** (bool)
    effects Returns true if *s* contains no elements; otherwise
    returns false.

fetch = **proc** (s: sequence[t], i: int) **returns** (t) **signals** (bounds)
    effects If $i < 1$ or $i > size(s)$, signals *bounds*. Otherwise
    returns the *i*th element of *s*.

bottom = **proc** (s: sequence[t]) **returns** (t) **signals** (bounds)
    effects If *s* is empty, signals *bounds*. Otherwise returns $s[1]$.

top = **proc** (s: sequence[t]) **returns** (t) **signals** (bounds)
    effects If *s* is empty, signals *bounds*. Otherwise returns
    $s[size(s)]$.

replace = **proc** (s: sequence[t], i: int, elem: t) **returns** (sequence[t])
    signals (bounds)
    effects If $i < 1$ or $i > high(s)$, signals *bounds*. Otherwise
    returns a new sequence containing the same elements as *s*,
    except that *elem* is in the *i*th position. For example,
    replace(sequence[int]$[2,5], 1, 6]) = [6, 5]

addh = **proc** (s: sequence[t], elem: t) **returns** (sequence[t])
    effects Returns a new sequence containing the same elements
    as *s* followed by one additional element, *elem*. If the result-
    ing sequence would have size > *int_max*, signals *failure*.

addl = **proc** (s: sequence[t], elem: t) **returns** (sequence[t])
    effects Returns a new sequence containing *elem* as the first
    element followed by the elements of *s*. If the resulting
    sequence would have *size* > *int_max*, signals *failure*.

remh = **proc** (s: sequence[t]) **returns** (sequence[t])
       signals (bounds)
      effects If *s* is empty, signals *bounds*. Otherwise returns a new
      sequence containing all elements of *s* except the last one.

reml = **proc** (s: sequence[t]) **returns** (sequence[t]) **signals** (bounds)
      effects If *s* is empty, signals *bounds*. Otherwise returns a new
      sequence containing all elements of *s* except the first one.

concat = **proc** (s1, s2: sequence[t]) **returns** (sequence[t])
      effects Returns a new sequence containing the elements of *s1*
      followed by the elements of *s2*. Signals *failure* if the size of
      the new sequence is $>$ *int_max*.

e2s = **proc** (elem: t) **returns** (sequence[t])
      effects Returns a one-element sequence containing *elem* as its
      only element.

a2s = **proc** (a: array[t]) **returns** (sequence[t])
      effects Returns a sequence containing the elements of *a* in the
      same order as in *a*.

s2a = **proc** (s: sequence[t]) **returns** (array[t])
      effects Returns a new array with low bound 1 and containing
      the elements of *s* in the same order as in *s*.

elements = **iter** (s: sequence[t]) **yields** (t)
      effects Yields the elements of *s* in order.

indexes = **iter** (s: sequence[t]) **yields** (int)
      effects Yields the indexes of *s* from 1 to *size(s)*.

equal = **proc** (s1, s2: sequence[t]) **returns** (bool)
      requires *t* has operation
            equal: **proctype** (t, t) **returns** (bool)
      effects Returns true if *s1* and *s2* have equal values as deter-
      mined by *t$equal*. This operation is equivalent to the fol-
      lowing body:

```
qt = sequence [t]
if qt$size(s1) ~= qt$size(s2) then return (false) end
for i: int in qt$indexes(s1) do
    if s1[i] ~= s2[i] then return(false) end
    end
return (true)
```

similar = **proc** (s1, s2: sequence[t]) **returns** (bool)
  requires *t* has operation
      similar: **proctype** (t, t) **returns** (bool)
  effects Returns true if *s1* and *s2* have similar values as
    determined by *t$similar*. *Similar* works in the same way as
    equal, except that *t$similar* is used instead of *t$equal*.

copy = **proc** (s: sequence[t]) **returns** (sequence[t])
  requires *t* has operation
      copy: **proctype** (t) **returns** (t)
  effects Returns a new sequence containing as elements copies
    of the elements of *s*. This operation is equivalent to the
    following body:
      qt = sequence[t]
      y: qt := qt$new()
      for e: t **in** qt$elements(s) **do**
          y := qt$addh(y, t$copy(e))
          end
      return (y)

fill_copy = **proc** (cnt: int, elem: t) **returns** (sequence[t])
      signals (negative_size)
  requires t has operation
      copy: **proctype** (t) **returns** (t)
  effects If *cnt* < 0, signals *negative_size*. Otherwise returns
    a new sequence containing *cnt* elements each of which is a
    copy of *elem*. This operation is equivalent to the following
    body:
      qt = sequence[t]
      if cnt < 0 **then signal** negative_size **end**
      x: qt := qt$new()
      for i: int **in** int$from_to(1, cnt) **do**
          x := qt$addh(x, T$copy(elem))
          end
      return (x)

end sequence

A.9.10 Records

record = **data type** [n₁: t₁, ..., nₖ: tₖ] **is** get_ , set_ , r_gets_r,
r_gets_s, equal, similar, similar1, copy, copy1

Overview

A record is a mutable collection of one or more named objects. The
names are called *selectors*, and the objects are called *components*.
Different components may have different types. A record type specification has the form

record [field_spec , ...]

where

field_spec **::=** name, ... : type_spec

Selectors must be unique within a specification, but the ordering and
grouping of selectors is unimportant. For example, the following name
the same type:

record[last, first, middle: string, age: int]

record[last: string, age: int, first, middle: string]

A record is created using a record constructor, such as

info${last: "Jones", first: "John", age: 32, middle: "J."}

(assuming that "info" has been equated to one of the above type
specifications; see section A.4.3). An expression must be given for
each selector, but the order and grouping of selectors need not resemble the corresponding type specification. Record constructors are
discussed in section A.6.8.

In the following definitions of record operations, let

rt = record[n₁: t₁, ..., nₖ: tₖ]

Operations

get_nᵢ = **proc** (r: rt) **returns** (tᵢ)
effects Returns the component of r whose selector is n_i. There
is a *get_* operation for each selector.

set_n = **proc** (r: rt, e: tᵢ)
modifies r
effects Modifies r by making the component whose selector is
n_i be e. There is a *set_* operation for each selector.

r_gets_r = **proc** (r1, r2: rt)
modifies *r1*
effects Sets each component of *r1* to be the corresponding
component of *r2*.

r_gets_s = **proc** (r: rt, s: st)
    modifies r
    effects Here st is a struct type whose components have the
    same selectors and types as rt. Sets each component of r to
    be the corresponding component of s.

equal = **proc** (r1, r2: rt) **returns** (bool)
    effects Returns true if $r1$ and $r2$ are the same record object;
    otherwise returns false.

similar = **proc** (r1, r2: rt) **returns** (bool)
    requires Each t_i has operation
        similar: **proctype** (t_i, t_i) **returns** (bool)
    effects Returns true if $r1$ and $r2$ contain similar objects for
    each component as determined by the $t_i\$similar$ operations.
    The comparison is done in sorted alphanumeric order; if
    any comparison returns false, false is returned immedi-
    ately.

similar1 = **proc** (r1, r2: rt) **returns** (bool)
    requires Each t_i has operation
        equal: **proctype** (t_i, t_i) **returns** (bool)
    effects Returns true if $r1$ and $r2$ contain equal objects for each
    component as determined by the $t_i\$equal$ operations. The
    comparison is done in sorted alphanumeric order; if any
    comparison returns false, false is returned immediately.

copy1 = **proc** (r: rt) **returns** (rt)
    effects Returns a new record containing the components of r
    as its components.

copy = **proc** (r: rt) **returns** (rt)
    requires Each t_i has operation
        copy: **proctype** (t_i) **returns** (t_i)
    effects Returns a new record obtained by performing $copy1(r)$
    and then replacing each component with a copy of the cor-
    responding component of r. Copies are obtained by calling
    the $t_i\$copy$ operations. Copying is done in sorted alphanu-
    meric order.

end record

A.9.11 Structs

struct = **data type** $[n_1: t_1, \ldots, n_k: t_k]$ **is** get_ , replace_ , s2r, r2s,
        equal, similar, copy

Overview

A struct is an immutable record. A struct type specification has the
form
        struct[field_ spec, ...]
where (as for records)
        field_ spec **: :** = name, ... : type_ spec
A struct is created using a struct constructor, which syntactically is
identical to a record constructor. Struct constructors are discussed in
section A.6.8.

In the following operation descriptions,
        st = struct$[n_1: t_1, \ldots, n_k: t_k]$

Operations

get_ n_i = **proc** (s: st) **returns** (t_i)
        effects Returns the component of s whose selector is n_i. There
            is a *get_* operation for each selector.

replace_ n_i = **proc** (s: st, e: t_i) **returns** (st)
        effects Returns a new struct object whose components are
            those of s except that component n_i is e. There is a
            *replace_* operation for each selector.

s2r = **proc** (s: st) **returns** (rt)
        effects Here *rt* is a record type whose components have the
            same selectors and types as *st*. Returns a new record object
            whose components are those of s.

r2s = **proc** (r: rt) **returns** (st)
        effects Here *rt* is a record type whose components have the
            same selectors and types as *st*. Returns a struct object
            whose components are those of r.

equal = **proc** (s1, s2: st) **returns** (bool)
  requires Each t_i has operation
    equal: **proctype** (t$_i$, t$_i$) **returns** (bool)
  effects Returns true if *s1* and *s2* contain equal objects for each
    component as determined by the $t_i\$equal$ operations. The
    comparison is done in sorted alphanumeric order; if any
    comparison returns false, false is returned immediately.

similar = **proc** (s1, s2: st) **returns** (bool)
  requires Each t_i has operation
    similar: **proctype** (t$_i$, t$_i$) **returns** (bool)
  effects Returns true if *s1* and *s2* contain similar objects for
    each component as determined by the $t_i\$similar$ operations.
    The comparison is done in sorted alphanumeric order; if
    any comparison returns false, false is returned immedi-
    ately.

copy = **proc** (s: st) **returns** (st)
  requires Each t_i has operation
    copy: **proctype** (t$_i$) **returns** (t$_i$)
  effects Returns a struct containing a copy of each component
    of *s*; copies are obtained by calling the $t_i\$copy$ operations.
    Copying is done in sorted alphanumeric order.

end struct

A.9.12 Oneofs

oneof = **data type** [n$_1$: t$_1$, ..., n$_k$: t$_k$] **is** make_ , is_ , value_ , o2v, v2o,
  equal, similar, copy

Overview

A oneof type is a *tagged, discriminated union*. A oneof is a labeled
object, to be thought of as "one of" a set of alternatives. The label
is called the *tag part*, and the object is called the *data part*. A oneof
type specification has the form
  oneof [field_ spec , ...]
where (as for records)
  field_ spec **: :** = name, ... : type_ spec
Tags must be unique within a specification, but the ordering and
grouping of tags is unimportant.

Although there are oneof operations for decomposing oneof objects, they are usually decomposed using the **tagcase** statement, which is discussed in section A.7.9.

In the following descriptions of oneof operations,

$$\text{ot} = \text{oneof}[n_1\colon t_1, \ldots, n_k\colon t_k]$$

Operations

make_n_i = **proc** (e: t_i) **returns** (ot)
> **effects** Returns a oneof object with tag n_i and value e. There is a *make_* operation for each selector.

is_n_i = **proc** (o: ot) **returns** (bool)
> **effects** Returns true if the tag of o is n_i; otherwise returns false. There is an *is_* operation for each selector.

value_n_i = **proc** (o: ot) **returns** (t_i) **signals** (wrong_tag)
> **effects** If the tag of o is n_i, returns the value of o; otherwise signals *wrong_tag*. There is a *value_* operation for each selector.

o2v = **proc** (o: ot) **returns** (vt)
> **effects** Here vt is a variant type with the same selectors and types as *ot*. Returns a new variant object with the same tag and value as o.

v2o = **proc** (v: vt) **returns** (ot)
> **effects** Here vt is a variant type with the same selectors and types as *ot*. Returns a oneof object with the same tag and value as v.

equal = **proc** (o1, o2: ot) **returns** (bool)
> **requires** Each t_i has operation
> > equal: **proctype** (t_i, t_i) **returns** (bool)
>
> **effects** Returns true if *o1* and *o2* have the same tag and equal values as determined by the *equal* operation of their type.

similar = **proc** (o1, o2: ot) **returns** (bool)
> **requires** Each t_i has operation
> > similar: **proctype** (t_i, t_i) **returns** (bool)
>
> **effects** Returns true if *o1* and *o2* have the same tag and similar values as determined by the *similar* operation of their type.

copy = **proc** (o: ot) **returns** (ot)
   requires Each t_i must have operation
      copy: **proctype** (t_i) **returns** (t_i)
   effects Returns a oneof object with the same tag as o and
   containing as a value a copy of o's value; the copy is made
   using the *copy* operation of the value's type.

end oneof

A.9.13 Variants

variant = **data type** [n_1: t_1, ..., n_k: t_k] **is** make_ , change_ , is_ ,
      value_ , v_gets_v, v_gets_0, equal, similar, similar1, copy,
      copy1

Overview

A variant is a mutable oneof. A variant type specification has the
form
   variant [field_spec , ...]
where (as for records)
   field_spec **::=** name, ... : type_spec
Although there are variant operations for decomposing variant
objects, they are usually decomposed using the **tagcase** statement,
which is discussed in section A.7.9.

In the following descriptions of variant operations,
   vt = variant[n_1: t_1, ..., n_k: t_k]

Operations

make_n_i = **proc** (e: t_i) **returns** (vt)
   effects Returns a new variant object with tag n_i and value e.
   There is a *make_* operation for each selector.

change_n_i = **proc** (v: vt, e: t_i)
   modifies v
   effects Modifies v to have tag n_i and value e. There is a
   *change_* operation for each selector.

is_n_i = **proc** (v: vt) **returns** (bool)
   effects Returns true if the tag of v is n_i; otherwise returns
   false. There is an *is_* operation for each selector.

value_ n_i = **proc** (v: vt) **returns** (t_i) **signals** (wrong_tag)
  effects If the tag of v is n_i, returns the value of v; otherwise
    signals *wrong_tag*. There is a *value_* operation for each
    selector.

v_gets_v = **proc** (v1, v2: vt)
  modifies *v1*
  effects Modifies *v1* to contain the same tag and value as *v2*.

v_gets_o = **proc** (v: vt, o: ot)
  modifies *v*
  effects Here *ot* is the oneof type with the same selectors and
    types as *vt*. Modifies *v* to contain the same tag and value
    as *o*.

equal = **proc** (v1, v2: vt) **returns** (bool)
  effects Returns true if *v1* and *v2* are the same variant object.

similar = **proc** (v1, v2: vt) **returns** (bool)
  requires Each t_i must have operation
    similar: **proctype** (t_i, t_i) **returns** (bool)
  effects Returns true if *v1* and *v2* have the same tag and sim-
    ilar values as determined by the *similar* operation of their
    type.

similar1 = **proc** (v1, v2: vt) **returns** (bool)
  requires Each t_i must have operation
    equal: **proctype** (t_i, t_i) **returns** (bool)
  effects Returns true if *v1* and *v2* have the same tag and equal
    values as determined by the *equal* operation of their type.

copy = **proc** (v: vt) **returns** (vt)
  requires Each t_i must have operation
    copy: **proctype** (t_i) **returns** (t_i)
  effects Returns a variant object with the same tag as *v* and
    containing as a value a copy of *v*'s value; the copy is made
    using the *copy* operation of the value's type.

copy1 = **proc** (v: vt) **returns** (vt)
  effects Returns a new variant object with the same tag as *v*
    and containing *v*'s value as its value.

end variant

A.9.14 Procedure and Iterator Types

Procedures and iterators are objects created by the CLU system. The type specification for a procedure or iterator contains most of the information stated in a procedure or iterator heading; a procedure type specification has the form

proctype ([type_spec , ...]) [returns] [signals]

and an iterator type specification has the form

itertype ([type_spec , ...]) [yields] [signals]

where

returns	::= **returns** (type_spec , ...)
yields	::= **yields** (type_spec , ...)
signals	::= **signals** (exception , ...)
exception	::= name [(type_spec , ...)]

The first list of type specifications describes the number, types, and order of arguments. The **returns** or **yields** clause gives the number, types, and order of the objects to be returned or yielded. The **signals** clause lists the exceptions raised by the procedure or iterator; for each exception name, the number, types, and order of the objects to be returned are also given. All names used in a **signals** clause must be unique, and none can be *failure*, which has a standard meaning in CLU. The ordering of exceptions is not important. For example, both of the following type specifications name the procedure type for string$substr:

proctype (string, int, int) **returns** (string)
            signals (bounds, negative_size)
proctype (string, int, int) **returns** (string)
            signals (negative_size, bounds)

In the following operation descriptions, t stands for a proctype or itertype.

Operations

equal = **proc** (x, y: t) **returns** (bool)
similar = **proc** (x, y: t) **returns** (bool)
        effects These operations return true if and only if x and y are
            the same module with the same parameters.

copy = **proc** (x: t) **returns** (y: t)
    effects Returns y such that $x = y$.

A.10 Input/Output

This section describes a set of standard "library" data types and procedures for CLU, provided primarily to support I/O. We do not consider this facility to be part of the language proper, but we feel the need for a set of commonly used functions that have some meaning on most systems. This facility is minimal because we wish it to be general, that is, to be implementable, at least in large part, under almost any operating system. The facility also provides a framework in which some other operations that are not always available can be expressed.

Some thought has been given to portability of programs, and possibly even data, but we expect that programs dealing with all but the simplest I/O will have to be written very carefully to be portable, and might not be portable no matter how careful one is.

We shall describe types for naming files, for providing access to text and image files, and for attaching calendar date and time to files. No type "file" exists, as will be explained in section A.10.3.

A.10.1 Files

Our notion of file is a general one that includes not only storage files (disk files) but also terminals and other devices (for example, tape drives). Each file will in general support only a subset of the operations described here.

There are two basic kinds of files, *text files* and *image files*. The two kinds of files may be incompatible. However, on any particular system, it may not be possible to determine which kind a given file is.

A text file consists of a sequence of characters and is divided into lines terminated by newline (\n) characters. A nonempty last line might not be terminated. By convention, the start of a new page is indicated by placing a newpage (\p) character at the beginning of the first line of that page.

A text file will be stored in the (most appropriate) standard text file format of the local operating system. As a result, certain control characters (such as NUL, CR, FF, ↑C, ↑Z) may be ignored when written. In addition, a system may limit the maximum length of lines and may add (remove) trailing spaces to (from) lines.

Image files are provided to allow more efficient storage of information than is provided by text files. Unlike text files, there is no need for image files to be compatible with any local file format; thus image files can be defined more precisely than text files.

An image file consists of a sequence of encoded objects. Objects are written and read using *encode* and *decode* operations of their types. (These in turn will call *encode* and *decode* on their components until built-in types are reached.) The objects stored in an image file are not tagged by the system according to their types. Thus if a file is written by performing a specific sequence of *encode* operations, it must be read back using the corresponding sequence of *decode* operations to be meaningful.

A.10.2 File Names

file_ name = **data type is** create, get_dir, get_ name, get_ suffix,
get_other, parse, unparse, make_output, make_temp, equal,
similar, copy

Overview

File names are immutable objects used to name files. The system file name format is viewed as consisting of four string components:

1. directory—specifies a file directory or device;

2. name—the primary name of the file (for example, "thesis");

3. suffix—a name normally indicating the type of file (for example, "clu" for a CLU source file);

4. other—all other components of the system file name form.

The *directory* and *other* components may have internal syntax. The *name* and *suffix* should be short identifiers. (For example, in the TOPS-20 file name "ps:⟨cluser⟩ref.lpt.3", the *directory* is "ps:⟨cluser⟩", the *name* is "ref", the *suffix* is "lpt", and the *other* is "3". In the UNIX path name "/usr/snyder/doc/refman.r", the *directory* is "/usr/snyder/doc", the *name* is "refman", the *suffix* is "r", and there is no *other*.)

A null component has the following interpretation:

1. directory—denotes the current "working" directory (for example, the "connected directory" under TOPS-20 and the "current directory" under UNIX—see also section A.10.6);

2. name—may be illegal, have a unique interpretation, or be ignored (for example, under TOPS-20, a null name is illegal for most directories, but for some devices the name is ignored);

3. suffix—may be illegal, have a unique interpretation, or be ignored (for example, under TOPS-20, a null suffix is legal, as in "⟨rws⟩foo");

4. other—should imply a reasonable default.

Operations

create = **proc** (dir, name, suffix, other: string) **returns** (file_name)
       signals (bad_format)
       effects Creates a new file name from its components; if any component is not in the form required by the underlying system, signals *bad_format*. In the process of creating a file name, the string arguments may be transformed—for example, by truncation or case-conversion.

get_dir = **proc** (fn: file_name) **returns** (string)
get_name = **proc** (fn: file_name) **returns** (string)
get_suffix = **proc** (fn: file_name) **returns** (string)
get_other = **proc** (fn: file_name) **returns** (string)
       effects These operations return string forms of the components of a file name. If the file name was created using the *create* operation, the strings returned may be different than those given as arguments to *create*—for example, they may be truncated or case-converted.

parse = **proc** (s: string) **returns** (file_name) **signals** (bad_format)
       effects This operation creates a file name given a string in the system standard file name syntax.

unparse = **proc** (fn: file_name) **returns** (string)
       effects This operation transforms a file name into the system standard file name syntax. We require that
           parse(unparse(fn)) = fn
           create(fn.dir, fn.name, fn.suffix, fn.other) = fn
       for all file names *fn*. One implication of this rule is that there can be no file name that can be created by *create* but not by *parse*; if a system does have file names that have no string representation in the system standard file name syntax, *create* must reject those file names as having a bad format. Alternatively, the file name syntax can be extended so that it can express all possible file names.

make_output =**proc** (fn: file_name, suffix: string)
                    returns (file_name) **signals** (bad_format)
              effects This operation is used by programs that take input
                    from a file and write new files whose names are based on
                    the input file name. The operation transforms the file name
                    into one that is suitable for an output file. The transforma-
                    tion is done as follows: (1) the suffix is set to *suffix*; (2) if
                    the old directory is not suitable for writing, it is set to null;
                    (3) the name, if null and meaningless, is set to "output".

make_temp = **proc** (dir, prog, file_id: string) **returns** (file_name)
                    signals (bad_format)
              effects This operation creates a file name appropriate for a
                    temporary file, using the given directory name (*dir*),
                    program name (*prog*), and file identifier (*file_id*). To be
                    useful, both *prog* and *file_id* should be short and alpha-
                    betic. The returned file name, when used as an argument
                    to *stream$open* or *istream$open* to open a new file for writ-
                    ing, is guaranteed to create a new file and will not over-
                    write an existing file. Further file name references to the
                    created file should be made using the name returned by the
                    stream or istream *get_name* operation.

equal = **proc** (fn1, fn2: file_name) **returns** (**bool**)
similar = **proc** (fn1, fn2: file_name) **returns** (**bool**)
              effects Returns true if and only if the two file_names will
                    unparse to equal strings.

copy = **proc** (fn: file_name) **returns** (file_name)
              effects Returns a file_name that is equal to *fn*.

end file_name

A.10.3 A File Type

Although files are the basic information-containing objects in this package, we do not introduce a type file. Our reason is that few systems provide an adequate representation for files. On many systems, the most reliable representation of a file (accessible to the user) is a channel (stream) to that file. However, this representation is inappropriate for a CLU file type, since possession of a channel to a file often implies locking that file. Another possible representation is a file name. However, file names are one level removed from files, via the file directory. As a result, the relationship of a file name to a file object is time-varying. Using file names as representations for files would imply that all file operations could signal *non_existent_file*.

For these reasons, operations related to file objects are performed by two clusters, *stream* and *istream*, and operations related to the directory system are performed by procedures.

Note that two opens for read with the same file name might return streams to two different files. We cannot guarantee anything about what happens to a file after a program obtains a stream to it.

A.10.4 Streams

stream = **data type is** open, primary_input, primary_output,
        error_output, can_read, can_write,
        reset, flush, close, abort, is_closed, is_terminal,
        getc, peekc, empty, getl, gets,
        putc, puts, putzero, putleft, putright, putspace,
        getc_image, putc_image, puts_image, gets_image,
        get_lineno, set_lineno, get_line_length,
        get_page_length, get_date, set_date,
        get_name, set_output_buffered, get_output_buffered,
        equal, similar, copy,
        create_input, create_output, get_contents, % string I/O
        get_buf, get_prompt, set_prompt, % terminal I/O
        get_input_buffered, set_input_buffered, % terminal I/O
        add_script, rem_script, unscript % scripting

Overview

Streams provide the means to read and write text files and to perform some other operations on file objects. The operations allowed on any particular stream depend upon the access mode. In addition, certain operations may have no effect in some implementations.

When an operation cannot be performed because of an incorrect access mode, implementation limitations, or properties of an individual file or device, the operation will signal *not_possible* (unless the description of the operation explicitly says that the invocation will be ignored). *End_of_file* is signaled by reading operations when there are no more characters to read.

Streams provide operations to connect streams to strings, to interact with a user at a terminal, and to record input/output in one stream on another. These operations are described in subsequent sections.

Operations

open = **proc** (fn: file_name, mode: string) **returns** (stream)
   signals (not_possible(string))
  effects Opens a stream to *fn* in the given *mode*. The possible access modes are "read", "write", and "append". If *mode* is not one of these strings, *not_possible*("bad access mode") is signaled. In those cases where the system is able to detect that the specified preexisting file is not a text file, *not_possible*("wrong file type") is signaled. If *mode* is "read", the named file must exist and a stream is returned upon which input operations can be performed. If *mode* is "write", a new file is created or an old file is rewritten. A stream is returned upon which output operations can be performed. Write mode to storage files should guarantee exclusive access to the file, if possible. If *mode* is "append" and if the named file does not exist, one is created. A stream is returned, positioned at the end of the file, upon which output operations can be performed. Append mode to storage files should guarantee exclusive access to the file, if possible.

primary_input = **proc** ( ) **returns** (stream)
  effects Returns the "primary" input stream, suitable for reading. This is usually a stream to the user's terminal, but may be set by the operating system.

primary_output = **proc** ( ) **returns** (stream)
  effects Returns the "primary" output stream, suitable for writing. This is usually a stream to the user's terminal, but may be set by the operating system.

error_output = **proc** ( ) **returns** (stream)
>   **effects** Returns the "primary" output stream for error messages, suitable for writing. This is usually a stream to the user's terminal, but may be set by the operating system.

can_read = **proc** (s: stream) **returns** (bool)
>   **effects** Returns true if input operations appear possible on *s*.

can_write = **proc** (s: stream) **returns** (bool)
>   **effects** Returns true if output operations appear possible on *s*.

getc = **proc** (s: stream) **returns** (char)
>   **signals** (end_of_ file, not_ possible(string))
>   **modifies** *s*
>   **effects** Removes the next character from *s* and returns it. Signals *end_of_file* if there are no more characters.

peekc = **proc** (s: stream) **returns** (char)
>   **signals** (end_of_ file, not_ possible(string))
>   **effects** This input operation is like *getc*, except that the character is not removed from *s*.

empty = **proc** (s: stream) **returns** (bool)
>   **signals** (not_ possible(string))
>   **effects** Returns true if and only if there are no more characters in the stream. It is equivalent to an invocation of *peekc*, where true is returned if *peekc* returns a character and false is returned if *peekc* signals *end_of_ file*. Thus in the case of terminals, for example, this operation may wait until additional characters have been typed by the user.

getl = **proc** (s: stream) **returns** (string)
>   **signals** (end_of_ file, not_ possible(string))
>   **modifies** *s*
>   **effects** Reads and returns (the remainder of) the current input line and reads but does not return the terminating newline (if any). This operation signals *end_of_ file* only if there are no characters and end-of-file is detected.

gets = **proc** (s: stream, term: string) **returns** (string)
      signals (end_of_file, not_possible(string))
    modifies *s*
    effects Reads characters until a terminating character (one in
      term) or end-of-file is seen. The characters up to the ter-
      minator are returned; the terminator (if any) is left in the
      stream. Signals *end_of_file* only if there are no characters
      and end-of-file is detected.

putc = **proc** (s: stream, c: char) **signals** (not_possible(string))
    modifies *s*
    effects Appends *c* to *s*. Writing a newline indicates the end of
      the current line.

putl = **proc** (s: stream, str: string) **signals** (not_possible(string))
    modifies *s*
    effects Writes the characters of *str* onto *s*, followed by a new-
      line.

puts = **proc** (s: stream, str: string) **signals** (not_possible(string))
    modifies *s*
    effects Writes the characters in *str* using *putc*. It should be
      somewhat more efficient than doing a series of individual
      *putc*s.

putzero = **proc** (s: stream, str: string, cnt: int)
      signals (negative_field_width, not_possible(string))
    modifies *s*
    effects Outputs *str*. However, if the length of *str* is less than
      cnt, outputs *cnt* − *length*(*str*) zeros before the first digit or
      period in the string (or at the end, if no such characters).

putleft = **proc** (s: stream, str: string, cnt: int)
      signals (negative_field_width, not_possible(string))
    modifies *s*
    effects Outputs *str*. However, if the length of *str* is less than
      cnt, outputs *cnt* − *length*(*str*) spaces after the string.

putright = **proc** (s: stream, str: string, cnt: int)
        signals (negative_ field_width, not_ possible(string))
        modifies s
        effects Outputs *str*. However, if the length of *str* is less than
            cnt, outputs *cnt* − *length*(*str*) spaces before the string.

putspace = **proc** (s: stream, cnt: int)
        signals (negative_ field_width, not_ possible(string))
        modifies s
        effects Outputs *cnt* spaces on *s*.

putc_image = **proc** (s: stream, c: char)
        signals (not_ possible(string))
        modifies s
        effects Like *putc*, except that an arbitrary character may be
            written and the character is not interpreted by the CLU
            I/O system. (An escape sequence consists of a special char-
            acter followed by a fixed number of arbitrary characters.
            These characters could be the same as an end-of-line mark,
            but they are recognized as data by their context. On a
            record-oriented system, such characters would be part of
            the data. In either case, writing a newline in image mode
            would not be interpreted by the CLU system as indicating
            an end-of-line.) Characters written to a terminal stream
            with this operation can be used to cause terminal-dependent
            control functions.

getc_image = **proc** (s: stream) **returns** (char)
        signals (end_of_ file, not_ possible(string))
        modifies s
        effects Provided to read escape sequences in text files, as
            might be written using *putc_ image*. Using this operation
            inhibits the recognition of end-of-line marks, where used.
            When reading from a terminal stream, the character is not
            echoed and is not subject to interpretation as an editing
            command.

puts_image = **proc** (s: stream, str: string)
        signals (not_possible(string))
      modifies *s*
      effects Writes the characters in *str* using *putc_image*. It should
      be somewhat more efficient than doing a series of individual
      *putc_image*s.

gets_image = **proc** (s: stream, term: string) **returns** (string)
        signals (end_of_file, not_possible(string))
      modifies *s*
      effects Reads characters until a terminating character (one in
      term) or end-of-file is seen. Using this operation inhibits
      the recognition of end-of-line marks, where used. When
      reading from a terminal stream, the characters read are
      not echoed and are not subject to interpretation as edit-
      ing commands. The characters up to the terminator are
      returned; the terminator (if any) is left in the stream. This
      operation signals *end_of_file* only if there are no characters
      and end-of-file is detected.

close = **proc** (s: stream) **signals** (not_possible(string))
      modifies *s*
      effects Attempts to terminate I/O and remove the association
      between the stream and the file. If successful, further use of
      operations that signal *not_possible* will signal *not_possible*.
      This operation will fail if buffered output cannot be writ-
      ten.

abort = **proc** (s: stream)
      modifies *s*
      effects Terminates I/O and removes the association between
      the stream and the file. If buffered output cannot be writ-
      ten, it will be lost, and if a new file is being written, it may
      or may not exist.

is_closed = **proc** (s: stream) **returns** (bool)
      effects Returns true if and only if the stream is closed.

is_terminal = **proc** (s: stream) **returns** (bool)
      effects Returns true if and only if the stream is attached to an
      interactive terminal (see below, page 403).

get_ lineno = **proc** (s: stream) **returns** (int)
   signals (end_of_ file, not_ possible(string))
   effects Returns the line number of the current (being or about
   to be read) line. If the system maintains explicit line numbers in the file, these line numbers are returned. Otherwise
   lines are implicitly numbered, starting with 1.

set_ lineno = **proc** (s: stream, num: int)
   signals (not_ possible(string))
   modifies s
   effects If the system maintains explicit line numbers in the
   file, sets the line number of the next (not yet started) line
   to *num*. Otherwise it is ignored.

get_ line_ length = **proc** (s: stream) **returns** (int)
   signals (no_ limit)
   effects If the file or device to which the stream is attached has
   a natural maximum line length, that length is returned.
   Otherwise *no_ limit* is signaled. The line length does not
   include newline characters.

get_ page_ length = **proc** (s: stream) **returns** (int)
   signals (no_ limit)
   effects If the device to which the stream is attached has a
   natural maximum page length, that length is returned.
   Otherwise *no_ limit* is signaled. Storage files will generally
   not have page lengths.

get_date = **proc** (s: stream) **returns** (date)
   signals (not_ possible(string))
   effects Returns the date of the last modification of the corresponding storage file.

set_date = **proc** (s: stream, d: date) **signals** (not_ possible(string))
   modifies s
   effects Sets the modification date of the corresponding storage
   file. The modification date is set automatically when a file
   is opened in "write" or "append" mode.

get_name = **proc** (s: stream) **returns** (file_name)
   signals (not_possible(string))
   effects Returns the name of the corresponding file. It may
   be different than the name used to open the file, in that
   defaults have been resolved and link indirections have been
   followed.

set_output_buffered = **proc** (s: stream, b: bool)
   signals (not_possible(string))
   modifies *s*
   effects Sets the output buffering mode. Normally output may
   be arbitrarily buffered before it is actually written out.
   Unbuffered output can be used on some systems to
   decrease the amount of information lost if the program
   terminates prematurely. For terminal streams, unbuffered
   output is useful in programs that output incomplete lines
   as they are working since it allows the user to watch the
   progress of the program.

get_output_buffered = **proc** (s: stream) **returns** (bool)
   effects Returns true if and only if output to the stream is
   being buffered.

equal = **proc** (s1, s2: stream) **returns** (bool)
similar = **proc** (s1, s2: stream) **returns** (bool)
   effects Returns true if and only both arguments are the same
   stream.

copy = **proc** (s: stream) **returns** (stream)
   effects Returns a stream equal to *s*.

String Input/Output
It is occasionally useful to be able to construct a stream that is not connected to a file, but instead simply collects the output text into a string. Conversely, it is occasionally useful to be able to take a string and convert it into an input stream so that it can be given to a procedure that expects a stream. String streams allow these functions to be performed. A string stream does not have a file name, a creation date, a maximum line or page length, or explicit line numbers. The following stream operations deal with string streams:

create_input = **proc** (s: string) **returns** (stream)
> **effects** An input stream is created that will return the characters in the given string. If the string is nonempty and does not end with a newline, an extra terminating newline will be appended to the string.

create_output = **proc** ( ) **returns** (stream)
> **effects** An output stream is created that will collect output text in an internal buffer. The text may be extracted using the *get_contents* operation.

get_contents = **proc** (s: stream) **returns** (string)
> **signals** (not_possible(string))
> **effects** Returns the text that has so far been output to *s*. Signals *not_possible* if the stream was not created by *create_output*.

Terminal I/O

Terminal I/O is performed via streams attached to interactive terminals. Such a stream is normally obtained using the *primary_input* and *primary_output* operations. A terminal stream is capable of performing both input and output operations. A number of additional operations are possible on terminal streams, and a number of standard operations have special interpretations.

Terminal input will normally be buffered so that the user can perform editing functions, such as deleting the last character on the current line, deleting the current line, redisplaying the current line, and redisplaying the current line after clearing the screen. Specific characters for causing these functions are not suggested. In addition, some means must be provided for the user to indicate end-of-file, so that a terminal stream can be given to a program that expects an arbitrary stream and reads it until end-of-file. The end-of-file status of a stream is cleared by the *reset* operation.

Input buffering is normally provided on a line basis. When a program first asks for input (using *getc*, for example), an entire line of input is read from the terminal and stored in an internal buffer. Further input is not taken from the terminal until the existing buffered input is read.

New input caused to be read by the *getbuf* operation will be buffered as a unit. Thus one can read in a large amount of text and allow editing of that entire text. In addition, when the internal buffer is empty, the *getc_image* operation will read a character directly from the terminal, without interpreting it or echoing it.

The user may specify a prompt string to be printed whenever a new buffer of input is requested from the terminal; the prompt string will be reprinted when redisplay of the current line is requested by the user. However, if new input is requested just when an unfinished line has been output to the terminal, that unfinished line is used instead as a prompt.

The routine *putc_ image* can be used to cause control functions, such as '\007' (bell) and '\p' (new-page or clear-screen). CLU does not guarantee the effect caused by any particular control character, but we recommend that the standard ASCII interpretation of control characters be supported wherever possible.

Terminal output may be buffered by the system up to one line at a time. However, the buffer must be flushed when new input is requested from the terminal. Terminal streams do not have modification dates, but they should have file names and implicit line numbers. The following additional operations are provided:

getbuf = **proc** (s: stream, str: string) **returns** (string)
        signals (end_of_ file, not_ possible(string))
      modifies *s*
      effects This operation is the same as *gets*, but for terminals
          with input buffering, the new input read by *getbuf* is
          buffered as a unit, rather than a line at a time, allowing
          input editing of the entire text.

get_ prompt = **proc** (s: stream) **returns** (string)
      effects Returns the current prompt string. The prompt string
          is initially empty (" "). The empty string is returned for
          nonterminal streams.

set_ prompt = **proc** (s: stream. str: string)
      modifies *s*
      effects Sets the string to be used for prompting to *str*. For
          nonterminal streams there is no effect.

get_ input_buffered = **proc** (s: stream) **returns** (bool)
      effects Returns true if and only if *s* is attached to a terminal
          and input is being buffered.

set_input_buffered = **proc** (s: stream, b: bool)
  signals (not_possible(string))
  modifies *s*
  effects Sets the input buffering mode. Only buffered terminal
  input is subject to editing.

Scripting
Streams provide a mechanism for recording the input and/or output from
one stream onto any number of other streams. This can be particularly
useful in recording terminal sessions. The following additional operations
are provided:

add_script = **proc** (s1, s2: stream) **signals** (script_failed)
  modifies *s1*
  effects Adds *s2* as a script stream of *s1*. All subsequent input
  from and output to *s1* will also be output to *s2*.
  Not_possible exceptions that arise in actually outputting
  to *s2* will be ignored. This operation signals *script-failed*
  if *s2* cannot be written to or if either stream is a direct or
  indirect script stream of the other.

rem_script = **proc** (s1, s2: stream)
  modifies *s2*
  effects Removes, but does not close, *s2* as a direct script
  stream of *s1*.

unscript = **proc** (s: stream)
  effects Removes, but does not close, all direct script streams
  of *s*.

end stream

A.10.5 Istreams

istream = **data type is** open, can_read, can_write, empty, reset, flush,
  get_date, set_date, get_name, close, abort, is_closed, equal,
  similar, copy

Overview

Istreams provide the means to read and write image files and to perform some other operations on file objects. The operations allowed
on any particular istream depend upon the access mode. In addition,
certain operations may be null in some implementations.

When an operation cannot be performed, because of an incorrect access mode, implementation limitations, or properties of an individual file or device, the operation will signal *not_possible* (unless the description of the operation explicitly says that the invocation will be ignored).

Actual reading and writing of objects is performed by *encode* and *decode* operations of the types involved. All of the built-in CLU types and type generators (except the routine type generators) and the file_name and date types, provide these operations. Designers of abstract types are encouraged to provide them also. The type specifications of the *encode* and *decode* operations for a type T are

encode = **proc** (c: T, s: istream)
  signals (not_possible(string))
decode = **proc** (s: istream) **returns** (T)
  signals (end_of_file, not_possible(string))

For parameterized types, *encode* will have a **where** clause requiring *encode* operations for all components, and *decode* will have a **where** clause requiring *decode* operations for all components.

The *encode* operations are output operations. They write an encoding of the given object onto the istream. The *decode* operations are input operations. They decode the information written by *encode* operations and return an object similar to the one encoded. If the sequence of *decode* operations used to read a file does not match the sequence of *encode* operations used to write it, meaningless objects may be returned. The system may in some cases be able to detect this condition, in which case the *decode* operation will signal *not_possible* ("bad format"). The system is not guaranteed to detect all such errors.

Operations

open = **proc** (fn: file_name, mode: string)
  signals (not_possible(string))
  effects The possible access modes are "read", "write", and "append". If *mode* is not one of these strings, not_possible("bad access mode") is signaled. In those cases where the system is able to detect that the specified pre-existing file is not an image file, *not_possible*("wrong file type") is signaled. If *mode* is "read", the named file must exist. If the file exists, an image stream is returned upon which *decode* operations can be performed. If *mode* is

"write", a new file is created or an old file is rewritten. An
image stream is returned upon which *encode* operations can
be performed. Write mode to storage files should guarantee
exclusive access to the file, if possible. If *mode* is "append"
and if the named file does not exist, one is created. An
image stream is returned, positioned at the end of the file,
upon which *encode* operations can be performed. Append
mode to storage files should guarantee exclusive access to
the file, if possible.

can_read = **proc** (s: istream) **returns** (bool)
    effects Returns true if *decode* operations appear possible
      on *s*.

can_write = **proc** (s: istream) **returns** (bool)
    effects Returns true if *encode* operations appear possible
      on *s*.

empty = **proc** (s: istream) **returns** (bool)
    effects Returns true if and only if there are no more objects in
      the associated file.

reset = **proc** (s: istream) **signals** (not_possible(string))
    effects Resets *s* so that the next input or output operation
      will read or write the first item in the file.

flush = **proc** (s: istream) **signals** (not_possible(string))
    effects Writes any buffered output to the associated file, if
      possible.

get_date = **proc** (s: istream) **returns** (date)
      signals (not_possible(string))
    effects Returns the date of the last modification of the corre-
      sponding storage file.

set_date = **proc** (s: istream, d: date) **signals** (not_possible(string))
    modifies *s*
    effects Sets the modification date of the corresponding storage
      file. The modification date is set automatically when a file
      is opened in "write" or "append" mode.

get_name = **proc** (s: istream) **returns** (file_name)
> **effects** Returns the name of the corresponding file. It may
> be different than the name used to open the file, in that
> defaults have been resolved and link indirections have been
> followed.

close = **proc** (s: istream) **signals** (not_possible(string))
> **modifies** s
> **effects** Attempts to terminate I/O and remove the association
> between s and the file. If successful, further use of opera-
> tions that signal *not_possible* will signal *not_possible*. This
> operation will fail if buffered output cannot be written.

abort = **proc** (s: istream)
> **modifies** s
> **effects** Terminates I/O and removes the association between
> the istream and the file. If buffered output cannot be writ-
> ten, it will be lost, and if a new file is being written, it may
> or may not exist.

is_closed = **proc** (s: istream) **returns** (bool)
> **effects** Returns true if and only if s is closed.

equal = **proc** (s1, s2: istream) **returns** (bool)
similar = **proc** (s1, s2: istream) **returns** (bool)
> **effects** Returns true if and only both arguments are the same
> istream.

copy = **proc** (s: istream) **returns** (istream)
> **effects** Returns a stream that is equal to s.

end istream

A.10.6 Miscellaneous Procedures

There are a number of miscellaneous procedures that are useful for in-
put/output:

working_dir = **proc** ( ) **returns** (string)
>   **effects** Returns the name of the current working directory.
>   The working directory is used by the I/O system to fill in
>   a null directory in a file name.

set_working_dir = **proc** (s: string) **signals** (bad_format)
>   **effects** Used to change the working directory. No checking of
>   directory access privileges is performed.

delete_file = **proc** (fn: file_name) **signals** (not_possible(string))
>   **effects** Deletes the specified storage file. An exception may be
>   signaled if the specified file does not exist, but an exception
>   will not be signaled solely because the file does not exist.
>   For example, an exception may be signaled if the specified
>   directory does not exist or if the user does not have access
>   to the directory.

rename_file = **proc** (fn1, fn2: file_name)
>   **signals** (not_possible(string))
>   **effects** Renames the file specified by *fn1* to have the name
>   specified by *fn2*. Renaming across directories and devices
>   may or may not be allowed.

user_name = **proc** ( ) **returns** (string)
>   **effects** Returns some identification of the user who is associ-
>   ated with the executing process.

now = **proc** ( ) **returns** (date)
>   **effects** Returns the current date and time (see the next
>   section).

e_form = **proc** (r: real, x, y: int) **returns** (string)
>   **signals** (illegal_field_width)
>   **effects** Returns a real literal of the form
> $$[-] i\_field\ [.f\_field\ ] e \pm x\_field)]$$
>   where *i_field* is x digits, *f_field* is y digits, and *x_field* is
>   *Exp_width* digits (see section A.9.4). If $y = 0$, the decimal
>   point and *f_field* are not present. If $r \neq 0.0$, the leftmost
>   digit of *i_field* is not zero. If $r = 0.0$, *x_field* is all zeros.
>   *Illegal_field_width* occurs if $x < 0$ or $y < 0$ or $x + y < 1$. If
>   necessary, x may be rounded off to fit the specified form.

f_form = **proc** (r: real, x, y: int) **returns** (string)
    signals (illegal_field_width, insufficient_field_width)
   effects Returns a real literal of the form
        [−] i_field.f_field
    where f_field is y digits. If $x > 0$, i_field is at least one
    digit, with leading zeros suppressed. If $x = 0$, i_field is not
    present. *Illegal_field_width* occurs if $x < 0$ or $y < 0$ or
    $x + y < 1$. If necessary, r may be rounded off to fit the
    specified form. *Insufficient_field_width* occurs if
    real$exponent(r) \geq x$ after any rounding.

g_form = **proc** (r: real, x, y: int) **returns** (string)
    signals (illegal_field_width, insufficient_field_width)
   effects If $r = 0.0$ or $-1 \leq$ real$exponent(r) < x$, the result
    returned is $f\_form(r, x, y)$. Otherwise the result is
    $e\_form(r, 1, x + y - Exp\_width - 3)$. *Illegal_field_width*
    occurs if $x < 0$ or $y < 0$ or $x + y < 1$. If necessary, r may be
    rounded off to fit the specified form. *Insufficient_field_width*
    occurs if $r \neq 0.0$ and $\sim (-1 \leq$ real$exponent(x) < y)$ and
    $(x + y < Exp\_width + 3)$ after any rounding.

A.10.7 Dates

date = **data type is** create, get_all, get_day, get_month, get_year,
       get_hour, get_minute, get_second

Overview

   Dates are immutable objects that represent calendar dates and times.

Operations

create = **proc** (day, month, year, hour, minute, second: int)
    returns (date) **signals** (bad_format)
   effects Creates the specified date. The ranges for the argu-
    ments are
        (1 .. 31), (1 .. 12), (1 ..), (0 .. 23), (0 .. 59), (0 .. 59),
    respectively; *bad_format* is signaled if an argument is out of
    range.

get_all = **proc** (d: date) **returns** (int, int, int, int, int, int)
   effects Returns the components in the same order as given to
    create.

get_day = **proc** (d: date) **returns** (int)
get_ month = **proc** (d: date) **returns** (int)
get_ year = **proc** (d: date) **returns** (int)
get_ hour = **proc** (d: date) **returns** (int)
get_ minute = **proc** (d: date) **returns** (int)
get_ second = **proc** (d: date) **returns** (int)
     effects Returns the specified component of *d*.

unparse = **proc** (d: date) **returns** (string)
     effects Returns a string representation of *d*—for example,
       "12 January 1978 01:36:59".

unparse_date = **proc** (d: date) **returns** (string)
     effects Returns a string representation of the date part of
       d—for example, "12 January 1978".

unparse_time = **proc** (d: date) **returns** (string)
     effects Returns a string representation of the time part of
       d—for example, "01:36:59".

lt = **proc** (d1, d2: date) **returns** (bool))
le = **proc** (d1, d2: date) **returns** (bool)
ge = **proc** (d1, d2: date) **returns** (bool)
gt = **proc** (d1, d2: date) **returns** (bool)
equal = **proc** (d1, d2: date) **returns** (bool)
     effects The obvious relational operations; if *d1* < *d2*, then *d1*
       occurs earlier than *d2*.

similar = **proc** (d1, d2: date) **returns** (bool)
     effects *d1* = *d2*.

copy = **proc** (d: date) **returns** (date)
     effects Returns a date equal to *d*.

end date

Appendix B
Implementation of the Formatter

Chapter 13 gave a specification and design of a program to do text formatting. This appendix contains an implementation of that program.

```
format = proc (ins, outs, errs: stream) signals (bad_arg(string))
    if ~stream$can_read(ins) then signal bad_arg ("input stream")
        elseif ~stream$can_write(errs)
            then signal bad_arg ("error stream")
        elseif ~stream$can_write(outs)
            then signal bad_arg ("output stream")
        end
    d: doc := doc$create(outs)
    while true do
        do_line(ins, d, errs)
        end except when all_done: end
    doc$terminate(d)
    stream$close(errs)
    stream$close(ins)
    end format
```

```
newline = '\n'
space = ' '
tab = '\t'
endword = " \t\n"   % end of word is space, tab, or newline

do_line = proc (ins: stream, d: doc, errs: stream) signals (all_done)
    c: char := stream$peekc(ins)
        except when end_of_file: signal all_done end
    if c = '.'
        then n: int := ins.lineno      % the number of the current line
            do_command(ins, d)
                except when error (why: string):
                    stream$putl(errs, int$unparse(n) || ": " || why)
                end
        else do_text_line(ins, d)
        end
    end do_line

do_command = proc (ins: stream, d: doc) signals (error(string))
    stream$getc(ins)    % skip the period
    cmd: string := stream$gets(ins, endword)  % read command
        except when end_of_file: signal error ("missing command") end
    stream$getl(ins)  % skip rest of line
        except when end_of_file: end
    if cmd = "nf" then doc$set_nofill(d)
        elseif cmd = "fi" then doc$set_fill(d)
        elseif cmd = "br" then doc$break_line(d)
        elseif cmd = " " then signal error ("missing command")
        else signal error ("'" || cmd || "' not a command")
        end
    end do_command
```

```
do_text_line = proc (ins: stream, d: doc)
    c: char := stream$getc(ins)
    if c = newline
        then doc$skip_line(d)
            return
        elseif c = space cor c = tab then doc$break_line(d)
        end
    while c ~= newline do
        if c = space then doc$add_space(d)
            elseif c = tab then doc$add_tab(d)
            else doc$add_word(d, string$c2s(c)|| stream$gets(ins, endword))
                except when end_of_file:
                    doc$add_word(d, string$c2s(c))
                    end
            end
        c := stream$getc(ins) except when end_of_file: break end
        end
    doc$add_newline(d)
    end do_text_line
```

doc = **cluster is** create, set_fill, set_nofill, add_newline, break_line,
        skip_line, add_word, add_space, add_tab, terminate

rep = record[line: line,	% the current line
fill: bool,	% true if in fill mode
from_right: bool,	% true if justify next line right to left
lineno: int,	% cnt of lines output on page so far
%	(not including header lines)
pageno: int,	% number of current output page
outs: stream]	% the output stream

chars_per_line = 60
lines_per_page = 50
left_margin_size = 10

% A typical doc object is a sequence of characters of all the lines
%   completed so far, followed by the characters in the current
%   line.
% The abstraction function is:
%   Characters in d.outs ‖ Characters in d.line

% The rep invariant is
%   can_write(r.outs)
%   & 1 < r.lineno <= lines_per_page
%   & 0 <= r.pageno
%   & r.page_no > 0 =>
%   The current page has the proper header &
%   r.line_no lines, not counting the header
%   & ∀ previous pages p (i.e., pages with numbers
%   1, ..., r.page_no − 1)
%   p has the proper header
%   & p has lines_per_page lines, not counting the header
%   & p is followed by a page break

create = **proc** (outs: stream) **returns** (**cvt**)
    return(rep${line: line$create( ),
                    fill: **true**,
                    from_right: **true**,
                    lineno: lines_per_page,
                    pageno: 0,
                    outs: outs})
        end create

```
add_word = proc (d: cvt, w: string)
    if d.fill cand line$length(d.line) > 0
        then if line$length(d.line) + string$size(w) > chars_per_line
            then line$justify(d.line, chars_per_line, d.from_right)
                d.from_right := ~d.from_right
                output_line(d)
            end
        end
    line$add_word(d.line, w)
    end add_word

add_space = proc (d: cvt)
    line$add_space(d.line)
    end add_space

add_tab = proc (d: cvt)
    line$add_tab(d.line)
    end add_tab

add_newline = proc (d: cvt)
    if ~d.fill
        then output_line(d)
        else line$add_space(d.line)
        end
    end add_newline

break_line = proc (d: cvt)
    if line$length(d.line) > 0 then output_line(d) end
    end break_line

skip_line = proc (d: cvt)
    break_line(up(d))
    output_line(d)  % output an empty line
    end skip_line

set_fill = proc (d: cvt)
    break_line(up(d))
    d.fill := true
    end set_fill
```

set_ nofill = **proc** (d: **cvt**)
   break_ line(**up**(d))
   d.fill := **false**
   end set_ nofill

terminate = **proc** (d: **cvt**)
   break_ line(**up**(d))
   stream$close(d.outs)
   end terminate

% internal routines of doc

output_ line = **proc** (d: **rep**)
   if d.lineno = lines_ per_ page
     then start_ page(d)
     else d.lineno := d.lineno + 1
     end
   if line$length(d.line) > 0
     then stream$putspace(d.outs, left_ margin_ size)
       line$output(d.line, d.outs)
       line$clear(d.line)
     end
   stream$putl(d.outs, " ") % output the end-of-line
   end output_ line

start_ page = **proc** (d: **rep**)
% modifies d
% effects outputs a page header on d.outs and prepares for
% the next page
     two_ lines = "\n\n"   % two empty lines
     page_break = '\p'
     d.from_ right := **true**
     d.lineno := 1
     d.pageno := d.pageno + 1
     if d.pageno > 1 **then** stream$putc(d.outs, page_break) **end**
     stream$puts(d.outs, two_ lines)
     stream$putspace(d.outs, left_ margin_ size)
     stream$puts(d.outs, "Page ")
     stream$putl(d.outs, int$unparse(d.pageno))
     stream$puts(d.outs, two_ lines)
     end start_ page

end doc

line = **cluster is** create, add_word, add_space, length, add_tab, justify,
output, clear

item = variant[word: string,
space: int,   % the int is the number of spaces
tab: int]   % the int is the width of the tab

items = array[item]

rep = record[length: int, right_tab: int, stuff: items]

max_tab_width = 8

% A typical line is a sequence of characters.
% The abstraction function is
%   if l.length = 0 then " "
%   else chars(l.stuff[1]) || ... || chars(top(l.stuff))
% where chars(item) is
%   if the item is a word, then the characters in the word;
%   otherwise n spaces, where n is the integer in the tab or
%   space item

% The rep invariant is
%   r.length = the sum of the ints in the tab and space items +
%   the sizes of the words
%   & r.right_tab >= 0
%   & low(r.stuff) = 1
%   & r.right_stuff > 0 => r.stuff[r.right_tab] is a tab
%   & there are no tabs higher in r.stuff
%   & there are no adjacent spaces in r.stuff
%   & the integer in each space and tab token is > 0
%   & the integer in each tab token is <= max_tab_width
%   & for each tab item in r.stuff,
%   count mod max_tab_width = 1
%   where count is the number of characters from the
%   beginning of r.stuff up to and including this tab

create = **proc** ( ) **returns** (**cvt**)
return (rep${length: 0, right_tab: 0, stuff: items$new( )})
end create

add_word = **proc** (l: **cvt**, w: string)
items$addh(l.stuff, item$make_word(w))
l.length := l.length + string$size(w)
end add_word

```
add_space = proc (l: cvt)
    l.length := l.length + 1
    tagcase items$top(l.stuff)
        tag space (cnt: int):
            item$change_space(items$top(l.stuff), cnt + 1)
            return
        others:
        end except when bounds: end  % handle empty array
    items$addh(l.stuff, item$make_space(1))
    end add_space

length = proc (l: cvt) returns (int)
    return (l.length)
    end length

add_tab = proc (l: cvt)
    cnt: int := max_tab_width − (l.length//max_tab_width)
    l.length := l.length + cnt
    items$addh(l.stuff, item$make_tab(cnt))
    l.right_tab := items$high(l.stuff)
    end add_tab

justify = proc (l: cvt, len: int, from_right: bool)
    tagcase items$top(l.stuff)        % remove trailing spaces
        tag space (cnt: int):
            items$remh(l.stuff)
            l.length := l.length − cnt
        others:
        end
    if l.length >= len then return end
    diff: int := len − l.length   % extra spaces
    first: int := find_first_justifiable_space(l)
        except when none: return end
    enlarge_spaces(l, first, diff, from_right)
    end justify
```

```
output = proc (l: cvt, outs: stream)
    for t: item in items$elements(l.stuff) do
        tagcase t
            tag word (s: string): stream$puts(outs, s)
            tag space, tab (cnt: int): stream$putspace(outs, cnt)
            end
        end
    end output

clear = proc (l: cvt)
    l.length := 0
    items$trim(l.stuff, 1, 0)
    l.right_tab := 0
    end clear

%  internal routines of line

find_first_justifiable_space = proc (l: rep) returns (int)
            signals (none)
% effects returns index in l.stuff of first justifiable space;
% signals none if no such space
        a: items := l.stuff
        if items$empty(a) then signal none end
        hi: int := items$high(a)
        i: int := l.right_tab + 1  % position after rightmost tab
        if i <= hi cand item$is_space(a[i])  % find first word after tab
            then i := i + 1 end  % there are no adjacent spaces
        % find first space after word
        while i <= hi cand ~item$is_space(a[i]) do
            i := i + 1
            end
        if i > hi then signal none end
        return (i)
        end find_first_justifiable_space
```

```
enlarge_spaces = proc (l: rep, first, diff: int, from_right: bool)
% modifies l
% effects distributes diff spaces between first and the last space
%     in l.stuff; extra spaces are added from the right if
%     from_right is true; otherwise from the left
    nspaces, last: int := count_spaces(l, first)
    if nspaces = 0 then return end
    by: int := 1
    if from_right
        then by := − 1
            first, last := last, first
        end
    neach: int := diff/nspaces   % amount to increase each space
    nextra: int := diff//nspaces   % leftovers to distribute
    for i: int in int$from_to_by(first, last, by) do
        tagcase l.stuff[i]
            tag space (cnt: int):
                cnt := cnt + neach
                if nextra > 0
                    then cnt := cnt + 1
                        nextra := nextra − 1
                    end
                item$change_space(l.stuff[i], cnt)
            others:
            end
        end
    l.length := l.length + diff
    end enlarge_spaces
```

```
count_spaces = proc (l: rep, idx: int) returns (int, int)
% effects returns a count of spaces subject to justification and the
%       index in l.stuff of the last such space
    cnt: int := 0
    for i: int in int$from_to(idx, items$high(l.stuff)) do
        tagcase l.stuff[i]
            tag space:
                cnt := cnt + 1
                idx := i
            others:
            end
        end
    return (cnt, idx)
    end count_spaces

end line
```

Appendix C
A Sequence of Programming Assignments

This appendix contains five programming problems representative of the sequence of assignments in our one-semester undergraduate laboratory course.

Problems 1 and 2 deal with trees and graphs. The first is handed out during the second week of the term and takes 1 week. It involves writing some relatively simple programs that use data abstractions and is intended to illustrate the utility of good abstractions. The second is a 1-week assignment in which the students implement the data abstractions used in the first problem.

Problem 3 is based on the formatter designed in chapter 13. The students are shown the implementation in appendix B and are given 3 weeks to make a set of modifications to it. The assignment is designed to give students some insight into the problems that arise in modifying programs built by others.

Problem 4 is the first assignment in which students are asked to do a substantial amount of design work. They are given 3 weeks to design and implement an interactive spelling corrector. The first phase of the assignment involves tightening up an intentionally vague requirements specification and designing an implementation. In the second phase they implement their design.

Problem 5 is a monthlong team programming project. Groups of two or three students are asked to design and implement a program similar to VisiCalc^{TM}. The project involves several intermediate due dates, as explained below.

Problem 1: Using Data Abstractions

In this assignment you are to write procedures for traversing trees and graphs. Your procedures will use directed graphs, trees, integer sets, and queues. These data abstractions are specified below, and their implementations will be provided online.

A *directed graph* (*digraph*) consists of a set of nodes together with a set of edges. An edge is an ordered pair of nodes. If the edge $\langle n_0, n_1 \rangle$ is in a directed graph g, n_1 is said to be a *successor* of n_0 in g.

A *path* from node n_0 to node n_m is a sequence, of length greater than 1, of nodes $n_0, \ldots, n_m$, such that for each $i = 0, \ldots, m - 1$, there is an edge from n_i to n_{i+1}. We say that node n_m is *reachable* from node n_0 if there exists a path from n_0 to n_m. A node may or may not be reachable from itself.

The *adjacency list* of a node n in g is the list of successors of n. The *adjacency-list representation* of a directed graph g is the collection of adjacency lists for each node of g.

A *breadth-first list* of the nodes in a directed graph g starting from a node n consists of n followed by each successor of n, followed by each of the successors of those nodes, and so on. All of the nodes reachable from n appear exactly once in the list.

For this problem, each node in a directed graph with n nodes will be labeled with a unique integer from 1 through n. These labels are the only means of referring to nodes.

A *tree* is a directed graph satisfying the following properties:

1. A tree has a distinguished node called the *root* that is not the successor of any node.

2. Every node in a tree other than the root is the successor of exactly one node.

3. Trees are connected; that is, all nodes in a tree are reachable from the root. Trees are acyclic; that is, there is no path from a node to itself.

When a directed graph is a tree, we use the word "child" in place of "successor."

The Tree Traversal Program

In the first part of the problem, you are to write a procedure that accepts a tree and the label of a node in the tree as input and prints a breadth-first list of that tree. The specification for this procedure is

> breadth_first_tree = **proc** (t: tree, n: int)
>   **requires** n is the label of a node in tree t.
>   **modifies** primary output.
>   **effects** Prints (on the primary output stream) a breadth-first list
>   in t starting at n.

You will be provided with implementations of the tree and queue data abstractions; their specifications are given in figures C.1 and C.2. You will find the queue abstraction handy for internal bookkeeping while traversing the tree.

tree = **data type is** create, add_node_and_edge, children, put

Overview

> A tree is an acyclic, connected directed graph with a distinguished *root* node. Trees are immutable.

Operations

> create = **proc** ( ) **returns** (tree)
> > **effects** Creates a new tree that contains exactly one node, labeled with the integer 1.

> add_node_and_edge = **proc** (t: tree, n: int) **returns** (tree, int)
> > **requires** n is the label of a node in t.
> > **effects** Creates a new tree containing the nodes and edges of t plus a new node and a new edge from n to that node. Returns the new tree and the unique integer label for the new node. Note that t is not modified.

> children = **proc** (t: tree, n: int) **returns** (sequence(int))
> > **requires** n is the label of a node in t.
> > **effects** Returns a sequence containing the label of each child of node n.

> put = **proc** (t: tree, s: stream)
> > **modifies** s
> > **effects** Outputs an adjacency-list representation of the tree t onto the stream s.

end tree

Figure C.1 Specification of the tree data type.

Traversing a Graph

You should now modify a copy of the procedure *breadth_first_tree* to satisfy the following specifications:

> breadth_first_graph = **proc** (g: digraph, n: int)
> > **requires** n is the label of a node in digraph g.
> > **modifies** primary output.
> > **effects** Prints (on the primary output stream) the breadth-first list in g starting at n.

We shall provide implementations of the intset and digraph data abstractions. The specification of digraph is given in figure C.3. Intsets are specified in figure 4.2; they may be useful for ensuring that each node is visited exactly once.

queue = **data type is** create, enq, deq, is_empty

Overview

A queue stores integers in a first-in/first-out discipline. Queues are mutable.

Operations

create = **proc** ( ) **returns** (queue)
    effects Creates and returns a new empty queue.

enq = **proc** (q: queue, n: int)
    modifies q
    effects Adds n to the back of q.

deq = **proc** (q: queue) **returns** (int)
    requires q is not empty.
    modifies q
    effects Removes the integer at the front of q and returns it.

is_empty = **proc** (q: queue) **returns** (bool)
    effects Returns true if q is empty; otherwise returns false.

end queue

Figure C.2 Specification of the queue data type.

Problem 2: Writing Data Abstractions

Your job now is to implement the tree, queue, and digraph abstractions that were provided in problem 1. Be sure to include the rep invariants and abstraction functions for your implementations.

Each of these abstractions has operations with requires clauses. These clauses should be removed and replaced with appropriate exceptions. Provide the amended specifications as part of your solution. (An implementation of the amended specification for intset is given in figure 5.4.)

Your solution should describe the tests that you ran. A required part of your testing is to show that your solutions to problem 1 work with your clusters. This is not sufficient, however, to serve as a complete demonstration that your clusters meet their specification.

Problem 3: Modification of a Text Formatter

This problem is intended to give you practice in modifying software written by another person. You will be required to read both the original program and its documentation and to turn in a revised program and revised

digraph = **data type is** create, add_node, add_edge, add_node_and_edge,
successors, put

Overview

Digraphs are immutable directed graphs.

Operations

create = **proc** ( ) **returns** (digraph)
effects Creates a new digraph that contains exactly one node. The
node is labeled with the integer 1.

add_node = **proc** (g: digraph) **returns** (digraph, int)
effects Creates a new digraph containing the nodes and edges of g
and one new node; returns the new graph and the unique integer
label for the new node. Note that g is not modified.

add_edge = **proc** (g: digraph, initial, final: int) **returns** (digraph)
requires *initial* and *final* are labels of nodes in g.
effects Creates a new digraph containing the nodes and edges of g
and an edge from *initial* to *final*; returns the new graph. If the
edge already exists, the new graph is indistinguishable from g.
Note that g is not altered in either case.

add_node_and_edge = **proc** (g: digraph, n: int) **returns** (digraph, int)
requires n is the label of a node in g.
effects Creates a new digraph containing the nodes and edges of g
plus a new node and an edge from n to that node. Returns the
new graph and the unique integer label for the new node. Note
that g is not modified.

successors = **proc** (g: digraph, n: int) **returns** (sequence[int])
requires n is the label of a node in g.
effects Returns a sequence listing the successors of n in g.

put = **proc** (g: digraph, s: stream)
modifies s
effects Outputs an adjacency-list representation of the digraph g
onto the stream s.

end digraph

Figure C.3 The graph abstraction.

documentation. The program you will be modifying is the text formatter
in appendix B.

You are to make the following modifications to the formatter:

1. Change the size of the left margin to 12 spaces. The left margin is
the number of spaces between the left-hand edge of the page and the first
position where text might be put.

2. Add the commands `.line width [+/-]N` and `.page_length [+/-]N`, which modify the line width and the number of lines per page, respectively, in the following fashion: If the integer argument N is preceded by a plus (or minus) sign, it is added to (or subtracted from) the previous value to give the new value; otherwise the new value simply becomes the argument N. (The initial value for the line width and page length are 60 and 50, respectively.) The command `.page_length` takes effect beginning with the page *following* the current page at the time of the command, unless the current page is empty. If the current page is empty, the command takes effect immediately.

3. Add a command `.title "text"` to specify the title of a text. The title should appear on the first line of every page's header. This command causes a line break.

4. Modify the printing of page numbers to make them appear on the right side on odd-numbered pages and on the left side on even-numbered pages. The page number need not appear on the same line as the title.

5. Add commands `.begin_quote N` and `.end_quote`, which are used to set off a long quotation from the rest of the text. The command `.begin_quote N` will cause a line break, print a blank line, and indent both margins by N spaces (the margins become wider; the text lines are shorter). The command `.end_quote` will cause a line break, print a blank line, and restore the margins to their previous values.

6. Add a command `.begin_paragraph` that marks the start of a paragraph. This command should cause a line break; then if there are less than three lines of text left on the page, a new page should be started. The first line of the paragraph should be indented five spaces from the left margin.

You should also design the following changes to the program, but do not implement them. Provide specifications and a list of the program changes that would be needed to implement the following commands:

1. Add a command `.need N` that causes the formatter to continue output on the current page only if the number of lines of text specified, N, will fit on the page. Otherwise this command starts a new output page. In either case the command causes a line break.

2. Add commands `.begin_table` and `.end_table`, which are used for setting off tables from the rest of the text. All lines between the `.begin_table` command and the `.end_table` command are saved until the top of the next page (or the current page if it is empty). If the current page is not empty, output continues as if everything between the `.begin_table` command and the `.end_table` command were not there. At the top of the next page (or

the current page if it is empty), the saved lines are formatted and printed immediately below the header. Then a blank line, a row of hyphens, and another blank line are printed before output continues in a normal fashion.

You will need two kinds of specification amendments for this problem: amendments to the requirements specification and amendments to the module specifications. Amendments to the module specifications will be needed as you modify the program. Amendments to the requirements specification will also be needed because the list of modifications given above doesn't take special cases or error conditions into account. You should complete the specification by deciding how those cases are handled. For instance, for commands that take a numerical argument, decide what to do if the argument is missing or invalid. Be consistent; if you decide to provide default values, for example, do so for all such commands. Also, decide how the table commands affect the formatting environment.

Your solution to the problem should include the following:

1. A description of your changes to the requirements specification. This should include a brief but complete description of which user errors are detected and how they are handled.

2. A description of your changes to the module specifications. The module specifications themselves should appear with the module listings.

3. Documentation of all changes to the design. In addition to the usual information, this section should include an analysis of the difficulty of each change. A key property of a good software design is that the resulting program is easy to extend and modify. By examining the difficulty of each change you will gain insight into how the design and implementation of a program affect the ease of modification.

4. Listings of all modules that you have modified. The changes should be clearly marked in the listings with comments. You should not list modules that did not change.

5. Output from test cases showing that the changes were successful. Include an overview, description, and rationale for your test cases.

6. Design documentation for the changes we asked you to design but not implement.

Problem 4: An Interactive Spelling Corrector

A spelling corrector is a program that takes a dictionary and a text file as input and corrects the spelling of words in the text file. The program operates by looking up each word in the text file in the dictionary. If a word is not in the dictionary, it may or may not be misspelled. In either case the program displays the word to the user with enough of the surrounding text to show the context in which the word was found. The user may take one of several corrective actions: retype a misspelled word, force the program to accept it as spelled, or have the word accepted as spelled and included in the dictionary. The output of the spelling corrector is the corrected text and the possibly updated dictionary.

The spelling corrector must have reasonably efficient performance for two reasons. First, the program does enough crunching that an inefficient approach makes it impractical to use; and second, since it is interactive, you have the user's peace of mind to consider. The potential bottleneck is the lookup operation on the dictionary. This operation is invoked once for each word in the input text. To satisfy the efficiency constraint, we require that the lookup algorithm be faster than linear in the number of entries of the dictionary. Examples of lookup schemes with acceptable efficiency are (1) binary search on a sorted list, (2) lookup in a sorted binary tree, and (3) hashing.

We shall supply a dictionary of several thousand words. The dictionary is an ASCII file, with all words in lowercase characters, separated by spaces, tabs, and newlines. The words are not in sorted order.

This statement of the requirements of the *spell* program is incomplete. You may need more information about the dictionary. You also have decisions to make that concern both the high-level facilities the program will provide and the detailed processing of input characters. Some suggestions follow:

1. *Character processing.* You will need to specify how your program treats punctuation such as quotation marks, words or acronyms with numbers in them, hyphens, standard abbreviations ("etc."), and so forth. For example, you could treat hyphens as delimiters, since most hyphenated words are compounds formed from recognizable subparts.

2. *User interface.* The user interface in an interactive program is important and should be designed with care. The program should be easy to use. Some ideas along that line are:

    a. Allow the user to specify the names for the source and destination files for both the text and dictionary.

    b. Let the user correct several text files without leaving the spell program.

    c. Prompt the user by listing the possible responses when a misspelled word is displayed.

d. Implement a help command that describes the effect of each of these responses in detail.

In addition to the basic checking described above, you may want to try some of the following extensions. These are not required. You should attempt one or both *only* if the required part of the assignment is in good shape.

1. Many spell programs assume that an unknown word is misspelled and try to divine the word the user intended by locating "near misses" in the dictionary. Candidates for the near misses can be generated by looking at the common kinds of spelling errors that people commit; letters are often inverted (as in "hte"), a word will contain one extra letter, or one letter too few, or one letter might be wrong. In this problem you have the option of implementing near-miss strategies for the first two kinds of errors; corrections for missing and incorrect letters require too much attention to efficiency details to be worthwhile here.

2. Another extension to a spell program is an automatic correction facility. Most people have collections of words that they consistently spell incorrectly. The spell program that has been described so far would ask the user for the proper spelling each time one of these words was encountered. This quickly gets tedious. With this extension, the user specifies the proper spelling for a bad word once, and indicates that the mapping should be recorded. From that time on, the program automatically makes the change without bothering the user again.

The solution to this problem should be submitted in in two parts. The first part (due halfway through the assignment) is a document describing your spelling corrector design. This document should include specifications for the modules you intend to implement, as well as refinements and clarifications to the problem specifications that you made during the design process. The second part should include program listings, test cases, and complete documentation for your spelling corrector.

Problem 5: TriviCalc (A Monthlong Project)

You are an engineer at a software/footware conglomerate, Softfeet, Inc. All the company's existing software is written in assembly language. However, in a major new policy decision, the company's directors have decreed that all software will be converted to a higher-level language. CLU has been chosen as the company's new *lingua franca*. Your arguments were instrumental in the company's new policy and in the choice of CLU.

Your division head calls you into the office. You know that he resents the attention you have received from his superiors during the move to CLU and is eager to put you in your place. He assigns you the task of rewriting Softfeet's most profitable program, TriviCalc. The original TriviCalc implementation took ten person-months from program design to a debugged first release. Since you have pointed out that it is easier to design and debug software in a higher-level language, your division head explains that he expects you and your programming team of one or two other engineers to complete the CLU program in four weeks.

The rest of this section contains information about due dates and what to hand in, a TriviCalc reference manual, some extensions to consider in your design, some hints about how to do the implementation, and the specification of a data abstraction that can be used in the implementation.

Due Dates

Before the whole project is due, you must prepare two design documents. The first, due in nine days, should contain refinements and clarifications to the TriviCalc specification and should describe a preliminary design for TriviCalc. Each team should hand in one document. The designs will be discussed that day. At the next class meeting, your teaching assistant will return your design with comments about how it might be improved. You will then have nine days to work on the design and prepare a revised design document. The revised designs will be discussed in class. The final writeup, including module listings, is due nine days after the revised design is due. You will lose points if the final design differs significantly from the revised design.

After the project is complete, each team must present their program. The presentation is limited to 25 minutes per group. During the first 20 minutes, the group must demonstrate the program's capabilities. During the last 5 minutes, the teacher will test the program.

Design Documentation

The preliminary and revised design documents should be as complete as possible. They must contain the following information:

1. Any amendments to the reference manual, including a list of all error messages. Be sure to justify any amendments to the functionality of TriviCalc.

2. A module dependency diagram.

3. An overview of the design.

4. A specification of each module.

5. Implementation. Write an overview of your implementation strategy that focuses on the order in which modules are to be implemented and tested and what stubs and drivers are to be used. Be sure to discuss both unit and integration testing. Point out which modules can be developed and tested in parallel.

6. List of integration test cases. Describe and justify a list of test cases that you will use to convince your teaching assistant that the program works completely. Try to make the test cases as concise and complete as possible.

7. Schedule and division of labor. Give a plausible timetable for completing the project. Indicate who will implement each module. The timetable should have well-defined milestones like completing a specification, finishing a stub implementation for other team members to use, finishing the implementation of some group of operations, or finishing the unit testing for some module.

TriviCalc Reference Manual

TriviCalc is a program that can be used as a scratchpad for problems involving arithmetic. The user of the program enters numbers or text into the *storage area* of the computer's memory. The data are then displayed on the terminal's screen. The user can also combine values that are already displayed on the screen to obtain new values, which are stored in the computer and displayed elsewhere on the screen. When this is done, the program remembers the relationships between the numbers, so that the calculation can be repeated on different values.

As an example, imagine two numbers displayed at places on the screen labeled A1 and B1. The user types a command that causes TriviCalc to add the value stored at A1 to that stored at B1 and store the result at C1. If the user later changes the value at A1, the value at C1 will also change so that it remains the sum of the values at A1 and B1.

User Interface

During the operation of TriviCalc, the display is divided into three parts. Figure C.4 contains a diagram of the display when the program is first begun.

The first line of the display, known as the *status line*, is used to display descriptions of the command state of the program. The line is divided in two: The left portion contains error messages. The right portion displays information about the contents of the current slot.

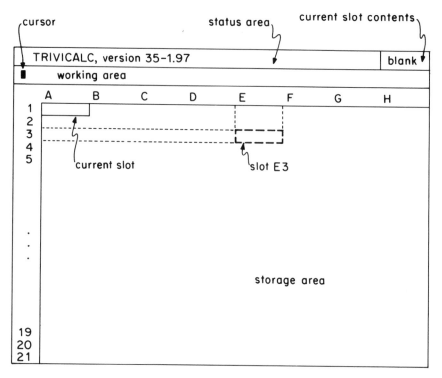

Figure C.4 TriviCalc initial display.

    The second line of the display is used as a *working area*. When the user types textual or numerical input, it is displayed here, pending action by some user-specified command. The text displayed in this area may be edited using a simple editor, described in detail in a later section.

    The remainder of the screen is occupied by what is known as the *storage area*. It is used to display textual representations of all the values stored in TriviCalc's internal storage. The storage area is divided into fields or "slots," which are eight characters wide and one line high. Each slot is named by a number describing its row and a letter describing its column, starting at the upper left-hand corner of the storage area. Thus the slot in this corner is named "A1," and the slot diagonally opposite it, in the lower right-hand corner of the storage area (and of the screen), is named "H21." Surrounding the storage area to the left and top is a border containing the name of each row and column.

    There is always one slot in the storage area that is considered to be the *current slot*. This slot is displayed specially, perhaps in reverse video, so as to be easily recognizable by the user. The initial current slot is A1. To make it easier to distinguish the different areas, the current slot contents, the working area, and the header lines for the storage area should also be displayed specially. (Reverse video is not shown in the figure.)

Description of Operation

TriviCalc's storage consists of a two-dimensional array whose elements are referred to by a letter in the range A–H describing one dimension and a number in the range 1–21 describing the other. A textual approximation to the contents of each storage element is displayed continuously in the storage area of the terminal display.

Elements of storage may be of three kinds: *blank, value,* or *comment.* A blank element has no value. A comment is a string of up to eight characters. When a comment occupies an element of storage, it has no effect on any other element of storage.

A value is a floating-point number. Values in storage may be related to each other by multiplication, division, addition, and subtraction. These relations are set by the user and may be changed at any time. Some values in storage will be entered by the user as constants or parameters; others will be derived as a result of one of the relations mentioned above. A derived value is said to *depend* on other values.

There are commands to change the contents of elements of storage. Subject to restrictions outlined later, the new contents of such an element need not be of the same type as the old. When a value is changed, the value of elements of storage that depend on that value may also change.

There are some general rules that apply to changing the contents of the storage area of TriviCalc. TriviCalc maintains the *consistency* of the storage area. Given the set of commands discussed above, it is possible to set up a configuration of values in storage that has no consistent interpretation. The TriviCalc command interpreter will not allow such a situation to occur. Instead, when a command would cause the storage area to become inconsistent, the command is not carried out and an error message is displayed.

The rules that the TriviCalc command interpreter follows to ensure the consistency of the storage area are:

1. Only elements of storage whose contents are values may be depended on by other elements.

2. When an element's contents are depended on by other elements, that element's type may not be changed.

3. When the contents of an element depend on those of other elements, changing the dependent element's contents causes the dependency to be broken.

4. A given element may not depend, even indirectly, on its own value.

The TriviCalc Command Processor

Except for the movement commands described below, all commands and their arguments are entered through the working-area editor. When a command is executed, the entire working area is submitted to the TriviCalc command processor, which either performs the command at once or rejects the command as an error. The TriviCalc command processor never modifies the contents of the working area of the display.

A full description of the operation of each command supported by the TriviCalc command processor follows. The notation ↑ ⟨character⟩ is used to denote holding down the CONTROL key and pressing the key labeled ⟨character⟩.

SAVE; FILE:*name*; The current state of the storage area is saved in a file named *name.tc*, where *name* is the FILE argument. The state is stored as an ASCII string containing a sequence of TriviCalc commands separated by line-feed characters. These commands should have the property that if they were executed in sequence, beginning with a blank storage area, they would generate the storage area in effect at the time the save is done.

LOAD; FILE:*name*; The current state of the storage area is discarded and then reloaded based on the contents of the file named *name.tc*. The file is assumed to be in the format produced by the SAVE command.

STORE-COMMENT; WITH:*string*; AT:*slot-address*;
The comment *string* is stored in the element of storage labeled by *slot-address*.

STORE-VALUE; WITH:*number*; AT:*slot-address*;
The value *number* is stored in the element labeled by *slot-address*. After this command has been completed, this element will not depend on any other elements.

BLANK; SLOT:*slot-address*;
The element labeled by *slot-address* becomes blank.

QUIT; The execution of TriviCalc is terminated and control is returned to the executive.

The following movement commands affect the position of the current slot of the storage area:

↑P   The current slot is changed to be the slot vertically above the current slot, if there is one. If the row of the current slot is 1, the command has no effect.

↑N   The current slot is changed to be the slot vertically below the current slot, if there is one. If the row of the current slot is 21, the command has no effect.

↑B   The current slot is changed to be the slot to the left of the current slot, if there is one. If the column of the current slot is A, the command has no effect.

↑F   The current slot is changed to be the slot to the right of the current slot, if there is one. If the column of the current slot is H, the command has no effect.

Binary operator commands take three arguments, each of which is a label of an element of the storage area. Their command relates the elements of storage labeled by the first two arguments to a result stored in the element of storage labeled by the third argument. If *slot3* labels an element of storage that is already the result of a relation, the old relation is discarded in favor of the new one. All these commands are of the form

    op; VALUE1:*slot1*; VALUE2:*slot2*; GIVING:*slot3*;

Op may have the following values:

ADD *or* +   The element of storage labeled *slot3* is related to the other two elements as (*slot1* + *slot2*).

SUBTRACT *or* −   The element of storage labeled *slot3* is related to the other two elements as (*slot1* − *slot2*).

MULTIPLY *or* ∗   The element of storage labeled *slot3* is related to the other two elements as (*slot1* ∗ *slot2*).

DIVIDE *or* /   The element of storage labeled *slot3* is related to the other two elements as (*slot1* ÷ *slot2*).

The Working-Area Editor

The working-area editor is a simple modeless editor with special functions to simplify the input of commands to the TriviCalc command processor. The editor maintains a cursor in the working area. Every keystroke is considered to be a command to the editor. All commands are atomic; they are either processed to completion immediately or halt in error, after doing nothing except possibly displaying an error message. Some keystrokes denote textual values (the characters, numerals, and punctuation keys). The command that is run by typing any of these keystrokes merely inserts the key's textual value at the cursor. These are known as the *textual input commands*.

Other keystrokes do not denote textual values. These are either special keys (such as the carriage return or delete) or are typed by holding down the CONTROL key and pressing some other key. These nontextual keystrokes are interpreted by the editor as commands that affect the text in the working area. A brief description of the nontextual commands follows.

↑L   Move the cursor to the left one position.

↑R   Move the cursor to the right one position.

↑D   Delete the character at the cursor, if there is one.

DELETE   Deletes the character to the left of the cursor, if there is one.

↑A   *Operator Adjust*: If the working area is of the form

                slot1 op slot2

            where *op* is one of the characters ∗, /, −, or + and *slot1* and *slot2* are strings, the contents of the working area are replaced with

                op; VALUE1:*slot1*; VALUE2:*slot2*; GIVING:%;

            Otherwise, if the contents of the working area represent a valid numerical value, the working area is interpreted as

                number

            and its contents are replaced by

                STORE-VALUE; WITH:*number*; AT:%;

            Otherwise the working area is interpreted as

                string

            and its contents are replaced by

                STORE-COMMENT; WITH:*string*; AT:%;

            Finally, the effect of a ↑K command with the cursor at the beginning of the working area, followed by a ↑E command is simulated. The effect of this is to replace the % character with the address of the current slot.

↑K   Search from the position of the cursor to the right for a %, wrapping around to the beginning of the working area if the end of the working area is reached. If a % is found, delete it and leave the cursor at its position. If none is found, do nothing.

↑E   The address of the current slot in the storage area is inserted at the cursor.

RETURN   Pass the contents of the working area to the command interpreter for immediate execution. If the command interpreter does not flag an error, delete the contents of the working area.

LINE-FEED   Same as RETURN, but leave the contents of the working area unaffected in all cases.

↑V   Insert a textual representation of the contents of the current slot into the working area at the cursor.

↑U   Delete the entire contents of the working area.

↑W   The working area must be of the form ⟨digit⟩⟨string⟩, where ⟨digit⟩ is between 1 and 9. The effect of the command is to cause the ⟨string⟩ to be saved at internal location ⟨digit⟩. When TriviCalc is first started up, the contents of the first four internal locations are the strings
                   1: SAVE; FILE:%;
                   2: LOAD; FILE:%;
                   3: QUIT;
                   4: BLANK; SLOT:%;
               Internal locations 5–9 contain the empty string initially.

↑X   The working area must be of the form ⟨digit⟩⟨string⟩, where ⟨digit⟩ is between 0 and 9 and ⟨string⟩ may be empty. The effect of the command is to cause the contents of the working area to be replaced by the string stored at internal location ⟨digit⟩. Then the effect of a ↑K followed by a ↑E is simulated, causing the % character to be replaced with the address of the current slot.

Every time a nontextual editing command causes the working area's contents to change, the contents are first saved in internal location 0. Entering ↑X when the working area is of the form 0⟨string⟩ causes the current contents of the working area and internal location 0 to be swapped.

Extensions

In your project writeup, in addition to the usual requirements, you should consider briefly (a paragraph or two each will suffice) the following extensions to TriviCalc. Do not implement any of these extensions (regardless of whether you have the time or not)! For each possible extension, indicate:

1. a more complete description of what the change would be;

2. whether or not it is a good idea (and why you think so);

3. how it would affect your code (what modules would need to be changed, and so forth); and

4. how many person-hours would be required to perform the extension.

The extensions are as follows:

1. Extending the size of the storage area to be a 100 × 100 array. (Consider addressing slots, screen management, and so forth.)

2. Allowing cyclic dependencies. (Be sure to say how this is to be "made sense of," since it is clearly meaningless in the current TriviCalc.)

3. Adding two multiple argument relations that relate an aggregate of elements of storage within the same row or column to some other element of storage not in the aggregate. One causes the resultant element to have the value of the sum of the aggregate, while the other causes the resultant element to have the value of the average of the aggregate. Their form is

SUM; FROM: *slot1*; THROUGH: *slot2*; GIVING: *slot3*;
AVERAGE; FROM: *slot1*; THROUGH: *slot2*; GIVING: *slot3*;

Slot1 and *slot2* must be in either the same row or the same column, and all elements of storage "between" the two selected elements must have contents of type Value. The sum or average is computed on all the slots in the row or column that are between *slot1* and *slot2*, inclusive.

Implementation Hints

This section contains some hints to help you in your implementation.

1. *Command Processing.* There are two main issues in the processing of commands. One is *parsing*—the process of breaking the command line up into the individual arguments. The second is *command interpretation*—the processing of the command. Experience has shown that a clear separation between the two is often a good idea.

2. *Displaying Values.* There is a procedure *g_form* described in appendix A that may be of some use.

3. *Screen Processing.* There are two common techniques for maintaining an up-to-date display in an interactive system. One is to have each command determine the changes to the display that should be made and cause those changes to occur. The other is for the command to register each change with a screen-manager module, which maintains a copy of what is to be displayed on the terminal screen when the command terminates. When the command is completed, the screen manager is notified and updates

the screen to reflect all the changes. Experience indicates that the latter technique is superior.

4. *Updating Storage.* The problem here is to update all the values that depend on an element that is being changed. Here are some hints on this problem:

a. One must be able to obtain the value of a storage element three ways: by slot address; as a predecessor to another element; and as a successor to another element.

b. Whenever a command modifies the storage area, it must do so *atomically*; this means that if there is an error during the modification required to process a command, the command should have no *net* effect. There are a number of ways to implement this kind of behavior. Two of them are (1) backing up (undoing the changes and reevaluating storage) and (2) shadow copy (performing the changes on a *copy* of storage and installing the new copy if and only if everything went well). The latter method is probably preferable.

c. There is a "correct" order in which to evaluate the new values of elements of storage that depend on other elements. It is crucial to evaluate all the elements of storage upon which a given element S depends (directly or indirectly) *before* evaluating S. This is not the only way to get the correct final result (in the event that there are no errors), since it would suffice to reevaluate all storage elements in a depth-first order, beginning from the changed element. In this case the values would eventually "settle down" to the correct ones. The problem is that this technique might produce inadvertent overflows, zero divides, and so forth, which would not happen were the elements evaluated in the *correct* order.

d. One way to determine the *correct* order of evaluation of elements of storage is to assign a *priority number* to each element of storage:

priority(S) = 1 + maximum (priorities of S's parents)

where a priority of 1 is used when S has no parents. If the elements of storage are assigned priority numbers in this way, the optimal order of evaluation is by increasing priority number. Note that when a given element of storage changes, one can determine just those other elements of storage that might also change. This means that it is not necessary to evaluate all the elements of storage with every command that is processed.

Table C.1 CLU literals for nontextual characters.

Character	CLU Literal
↑A	'\001'
↑B	'\002'
↑D	'\004'
↑E	'\005'
↑F	'\006'
↑K	'\013'
↑L	'\014'
↑N	'\016'
↑P	'\020'
↑R	'\022'
↑U	'\025'
↑V	'\026'
↑W	'\027'
↑X	'\030'
DELETE	'\177'
RETURN	'\r'
LINE-FEED	'\n'

5. *Support Software.* We shall provide an implementation of an abstraction that will help you in interfacing to terminals. The *screen* data type provides a primitive interface to the screen of a terminal such as a VT-100; its specification is given in figure C.5. The CLU literals for the nontextual characters used in TriviCalc are given in table C.1. In reading from the terminal, use stream$*getc_image* (this avoids the echo of the character being read).

screen = **data type is** create, clear, get_height, get_width, putc, puts, flush,
beep, cursor_forward, cursor_backward, cursor_up, cursor_down,
index, reverse_index, set_cursor_pos, special_modes, erase_to_eol,
erase_to_bottom

Overview

A screen is an abstract presentation of the primary output device; it can be
used only if the primary output device is supported. In addition, only one
screen object should be created within a program.

A screen is a matrix of characters, organized by lines and columns, together
with a cursor. The characters can be any ASCII character whose corresponding code is in the range 32–126 inclusive. Note that this does not
include the tab and newline characters. A clear screen is a screen where every character in the matrix is the blank character. The lines are numbered
from the top down starting with 1, and the columns from left to right also
starting with 1. The screen height is equal to the number of lines, and the
screen width is equal to the number of columns.

At any given time, the cursor points to one of the characters in the matrix.
Its position can be specified by its line and column. The cursor is special in
that most operations that modify the screen do so relative to the cursor's
current position. There are operations for moving the cursor around the
screen. In addition, there are operations that cause the screen to scroll up
or down. When the screen is scrolled up, each line of the screen is moved
up by one line. The top line "disappears" off the top of the screen, and a
blank line is written at the bottom. Scrolling down is the reverse of scrolling
up.

Characters output to the screen with the *puts* and *putc* operations are
buffered by the system until one of the following occurs: (1) A character is
read from the primary input device, or (2) an operation of the screen cluster is called that implicitly flushes the buffer. An explicit call to the *flush*
operation is needed only when output to the screen must be displayed while
the using program is writing characters to the screen and not reading from
the primary input device. An example would be displaying the status of the
program while it is performing some computation.

Operations

create = **proc** ( ) **returns** (scrn: **cvt**) **signals** (error(string))
　　　　modifies *primary_output*
　　　　effects Creates and returns a screen object for the primary output
　　　　　　device. Clears screen and moves cursor to home position (upper
　　　　　　left corner of screen). Signals *error*("unsupported terminal type")
　　　　　　if the primary output device is not supported.

Figure C.5 Specification of the screen abstraction.

clear = **proc** (scrn: **cvt**)
    modifies *scrn*
    effects Clears the screen and moves the cursor to the upper left-hand corner of screen (cursor position [1,1]).

get_height = **proc** (scrn: **cvt**) **returns** (height: int)
    effects *height* = number of lines that can be displayed on the screen.

get_width = **proc** (scrn: **cvt**) **returns** (width: int)
    effects *width* = number of columns that can be displayed on the screen.

putc = **proc** (scrn: **cvt**, ch: char)
    requires *ch* must be either a printing character (ASCII code 32–126 inclusive), a carriage return, or a newline character.
    modifies *scrn*
    effects Depends on the type of *ch* as follows:

    ch is an ASCII printing character with code in the range 32–126 inclusive (note that this includes the space character). The character is displayed at the current cursor position. If the cursor is not in the last column, it is moved one column to the right. Otherwise it remains in the last column.

    ch is a carriage return (\r). The cursor is moved to the first column of the line it is currently on (if the starting cursor position is [x,y], its final position is [x,1]).

    ch is a newline (\n). If the cursor is not on the last line of the screen, it is moved to the next line down on the screen, but remains in the same column (if the starting cursor position is [x,y], its final position is [x+1,y]). If the cursor is on the last line of the screen, the screen is scrolled up one line and the cursor position remains unchanged. This operation also flushes any output that may have been buffered. Note that this case is the same as the operation *reverse_index*.

puts = **proc** (scrn: **cvt**, str: string)
    requires Every character in *str* must be either a printing character (ASCII code 32–126 inclusive), a carriage return, or a newline character.
    modifies *scrn*
    effects For each character *c* in *str*, do *putc(scrn, c)*. Note that it is more efficient to use this operation on a string than it is to use the *putc* operation explicitly on each character of the string.

Figure C.5 (continued)

flush = **proc** (scrn: **cvt**)
> **effects** Flushes any buffered output to the screen.

beep = **proc** (scrn: **cvt**)
> **modifies** *scrn*
> **effects** Flushes buffered output and sounds beep tone on terminal.

cursor_forward = **proc** (scrn: **cvt**, n: int)
> **modifies** *scrn*
> **effects** Moves the cursor *n* columns to the right. The cursor remains on the same line. Once the cursor has reached the rightmost column of the line, it stops moving, regardless of how large *n* is.

cursor_backward = **proc** (scrn: **cvt**, n: int)
> **modifies** *scrn*
> **effects** Moves the cursor *n* columns to the left. The cursor remains on the same line. Once the cursor has reached the leftmost column of the line, it stops moving, regardless of how large *n* is.

cursor_up = **proc** (scrn: **cvt**, n: int)
> **modifies** *scrn*
> **effects** Moves the cursor *n* lines up. The cursor remains in the same column. Once the cursor has reached the top line, it stops moving, regardless of how large *n* is. No scrolling takes place.

cursor_down = **proc** (scrn: **cvt**, n: int)
> **modifies** *scrn*
> **effects** Moves the cursor *n* lines down. The cursor remains in the same column. Once the cursor has reached the bottom line, it stops moving, regardless of how large *n* is. No scrolling takes place.

index = **proc** (scrn: **cvt**)
> **modifies** *scrn*
> **effects** Moves the cursor to the next line (down) on the screen. The cursor remains in the same column. If the cursor is already on the last line when this procedure is called, the screen is scrolled up one line and the cursor position remains unchanged. Buffered output is flushed.

reverse_index = **proc** (scrn: **cvt**)
> **modifies** *scrn*
> **effects** Moves the cursor to the previous line (up) on the screen. The cursor remains in the same column. If the cursor is already on the first line when this procedure is called, he screen is scrolled down one line and the cursor position remains unchanged. Buffered output is flushed.

Figure C.5 (continued)

set_cursor_pos = **proc** (scrn: **cvt**, x, y: int) **signals** (bounds)
>   **modifies** *scrn*
>   **effects** Moves the cursor to line *x* and column *y* of the screen if this position exists. Signals *bounds* otherwise.

special_modes = **proc** (scrn: **cvt**, reverse_video, underline: bool)
>   **modifies** *scrn*
>   **effects** Puts the terminal into reverse-video mode or underline mode, depending on the values of the two arguments. Characters that were displayed on the screen before a mode is set remain unchanged. All characters displayed on the screen after a mode is set are displayed with the new mode. If the terminal is not capable of the mode selected, this procedure has no effect.

erase_to_eol = **proc** (scrn: **cvt**)
>   **modifies** *scrn*
>   **effects** Erases from cursor to column screen$*width* inclusive. Flushes any buffered output.

erase_to_bottom = **proc** (scrn: **cvt**)
>   **modifies** *scrn*
>   **effects** Erases from cursor to last column of last line inclusive. Flushes any buffered output.

end screen

Figure C.5 (continued)

Index